International Conference on Materials Research and Innovation (ICMARI 2019)

IOP Conference Series: Materials Science and Engineering
Volume 773

Bangkok, Thailand
16 - 18 December 2019

Editors:

Wirasak Smitthipong
Saree Phongphanphanee
Jirasak Wong-ekkabut

ISBN: 978-1-7138-3503-5
ISSN: 1757-8981

Printed from e-media with permission by:

Curran Associates, Inc.
57 Morehouse Lane
Red Hook, NY 12571

CURRAN ASSOCIATES INC. proceedings.com

Some format issues inherent in the e-media version may also appear in this print version.

Institute of Physics
Dirac House, Temple Back
Bristol BS1 6BE UK

Phone: 44 1 17 929 7481
Fax: 44 1 17 920 0979

techtracking@iop.org

Additional copies of this publication are available from:

Curran Associates, Inc.
57 Morehouse Lane
Red Hook, NY 12571 USA
Phone: 845-758-0400
Fax: 845-758-2633
Email: curran@proceedings.com
Web: www.proceedings.com

TABLE OF CONTENTS

IOP Conf. Series: Materials Science and Engineering **773** (2020) 011001 doi:10.1088/1757-899X/773/1/011001

IOP PROCEEDING

Organized by

1. Specialized center of Rubber and Polymer Materials in agriculture and industry (RPM), Faculty of Science, Kasetsart University, Bangkok, Thailand.
2. National Research Council of Thailand (NRCT).
3. Thailand Science Research and Innovation (TSRI).
4. Kasetsart University Research and Development Institute (KURDI).
5. Faculty of Architecture, Kasetsart University.

IOP Conf. Series: Materials Science and Engineering **773** (2020) 011001 doi:10.1088/1757-899X/773/1/011001

Preface

The 2nd International Conference on Materials Research and Innovation (2nd ICMARI) will be held on 16th – 18th December 2019 in Bangkok, Thailand. The conference will be conducted in frontier research on materials research and Innovation including Rubbers and Composites, Biomaterials, Materials of Energy and Environmental Applications, Computational Model and Simulations, Industrial Innovation, and Special Advanced Materials. This 2nd ICMARI was announced in order to created and build up collaboration network and sharing experience among researchers in ASEAN and around the world. It is a great opportunity for all participants to exchange knowledge and strengthen the research collaboration. We hope that the participants will be fruitful by plenary and invited speakers in knowledge, application, novel technology in the field of materials research and innovation.

Thank you all for participating,

ICMARI 2019 Organizing Committee

IOP Conf. Series: Materials Science and Engineering **773** (2020) 011002 doi:10.1088/1757-899X/773/1/011002

Organizing Committee

Chairman

Asst. Prof. Dr. Wirasak Smitthipong

Committee (Inside KU)

Dr. Saree Phongphanphanee

Dr. Nattasamon Petchsang

Asst. Prof. Dr. Chanapa Kongmark

Asst. Prof. Dr. Decha Dechtrirat

IOP Conf. Series: Materials Science and Engineering **773** (2020) 011002 doi:10.1088/1757-899X/773/1/011002

Committee (inside KU)

Asst. Prof. Dr. Pongthep Prajongtat

Assoc. Prof. Dr. Sutee Boonchui

Assoc. Prof. Dr. Jirasak Wong-ekkabut

Asst. Prof. Dr. Chomdao Sinthuvanich

Asst.Prof.Dr.Wirunya Keawwattana

Asst.Prof. Dr. Kiadtisak Saenboonruang

Assoc.Prof.Dr. Patthra Pengthamkeerati

Asst. Prof. Dr. Jukkrit Mahujchariyawong

Assoc. Prof. Dr. Nanthiya Hansupalak

Asst. Prof.Dr. Parames Kamhangrittirong

Ms. Ampaiwan Taweetanyaluk

Ms. Nipasiri Tapchai

IOP Conf. Series: Materials Science and Engineering **773** (2020) 011002 doi:10.1088/1757-899X/773/1/011002

Committee (outside KU)

Mr. Payab Namprasert (Thailand Science Research and Innovation)

Mrs. Pannee Jirabunpaisit (Thailand Science Research and Innovation)

Mr. Nakorn Tangavirapat (Rubber Authority of Thailand)

Dr. Orasa Onjun (Department of Science Service)

Dr. Wilaiporn Chetanachan (Siam Cement Group Pub Co.. Ltd.)

IOP Conf. Series: Materials Science and Engineering **773** (2020) 011002 doi:10.1088/1757-899X/773/1/011002

Secretary

Ms. Ratchadathip Purnajoti

Ms. Sukontip Suethao

Ms. Thridsawan Prasopdee

Ms. Nantikan Phuhiangpa

Mr. Khwanchat Promhuad

Mr. Khwanchai Buaksuntear

Ms. Siriwan Jansinak

Ms. Nisakorn Paetrat

Editors of Proceeding

IOP Conf. Series: Materials Science and Engineering **773** (2020) 011002 doi:10.1088/1757-899X/773/1/011002

Asst. Prof. Dr. Wirasak Smitthipong	Chair Editor
Dr. Saree Phongphanphanee	Editor
Assoc. Prof. Jirasak Wong-ekkabut	Editor

International Technical Committee

Prof. Sadhan C. Jana	University of Akron	USA
Prof. Arkadiusz Chworos	Polish Academy of Sciences	Poland
Prof. Katsura Nishiyama	Meijo University	Japan
Assoc. Prof. Li Jia	University of Akron	USA
Asst. Prof. Dr. Wirasak Smitthipong	Kasetsart University	Thailand
Prof. Koichiro Mitsuke	Josai University	Japan
Assoc. Prof. Norio Yoshida	Kyushu University	Japan
Dr. Saree Phongphanphanee	Kasetsart University	Thailand
Prof. Takaomi Kobayashi	Nagaoka University of Technology	Japan
Dr. Nattasamon Petchsang	Kasetsart University	Thailand
Assoc. Prof. Dr. Sutee Boonchui	Kasetsart University	Thailand
Dr. Karine MOUGIN	Institut de Science des Materiaux de Mulhouse	France
Assoc. Prof. Dr. Jirasak Wong-ekkabut	Kasetsart University	Thailand
Dr. Darshil U. Shah	University of Cambridge	UK

IOP Conf. Series: Materials Science and Engineering **773** (2020) 011003 doi:10.1088/1757-899X/773/1/011003

Peer review statement

All papers published in this volume of *IOP Conference Series: Materials Science and Engineering* have been peer reviewed through processes administered by the proceedings Editors. Reviews were conducted by expert referees to the professional and scientific standards expected of a proceedings journal published by IOP Publishing.

IOP Conf. Series: Materials Science and Engineering **773** (2020) 012001 doi:10.1088/1757-899X/773/1/012001

Facile preparation of cellulose nanofibers prepared by TEMPO-mediated oxidation

S Tanpichai[1, 2]

[1] Learning Institute, King Mongkut's University of Technology Thonburi, Bangkok, 10140, Thailand.
[2] Cellulose and Bio-based Nanomaterials Research Group, King Mongkut's University of Technology Thonburi, Bangkok, 10140, Thailand.

Email: supachok.tan@kmutt.ac.th

Abstract. Cellulose nanofibers (CNFs) with width of 20 nm and lengths of up to several μm were fully disintegrated from water hyacinth (*Eichhornia crassipes*) with aids of the 2, 2, 6, 6-tetramethylpiperidine-1-oxyl (TEMPO)-mediated oxidation and mild-mechanical treatment. TEMPO-oxidized CNFs were characterized by scanning electron microscopy (SEM), Fourier-transform infrared (FTIR) spectroscopy and thermogravimetric analysis (TGA). FTIR reveals the conversion of C-6 hydroxyl groups to sodium carboxylate groups, and lower thermal degradation was obtained from the TEMPO-oxidized CNFs in comparison to untreated cellulose fibers. The as-prepared TEMPO-oxidized CNFs might be possibly used in packaging and composite applications.

1. Introduction

Due to outstanding features such as low density, low thermal expansion, superior mechanical properties, biodegradability, biocompatibility and a large surface-to-volume ratio, cellulose nanofibers (CNFs) have gained much attention in applications such as nanocomposites, biomaterials, paper packaging, electronic devices and filtration membranes [1-4]. Many attempts have been made to extract CNFs from low-value agricultural wastes, for instance, banana [5], bamboo [6, 7], oil palm [8], pineapple leaf [9], wheat straw [7] and residues from industries such as tobacco [10], carrot [11] and barley [11] in order to use these materials in a greater beneficial way than landfills or animal feedstocks. In a tropical country like Thailand, water hyacinth (*Eichhornia crassipes*) grows rapidly on water surface in open areas, causing environmental problems for fish populations, irrigation and water transportation [12, 13]. It has been reported that water hyacinth contains high cellulose and low lignin contents [12-14].

Although CNFs can be disintegrated using mechanical treatment such as high-pressure homogenization [15], sonication [16, 17], high-speed blending [18, 19] and grinding [2], these processes require high demand of energy [20]. It has been reported that the high energy consumption of 700 – 1400 MJ is required to prepare 1 kg of CNFs [21], in comparison to that of ~40 MJ kg^{-1} for flex fiber production [22]. Therefore, chemical or enzymatic treatments have been used to reduce the time and energy consumption [23]. For example, the reduction in energy consumption to less than 7 MJ kg^{-1} has been reported with an aid of 2, 2, 6, 6-tetramethylpiperidine-1-oxyl (TEMPO)-mediated oxidation to prepare nanofibers [21]. The TEMPO mediated oxidation has been investigated to chemically modify primary hydroxyl groups at the C-6 position of D-glucose units in cellulose chains into carboxylic groups [24, 25]. With assistance of the mild mechanical disintegration, TEMPO-oxidized CNFs could be finally prepared.

Here, water hyacinth (*Eichhornia crassipes*) was used as a raw material to prepare CNFs with the TEMPO-mediated oxidation process with a high content of sodium hypochlorite (NaOCl) to skip chemical pre-treatment steps. Then, the mild mechanical treatment was introduced to disintegrate CNFs. Morphology, chemical groups and thermal properties of the CNFs were investigated.

2. Materials and methods

2.1. Materials

Water hyacinth fibers (*Eichhornia crassipes*) were purchased from a small-medium enterprise called Ban Pak Top Chawa, Thailand. 2,2,6,6-tetramethylpiperidine (TEMPO) was supplied from Sigma Aldrich while sodium bromide (NaBr), sodium hypochlorite (NaClO), sodium hydroxide (NaOH) and ethanol were provided by Polysci Service Co., Ltd. All chemicals and solvents were used as received without purification.

2.2. Preparation of TEMPO-oxidized cellulose nanofibers

Briefly, 5 g of the dried water hyacinth fibers were dispersed in 500 ml of distilled water, and TEMPO (0.08 g), NaBr (0.50 g) and NaOCl (30 mmol g^{-1}) were wisely dropped in the fiber suspension. The suspension was continuously stirred at 25 ± 3 °C for 180 min. Then, 20 ml of ethanol was added to stop the reaction, and the TEMPO-oxidized fibers were washed for several times until neutral. Subsequently, the TEMPO-oxidized fibers were disintegrated using s mixer grinder (MX-AC400, PANASONIC, Thailand) for 15 min. The as-prepared suspension of the TEMPO-oxidized cellulose nanofibers (CNFs) was stored in a refrigerator at 5 °C before use.

2.3. Characterizations of TEMPO-oxidized cellulose nanofibers

Morphology of the TEMPO-oxidized CNFs was investigated using a scanning electron microscope (JSM-6610 LV, JEOL Ltd., Japan) with an acceleration voltage of 20 kV. Before each observation, a sample was coated with thin layers of gold using a sputter coater (108 auto, Cressington Scientific Instruments Ltd., UK) for 100 s.

Chemical functional groups of the original water hyacinth fibers and TEMPO-oxidized CNFs were investigated using a Fourier-transform infrared (FTIR) spectrometer (Nicolet iS5, Thermo Fisher Scientific Inc., USA) equipped with an attenuated total reflectance (ATR) mode. The measurements were recorded between 600 – 4000 cm^{-1} with a resolution of 4 cm^{-1} and an overall scan of 32.

Thermogravimetric analysis (TGA) was performed using a thermal analyzer (TG-209 F3, NETZSCH-Gerätebau GmbH., Germany). ~10 mg of the untreated fibers and TEMPO-oxidized CNFs were measured with a heating rate of 10 °C min^{-1} under nitrogen atmosphere, and the measurements were recorded until 800 °C.

3. Results and discussion

Figure 1(a) and (b) present the fiber suspension before and after the TEMPO-mediated oxidation process. Obviously, the color of fibers was changed from brownish to white after the oxidation treatment. This change resulted from the high concentration of NaOCl (30 mmol g^{-1}) as an oxidant used in this work to eliminate lignin, pectin and hemicellulose. Therefore, the use of the TEMPO-mediated oxidation process with a high amount of NaOCl could skip the pre-treatment steps to purify fibers such as alkaline or bleaching treatment [26, 27]. The scanning electron microscopy (SEM) image reveals the full disintegration of CNFs after the TEMPO-mediated oxidation process and high speed blending for 15 min. The TEMPO-oxidized CNFs with widths of 20 nm and lengths of more than 1 μm were observed. The similar morphology of the TEMPO-oxidized CNFs has been previously prepared from bamboo [20] and biomass residues such as soybean hull, wheat straw and softwood flour [28].

FTIR spectra of the raw and TEMPO-mediated oxidized fibers are shown in Figure 2(a). The peaks located at ~3400 and 2800, corresponding to O-H stretching of the hydroxyl group and C-H stretching of CH_2 in the cellulose chains, were observed from both untreated and treated fibers [20]. After the treatment of the TEMPO-mediated oxidation process, the higher intensity of these peaks was observed. This might be due to the removal of lignin and hemicellulose from the fibers. This has been well-agreed with a previous investigation where with increasing NaClO concentrations, higher contents of the hemicellulose and lignin were removed [20]. Also, the higher intensity of the peak located at 1600 cm^{-1} corresponding to COONa stretching vibration, was observed from the spectrum of the chemical-treated fibers, confirming the successful conversion of C6 hydroxyl groups to sodium carboxylate groups induced by the TEMPO-mediated oxidation process [20, 28]. Also, the disappearance of the peak located at 1730 cm^{-1}, attributing to C-O stretching of the aryl group in lignin, was observed [28].

Figure 1. Appearance of the suspension of the water hyacinth fibers (a) before and (b) after the TEMPO-mediated oxidation process, and (c) scanning electron microscopy (SEM) image of the TEMPO-oxidized Cellulose nanofibers (CNFs) extracted from water hyacinth fibers.

Thermogravimetric (TG) curves of the untreated water hyacinth fibers and TEMPO-oxidized CNFs are presented in Figure 2(b).The onset degradation temperature of the TEMPO-oxidized CNFs with sodium carboxylate groups was 237.1 °C while that of the untreated fibers was 285 °C. The significant decrease in the onset degradation temperature of the TEMPO-oxidized CNFs was caused by the degradation of the sodium carboxylate groups generated during the TEMPO-oxidation process [24]. The higher residue content at 800 °C was obtained from the TEMPO-oxidized CNFs (25.4%) in comparison with that of the untreated fibers (22.4 %). This higher ash content was due to sodium carboxylate groups. It has also been reported that the addition of sodium on the cellulose surface may increase the ash content [29].

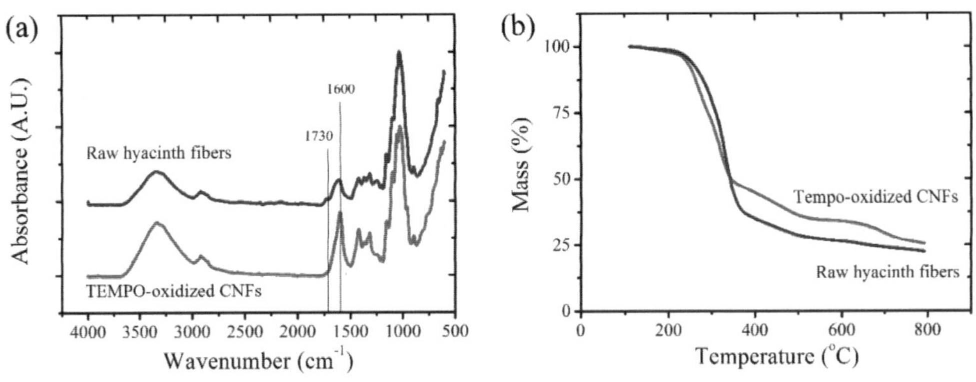

Figure 2. (a) Fourier-transform infrared (FTIR) spectra and (b) thermogravimetric (TG) curves of raw water hyacinth fibers and TEMPO-oxidized Cellulose nanofibers (CNFs).

4. Conclusions
TEMPO-oxidized CNFs with widths of 20 nm and lengths of up to several μm were successfully prepared from water hyacinth fibers with the assistance of the mild mechanical disintegration. The TEMPO-mediated oxidation process used in this work was adequate to purify fibers without any introduction of the pre-treatment steps. With the conversion of C6 hydroxyl groups to sodium carboxylate groups, the significant decrease in the thermal degradation temperature of the TEMPO-oxidized CNFs was observed.

5. Acknowledgements
The author would like to thank Learning Institute, King Mongkut's University of Technology Thonburi for financial support.

6. References
[1] Tanpichai S, Quero F, Nogi M, Yano H, Young R J, Lindstrom T, Sampson W W and Eichhorn S J 2012 *Biomacromolecules* **13** 1340
[2] Abe K, Iwamoto S and Yano H 2007 *Biomacromolecules* **8** 3276
[3] Varanasi S, Henzel L, Sharman S, Batchelor W and Garnier G 2018 *Carbohydr. Polym.* **184** 307
[4] Tanpichai S, Witayakran S and Boonmahitthisud A 2019 *J. Environ. Chem. Eng.* **7** 102836
[5] Khawas P and Deka S C 2016 *Carbohydr. Polym.* **137** 608
[6] Chang F, Lee S-H, Toba K, Nagatani A and Endo T 2012 *Wood. Sci. Technol.* **46** 393
[7] Chen W S, Yu H P, Liu Y X, Hai Y F, Zhang M X and Chen P 2011 *Cellulose* **18** 433
[8] Fahma F, Iwamoto S, Hori N, Iwata T and Takemura A 2010 *Cellulose* **17** 977
[9] Cherian B M, Leao A L, de Souza S F, Costa L M M, de Olyveira G M, Kottaisamy M, Nagarajan E R and Thomas S 2011 *Carbohydr. Polym.* **86** 1790
[10] Tuzzin G, Godinho M, Dettmer A and Zattera A J 2016 *Carbohydr. Polym.* **148** 69
[11] Berglund L, Noël M, Aitomäki Y, Öman T and Oksman K 2016 *Ind. Crops. Prod.* **92** 84
[12] Abral H, Kadriadi D, Rodianus A, Mastariyanto P, Ilhamdi, Arief S, Sapuan S M and Ishak M R 2014 *Mater. Design.* **58** 125
[13] Sundari M and Ramesh A 2012 *Carbohydr. Polym.* **87** 1701
[14] Tanpichai S, Biswas S K, Witayakran S and Yano H 2019 *ACS. Sustain. Chem. Eng.* **7** 18884
[15] Wang Y H, Wei X Y, Li J H, Wang F, Wang Q H, Chen J C and Kong L X 2015 *Fiber. Polym.* **16** 572
[16] Tanpichai S, Sampson W W and Eichhorn S J 2012 *Compos. Pt. A-Appl. Sci. Manuf.* **43** 1145
[17] Tanpichai S, Sampson W W and Eichhorn J S 2014 *Compos. Pt. A-Appl. Sci. Manuf.* **65** 186
[18] Uetani K and Yano H 2011 *Biomacromolecules* **12** 348
[19] Tanpichai S and Wootthikanokkhan J 2018 *Sci. Eng. Compos. Mater.* **25** 395
[20] Chitbanyong K, Pitiphatharaworachot S, Pisutpiched S, Khantayanuwong S and Puangsin B 2018 *Bioresources* **13** 4440
[21] Isogai A, Saito T and Fukuzumi H 2011 *Nanoscale* **71** 71
[22] Dissanayake N P J, Summerscales J, Grove S M and Singh M M 2009 *J. Nat. Fibers.* **6** 331
[23] Fujisawa S, Okita Y, Fukuzumi H, Saito T and Isogai A 2011 *Carbohydr. Polym.* **84** 579
[24] Fukuzumi H, Saito T, Okita Y and Isogai A 2010 *Polym. Degrad. Stab.* **95** 1502
[25] Tang Z, Li W, Lin X, Xiao H, Miao Q, Huang L, Chen L and Wu H 2017 *Polymers* **9** 421
[26] Yuwawech K, Wootthikanokkhan J and Tanpichai S 2018 *Mat. Renew. Sustain. Energy.* **7** 21
[27] Tanpichai S, Witayakran S, Srimarut Y, Woraprayote W and Malila Y 2019 *J. Mater. Res. Technol.* **8** 3612
[28] Sinclair A, Jiang L, Bajwa D, Bajwa S, Tangpong S and Wang X 2018 *J. Appl. Polym. Sci.* **135** 46304
[29] Faradilla R H F, Lee G, Arns J-Y, Roberts J, Martens P, Stenzel M H and Arcot J 2017 *Carbohydr. Polym.* **174** 1156

Surface plasmon resonance sensor based on core-shell metal nanorods

Y-F C Chau

Centre for Advanced Material and Energy Sciences, Universiti Brunei Darussalam, Tungku Link, Gadong BE1410, Negara Brunei Darussalam.

Email: chou.fong@und.edu.bn

Abstract. A surface plasmon resonance (SPR) sensor based on core-shell metal nanorod arrays for ultra-sensitive refractive index sensing applications is proposed using three-dimensional finite element method. Simulation results show that the absorptance (A) of the proposed structure can be reached A=90.0% which is about 3.5 times enhancement compared to its solid counterpart (A=25.57%). Result shows that the sensitivity can be obtained as high as 700.00 nm/RIU (RIU is the refractive index unit). The proposed SPR sensor could be a desirable candidate for applications in nanophotonic devices.

1. Introduction

Surface plasmon resonance (SPR) sensor have been attracted great attention for their fascinating properties due to their behaviour of absorbing and enhancing electromagnetic (EM) waves in nanophotonic field [1-6]. Recently, many research groups have made their efforts to enhance the absorption in plasmonic nanostructures for refractive index (RI) sensing applications [7,8]. SPR sensors require being designed with simple and tunable in near infrared if they are to be successfully implemented in sensor applications. Many SPR RI sensors using core-shell nanostructures have been proposed [9,10]. However, the RI sensitivity of the SPR sensor based on core-shell nanorods connected with crossing veins has been investigated less before.

In this paper, a new type of SPR sensor based on silver-shell nanorod arrays connected with crossing veins, which provides a connected resonance bridge between silver-shell nanorods and veins compared to its counterpart without the crossing veins, has been proposed and investigated using three-dimensional (3-D) finite element method (FEM) and the numerical results of coupled silver nanorods without the crossing veins are included for comparison.

2. Simulation method and model

The numerical simulation of the proposed structure was performed using 3-D FEM (COMSOL Multiphysics [11]). The silver permittivity data is obtained from Ref. [12] and the absorptance (A) is calculated from 1-reflectance (R)-transmittance (T). The sensitivity (S) can be calculated as $S=\Delta\lambda/\Delta n$ [13-15], where $\Delta\lambda$ is the shift of resonant peak wavelength of absorptance and Δn is the RI difference. Figs. 1(a) and (b) depict the unit cell of proposed structures without and with crossing veins, respectively. The case with veins is composed of four silver nanorods connected by silver crossing veins at central part placed on the surface of a uniform silver film. The reason why the proposed sensors are based on core-shell structures is due to the capacitance and inductance effects can be simultaneously induced in the designed sensor. The structural parameters of the proposed SPR sensor are denoted as period (*P*), height (*h*), length (*w*), gap distance (*g*), vein thickness (*t*) and the thickness of bottom Ag film (*s*), respectively.

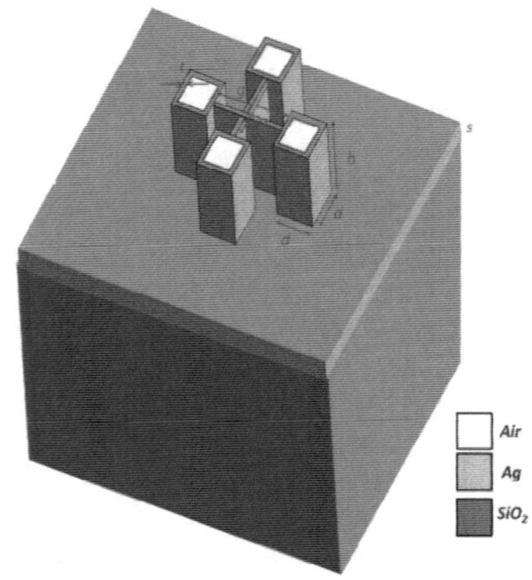

Figure 1. A truncated view of the proposed SPR sensor structure (a) without and (b) with crossing veins.

3. Results and discussion

Figure 2 shows the optical spectra of the designed structures without (dashed lines) and with (solid lines) the crossing veins. The values of *P*, *h*, *d*, *g*, *t* and *s* are fixed at 470nm, 100 nm, 60 nm, 80 nm, 10 nm and 100 nm, respectively. As shown in Figs. 2, the difference of peak/dip of absorptance/reflectance/transmittance spectra can be attributed to that of different SPR effect generated in the proposed structure. A remarkable peak/dip occurred at 780 nm for the case with crossing veins and a much weaker one at 1010 nm for the case without crossing veins. There is a huge difference between the two cases because of the vein effect has a dominant influence on plasmon resonance in the proposed structure. Note that the absorptance (A) of the case with veins can be reached A=90.00% which is about 3.5 times enhancement compared to its solid counterpart without veins (A=25.57%).

The electric field intensity ($|\boldsymbol{E}|$, V/m) distributions of the unit-cell structures at resonances for the cases without crossing veins and with veins are displayed in Figs. 3(a) and (b), respectively. It can be seen that the $|\boldsymbol{E}|$ profile of the case with veins is much higher than that of the case without veins, which is strongly confined in the gap vicinity of Ag nanorods and the surface of veins. In addition, an enhanced distribution of $|\boldsymbol{E}|$ can be extended around their outer sides. The gap and edge enhancement profiles in the case with crossing veins indicate that the hybrid surface, cavity and gap modes were stimulated by the incident EM waves coupled with the proposed structure.

To investigate the difference of the proposed structures without and with crossing veins, the surrounding RI of n=1.0, 1.20 and 1.40 are examined as shown in Figs. 4(a) and Figs. 4(b), respectively. There is a remarkable redshift of each mode when the RI is varied from 1.0 to 1.4. The calculated sensitivity can be achieved up to 700.00 nm/RIU. The magnitude of absorptance of the cases without veins is much less than that of the cases with veins because of the cases without veins possesses the less gap and cavity modes among Ag nanorods [16-18]. The absorptance (A) in the cases with crossing veins will be beneficial for plasmonic sensing application. The physical inference for these redshifts of the cases with crossing veins is due to the effective increase in capacitance and inductance of the resonant structure [19] lead to the increase in the RI of the surrounding media.

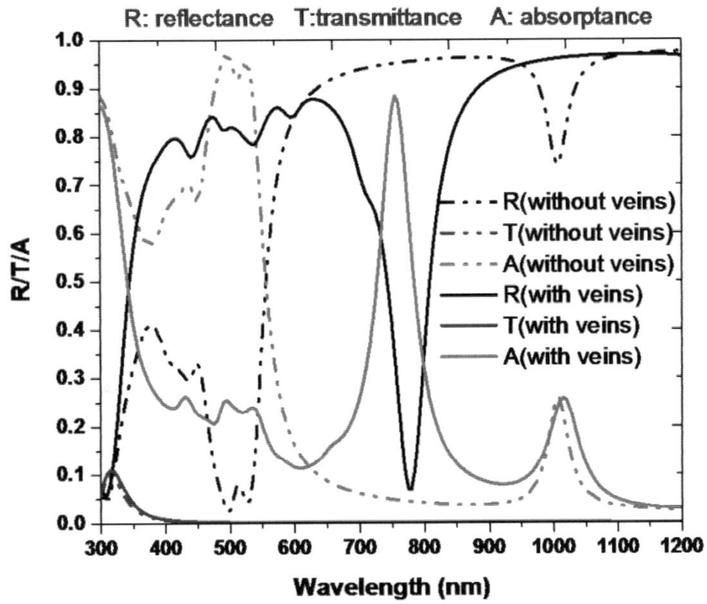

Figure 2. Reflectance (R), Transmittance (T) and Absorptance (A) spectra of the designed SPR sensors without and with the connected crossing veins.

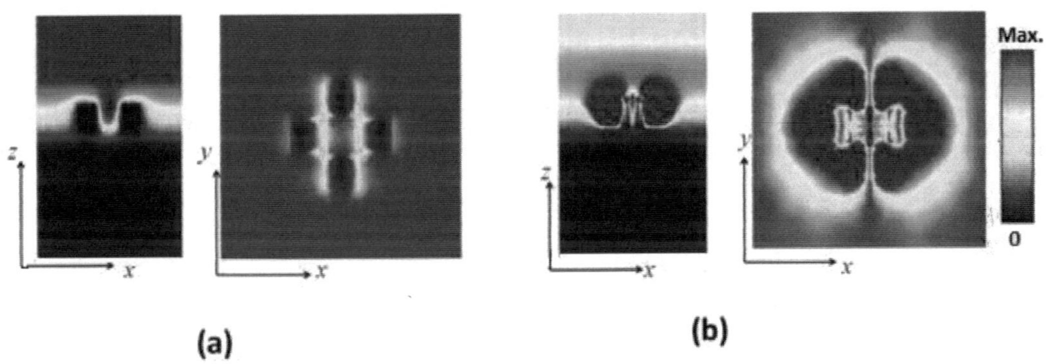

Figure 3. Electric field intensity (|$\boldsymbol{E}$|, V/m) for the case (a) without crossing veins (at λ_{res}=1010 nm) and (b) with crossing veins (at λ_{res}=780 nm), respectively.

4. Conclusions

In this paper, a new type of SPR sensor based on core-shell silver nanorod arrays with the crossing veins and compared the optical performance with its counterpart without the crossing veins is proposed and investigated using 3-D FEM. Results show that a huge difference of absorptance between the case with and without crossing veins can be found because of the vein effect has a dominant influence on plasmon resonance in the proposed structure. The absorptance of the case with crossing veins is nearly perfect which is much higher than the case without veins. The calculated sensitivity can be achieved up to 700.00 nm/RIU. The proposed structure could be a promising candidate for applications in the SPR sensors and the nanophotonic devices.

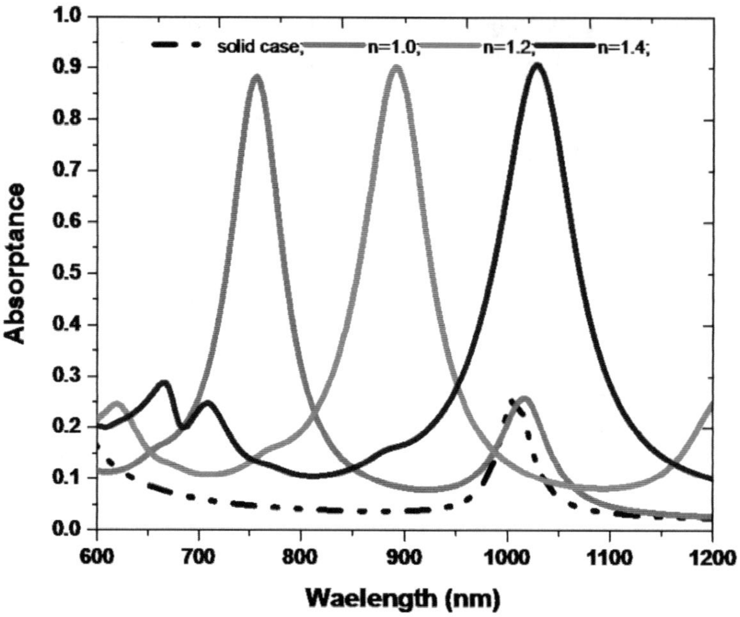

Figure 4. Absorptance spectrum of the case with veins by varying the surrounding RI of n=1.0, 1.2 and 1.4, respectively. Where black dashed line denotes the case without veins.

5. References

[1] Lin CT Chang MN et al. 2016 *Electrochimica Acta* **192** 15.
[2] Chau YF, Yeh HH and Tsai DP 2010 *J. Electromagnet. Wave* **24** 1621.
[3] Gao H, Zhou D, Cui W, Liu Z, Liu Y, Jing Z and Peng W 2019 *J. Opt. Soc. Am. A.* **36** 264.
[4] Liu Z, Liu G, Fu G, Liu X and Wang Y 2016 *Opt. Express* **24** 5020.
[5] Liu Z, Tang P, Liu X, Yi Z, Liu G, Wang Y and Liu M 2019 *Nanotechnology* **30** 305203.
[6] Chau YF et al. 2013 *J. Nanopart. Res.* **15** 1424.
[7] Wu D, Liu Y, Li R, Chen L, Ma R, Liu C and Han Ye HI 2016 *Nanoscale Res. Lett.* **11** 483.
[8] Yong Z, Zhang S, Gong C and He S 2016 *Sci. Rep.* **6** 24063.
[9] Luo S, Zhao J, Zuo D and Wang X 2016 *Opt. Express* **24** 9288.
[10] Chen Y, Li X, Luo X, Maier SA and Hong M. 2015 *Photon. Res.* **3** 54.
[11] COMSOL Multiphysics Reference Manual. http://www.comsol.com/.
[12] Chau YF, Jiang ZH, Li HY, Lin GM, Wu FL, Lin WH 2011 *Prog. Electromagn. Res. B* **28** 183.
[13] Chau YF, Yang TJ and Lee WD 2004 *Appl. Opt.* **43** 6656.
[14] Kumara N et al. 2016 *J. Opt.* **18** 115003
[15] Lin JM and Chau YF 1995 *J. Guid. Control Dyn.* **18** 637.
[16] Chou YF, Yeh HH and Tsai DP 2009 *Phy. Plasmas* **16** 022303.
[17]Chen MW, Chau YF and Tsai DP 2008 *Plasmonics* **3** 157.
[18] Shen L, Yang TJ and Chau YF *Appl. Phys. Lett.* **90** 251909.
[19] Huang CP, Yin XG, Huang H and Zhu YY 2009 *Opt. Express* **17** 6407.

Acknowledgement
This work was supported by Universiti Brunei Darussalam (UBD/RSH/1.9/FICBF(b)/2019/006).

Growth of graphene on stainless steel by chemical vapor deposition using soybean oil as a carbon source

A Ruammaitree

Department of physics, faculty of science and technology, Thammasat university, Thailand, 12120.

Email: u4605070@hotmail.com

Abstract. Graphene is a 2-dimension material which consists of carbon atoms arranged in a flat honeycomb lattice. Graphene has many outstanding properties such as high electron mobility, superb strength, and great flexibility. Although graphene film can be fabricated by economical method such as chemical vapor deposition, the utilization of flammable gas (such as methane, acetylene) in the graphene growth process is still dangerous therefore the safe flammable gas storage and gas line installation are needed. Soybean oil is available vegetable oil which contains carbon, hydrogen and oxygen. Utilization of soybean oil as a carbon source instead of hydrocarbon gas for graphene growth by chemical vapor deposition can avoid an explosion of flammable gas and save the installation cost of gas system. However, the presence of oxygen in soybean oil may obstruct the growth of graphene by chemical vapor deposition. Fortunately, annealing stainless steel with carbon at high temperature can eliminate oxygen from the surface. In this study, the growth of graphene on stainless steel by chemical vapor deposition using soybean oil as a carbon source has been study. The presence and quality of graphene are investigated by Raman spectroscopy. The morphology and element analysis are measured using scanning electron microscope and energy-dispersive X-ray spectroscopy, respectively.

1. Introduction

Graphene is 2-dimensional material which consists of carbon atoms arranged in flat honeycomb lattice. Graphene has considerably attracted researcher's attentions due to its exotic properties such as, high flexibility [1-2], large specific surface area [3-4], high electron mobility [5-6] and great mechanical properties [7]. Moreover, the coating graphene film on metal can increase corrosion resistance [8-9], electrical conductivity [8] and surface hardness [10] of the metal. Graphene film can be fabricated by many methods such as annealing SiC [11-15], chemical vapor deposition (CVD) [16-18] and so on. CVD is a simple and economical method for graphene growth on large-scale production. Generally, hydrocarbon gas (such as methane or acetylene) is preferred to use as a carbon source in the CVD process for the growth of graphene film on metal. However, methane and acetylene gases are flammable gases therefore the safe gas storage and gas line installation are needed. Soybean oil is carbon-rich liquid which is available and cheap. Thus, the utilization of soybean oil as a carbon source instead of hydrocarbon gas for graphene growth by CVD can avoid an explosion of flammable gas and save the installation cost of gas piping system. In this report, the growth of graphene on stainless steel by CVD using soybean oil as a carbon source is studied. The effect of soybean oil quantity on the growth of graphene film on stainless steel is emphasized.

2. Experimental

A s304 stainless-steel sheet with a size of 30×40×0.85 mm^3 was precleaned by ultrasonic sonication in acetone for 5 minutes and then put into a quartz tube. Thereafter, the quartz tube was first evacuated until the base pressure in the quartz tube reached ~3 Pa and then the stainless-steel plate was heated to the graphene growth temperature (900°C) by furnace. At the growth temperature, the soybean oil with the quantities of 1-25 ml was heated at 400°C to evaporate soybean oil. The vapor of the soybean oil was introduced into the quartz tube for 30 minutes. After that the sample was cooled down rapidly to room temperature.

The formation of graphene film on stainless-steel plates was confirmed by Raman spectroscopy measured at room temperature with a 50x objective and a 532 nm laser. The laser beam size is ~1 μm in diameter. Exposure time and accumulations are set at 10 seconds and 10 accumulations. Surface morphology and element analysis were investigated by field-emission scanning electron microscope (FE-SEM) equipped with an energy dispersive X-ray spectroscope (EDX) at incident beam of 15 keV.

3. Results and discussion

Figure 1 (a)-(f) show the morphology of the samples annealed at 900 °C for 30 minutes under the vapor of soybean oil quantities of 1 ml, 5 ml, 10 ml, 15 ml, 20 ml, and 25 ml, respectively. In case of 1 ml, 5 ml and 10 ml samples (Figure 1 (a)-(c)), the surface roughness increases with increase in the quantity of soybean oil. For 15 ml sample (Figure 1 (d)), many islands appear on the sample surface and they disappear when the samples annealed under the vapor of soybean oil quantities of 20 ml and 25 ml as shown in Figure 1 (e) and (f), respectively. Figure 1 (g) displays Raman spectra of the samples annealed at 900 °C for 30 minutes under vapor of various quantity of soybean oil. In general, graphene has 2 main specific peaks in the Raman spectra at ~1580 cm^{-1} (G band) and ~2700 cm^{-1} (2D band). G band and 2D band are originated from E_{2g} vibrational mode and second-order two-phonon mode, respectively [19]. However, The D band at ~1350 cm^{-1} is the third specific peak of Raman spectra of graphene. The D band will appear where the symmetry of graphene structure is broken by edge or deflect [20]. In the case of 1 ml, 5 ml and 10 ml samples, there is no significant peak appears in the Raman spectra while the Raman spectrum of 15 ml sample displays D band and G band at 1359 cm^{-1}, 1592 cm^{-1}, respectively. The absence of 2D band reveals that there is no graphene on the surface of the 15 ml sample. The D and G bands of the 15 ml sample may be originated from amorphous carbon on its surface. For 20 ml and 25 ml samples, the G and 2D bands obviously appear in the Raman spectra at ~1572 cm^{-1} and ~2700 cm^{-1}, respectively implying that there is graphene film on the surfaces. Figure 1 (h) shows the relations between soybean oil quantity and weight% of carbon and oxygen on the surface of the samples. This graph displays the weight% of oxygen reduces with increase in soybean oil quantity. Conversely, the weight% of carbon increases with increase in the soybean oil quantity. Generally, oxygen is an obstacle to graphene growth while carbon is a basic element of graphene structure therefore the low-oxygen and high-carbon environments are excellent condition for graphene growth on stainless-steel substrates. It is in a good agreement with Raman spectra in Figure 1 (g) which shows that the graphene can be grown on the samples annealed under soybean oil quantities of 20 ml and 25 ml.

Carbothermal reduction is the process for oxide reduction by annealing metal oxide at high temperature with reducing-agent carbon. The reducing-agent carbon may be gaseous CO and CO_2 in accordance with the following reactions [21]:

$$MO(s) + CO \rightarrow M(s) + CO_2$$

where M is metal. Stainless steel is iron-rich alloy, Liu et al. show a gas-phase equilibrium diagram of iron oxide which displays the temperature more than 600 K with large amount of gaseous carbon can reduce the iron oxide [22]. The reaction of reduction of iron oxide is displayed step by step as follows.

$$3Fe_2O_3+CO=2Fe_3O_4+CO_2$$
$$Fe_3O_4+CO=3FeO+CO_2$$
$$FeO+CO=Fe+CO_2$$

In this experiment, the vapour of soybean oil is used as metal vapor with carbon for reduction of metal oxide on the stainless-steel surface at 900 °C. The results show that the oxides can be eliminated from the surface at high temperature if the quantity of soybean oil is much enough therefore annealing stainless steel at high temperature with enough soybean oil can grow graphene on its surface.

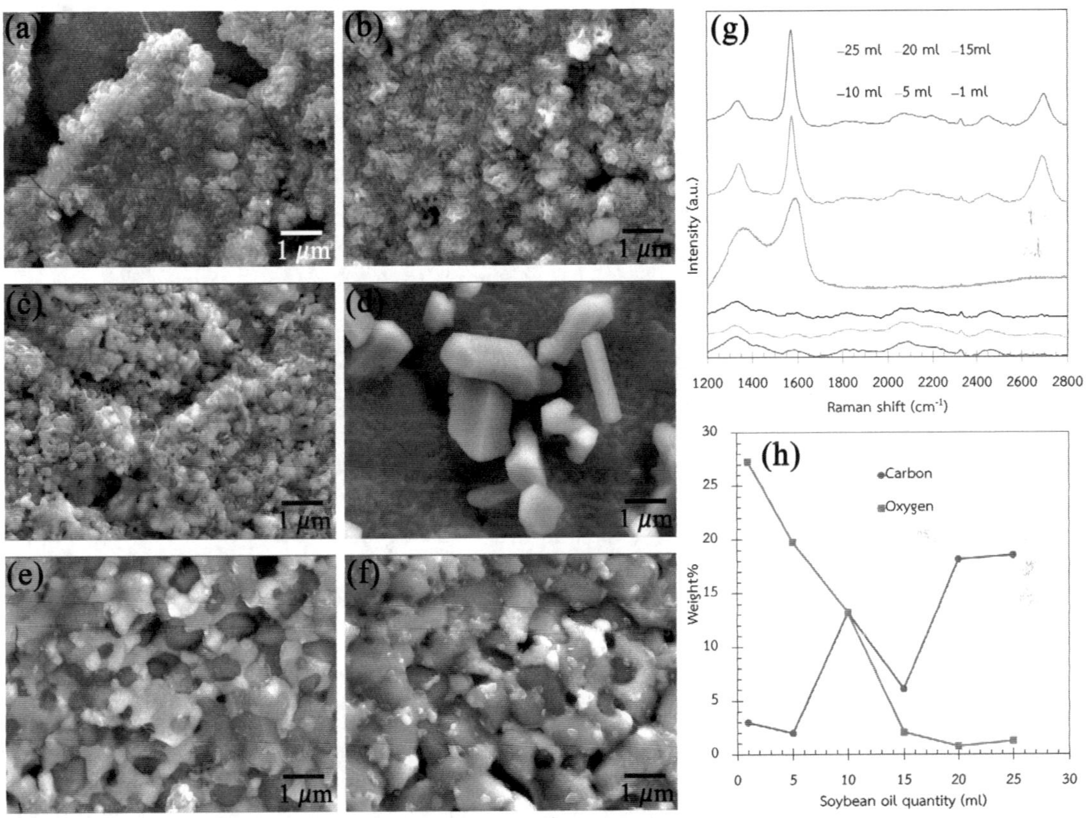

Figure 1. FE-SEM images measured on the surface of the samples annealed at 900 °C for 30 minutes using soybean oil quantities of (a) 1 ml, (b) 5 ml, (c) 10 ml, (d) 15 ml, (e) 20 ml and (f) 25 ml as a carbon source. (g) Raman spectra of samples annealed at 900°C for 30 minutes under the vapor of soybean oil quantities of 1-25 ml. (h) The relations between soybean oil quantity and weight% of carbon (blue) and oxygen (red).

4. Conclusions

In this report, the growth of graphene on stainless steel by chemical vapor deposition using soybean oil as a carbon source was demonstrated. The annealing temperature and growth time were set at 900°C and 30 minutes, respectively. In addition, the quantity of soybean oil was varied from 1 ml to 25 ml. We found that annealing stainless steel at high temperature (900 °C) with enough amount of carbon can eliminate oxygen which is an obstacle to graphene growth on stainless-steel surface. In this research, we can grow graphene film on stainless steel by CVD using soybean oil quantities of 20 ml to 25 ml as a carbon source.

5. References

[1] Sahoo M, Wang J-C, Nishina Y, Liu Z, Bow J-S and Lai C-S 2020 *Applied Surface Science* **499** 143839.
[2] https://www.foxnews.com/tech/5-ways-graphene-material-will-change-gadgets-forever: April 2014.
[3] Li X, Chen Y, Cheng Z, Jia L, Mo S and Liu Z 2014 *Applied Energy* **130** 824.
[4] Zhou L, Yang Z, Yang J, Wu Y, Wei D 2017 *Chemical Physics Letters* **677** 7.
[5] Hass J, Heer W-A and Conrad E-H 2008 *J. Phys: Condens. Matter* **20** 323202.
[6] Yadav V-K, Mir S-H and Singh J-K 2019 *Physica B: Condensed Matter* **571** 291.
[7] Lee C, Wei X, Kysar J-W and Hone J 2008 *Science* **321** 385.
[8] Ruammaitree A, Phokharatkul D, Nuntawong N and Wisitsoraat A 2018 *Surface Review and Letters* 1840003 (6 pages).
[9] Xue H, Wand T, Guo H, Fan X, Zhu Z, Pana X and He J 2014 *RSC Adv.* **4** 57724.
[10] Ruammaitree A, Phokharatkul D and Wisitsoraat A 2018 *Solid State Phenomena* **283** 173.
[11] Ruammaitree A, Nakahara H and Saito Y 2014 *Surf. Interface Anal.* **46** 1156.
[12] Sun C, Cai W, Hong R, Wu J, Chen X, Cai J, Zhang F, Wu Z 2020 *Journal of Physics and Chemistry of Solids* **137** 109224.
[13] Ruammaitree A, Nakahara H, Akimoto K, Soda K and Saito Y 2013 *Applied Surface Science* **282** 297.
[14] Ruammaitree A, Nakahara H and Saito Y 2014 *Applied Surface Science* **307** 136.
[15] Hu H, Ruammaitree A, Nakahara H, Asaka K and Saito Y 2012 *Surf. Interface Anal.* **44** 793.
[16] Kokmat P, Donnuea N, Nuntawong N, Wisitsoraat A and Ruammaitree A 2019 *IOP Conf. Series: Materials Science and Engineering* **526** 012008.
[17] Lee S, Park W-K, Yoon Y, Baek B, Yoo J-S, Kwon S-B, Kim D-H, Hong Y-J, Kang B-K, Yoon D-H and Yang W-S 2019 *Materials Science & Engineering B* **242** 63.
[18] Zan R, Altuntepe A 2020 *Journal of Molecular Structure* **1199** 127026.
[19] Allen M-J, Tung V-C and Kaner R-B 2010 *Chem. Rev.* **110** 132.
[20] Machac P and Hrebicek T 2017 *J Mater Sci: Mater Electron* **28** 12425.
[21] L'vov B V 2000 *Thermochimica Acta* **360** 109.
[22] Liu B, Zhang Y, Lu M, Su Z, Li G and Jiang T 2019 *Minerals Engineering* **131** 286.

Acknowledgement

This work was supported by Research Fund of Thammasat University (contract number 5/2562) and Biodiversity-based economy development office (public organization) (contract number 8/2562). The authors thank Center of Scientific Equipment for Advanced Research, Thammasat University for SEM and EDX measurements.

IOP Conf. Series: Materials Science and Engineering **773** (2020) 012004 doi:10.1088/1757-899X/773/1/012004

The analysis of a chilled beef supply chain for developing strategic improvement

R J Neisyafitri and P Ongkunaruk*

Department of Agro-Industrial Technology, Faculty of Agro-Industry, Kasetsart University, 50 Ngam Wong Wan Rd. Lad Yao, Chatuchak, Bangkok 10900, Thailand

*Corresponding author: pornthipa.o@ku.ac.th

Abstract. The growth of population and economics are factors that affect the increasing demand for livestock products in Indonesia. Beef and chicken are the most consumed meat in Indonesia with an average meat consumption of 2.72 kg per capita per year and there is an increasing trend. Currently, the domestic production of beef is only about 45% percent of the demand, the rest of the demand will be fulfilled by importing live cattle and frozen meat, mainly from Australia. The beef supply chain is quite complex and long. The available infrastructures still need to be improved in order to deliver the beef quickly and handled under appropriate cold chain. Our objectives are to identify the business process of the abattoir in the chilled beef product cold chain, to analyze the problem, and to propose some recommendations for their strategic improvement. The observation and in-depth interviews are conducted. Then, the result is analyzed by using Integration Definition for Function Modelling (IDEF0) to figure out the current situation and the activities in the abattoir. Based on the result, it is found that inventory management and temperature controlling are the major issues in the chilled beef cold chain that need to be monitored and controlled. This suggestion would be able to improve the efficiency of a chilled beef cold chain in Indonesia.

1. Introduction

The growth of population and economics are the factors that affect the increasing demand for livestock products in Indonesia. The most consumed meat in Indonesia is beef and chicken with an average meat consumption of 2.72 kg per capita per year and there is an increasing trend [1]. Currently, the domestic production of beef is only about 45% of the demand, the rest of the demand will be fulfilled by importing live cattle and frozen meat, mainly from Australia [2]. The beef supply chain is quite complex and long. Currently, the high demand for fresh beef in Indonesia is hot carcass, the beef that is not stored in the chiller. Indonesian consumer considers the hot carcass as the fresh meat. The butchers were just put the carcass exposed to the air in the stall. This handling may lead to temperature abuse since the ideal temperature for storing fresh meat is range between 1°C to 3°C. The temperature abuse will affect the beef quality during distribution and may cause defects before the use-by date is reached, leading to food waste and economic losses. The increasing consumer awareness towards food safety in Indonesia also affects the demand for the high quality of beef. However, the available infrastructures still need to be improved in order to deliver the beef quickly and handled under the appropriate cold chain to maintain the beef quality. Other than that, there are restrictions due to the limitation of knowledge of the workers about the cold chain system [3]. Therefore, cold chain management is very important and becomes a challenge to maintain the freshness and the quality of the beef until it reaches the final consumer [4].

This study is focused on the implementation of the cold chain from the abattoir to the final consumer. The objectives of this research are to identify the business process and risks of the chilled beef supply chain in Indonesia. A recommendation can thus be developed to improve the efficiency of chilled beef product cold chain in Indonesia.

2. Methodology

The in-depth interview is conducted with the abattoir manager in Bogor, West Java, Indonesia by using a questionnaire to figure out the structures and the activities of a chilled beef cold chain. Then, the business process is analyzed by using Integration Definition for Function Modelling (IDEF0). An IDEF0 model contains information about how an organization running the business process activities [5]. IDEF0 is a modeling technique that combines graphics and texts into a systematic way to give the understanding, support the analysis, specify the requirements, or support system-level design and integration activities [6]. The model is constructed from two main things, the function box that represents activities, processes, and transformations also the arrows that represent data and objects related to the functions. There are four types of arrow including input, output, control, and mechanism [7]. The input is an arrow that entering the left side of the box represents the inputs that are transformed by the function for producing outputs. Second, the output is the arrow that leaving the box on the righthand side, it represents the outputs produced by the function. Third, control is the arrow that entering the box on the top represents controls that specify the required condition for a function to produce correct outputs. Last, a mechanism is the arrow that is connected to the bottom of the box, it represents the resource that is needed to perform the function [8]. Moreover, there is a solid line that represents the current activities and dashed line that shows the improvement that is suggested for improving the efficiency of chilled beef product cold chain in Indonesia.

3. Results and Discussion

3.1. Background of an abattoir

A case study company is a service provider specialized in cattle slaughter and beef production. The goal of the company is to produce a final product in the form of chilled beef from the high quality of cattle and meet the national standard of Indonesia. This company is one of the modern slaughterhouses in Indonesia that produces chilled beef, frozen beef, and portioning product based on the demand from the customer. Meanwhile, the side products such as meat offal, fat, tail, skin, and the bone will be distributed to some local business enterprises. The cattle that will be slaughtered are imported Australian cattle and have been fattened in Lampung Province for 100 days. There are about 150-180 cattle slaughtered a week on average. Then, the product will be shipped to Java and Bali Island. The company serves Indonesian markets including high-end restaurants, hotels, and supermarkets across the country since 2005. Previously, the company is owned under the supervision of Australian Company, but since August 2018 the Indonesian government proposed acquisition for this company. In order to distribute the products, the company has a refrigerated container for serving the local customer and rent the refrigerated truck from the third-party logistics provider in order to support its distribution.

3.2. A generic Indonesian chilled beef cold chain

The business process in the organizational level involved in the chilled beef product cold chain in Indonesia from upstream to downstream is analyzed by using IDEF0 level 0. In Figure 1, it is shown that the feedlot imported cattle from Australia according to the trader order. The feedlot is in Lampung Province, the cattle will be fattened within 100 days. When the cattle are ready to be slaughtered, they are transported by a truck with a capacity of 16 cattle per shipment (8 tons). The shipment takes 24 hours until it reaches the abattoir in Bogor, West Java. In the abattoir, the cattle are kept at the crib for a night. In the morning before the slaughtering, the cattle are being cleaned and inspected by a veterinarian (antemortem). After that, the hot carcass will be kept at the chiller for about 24 h. Then, the chilled carcass temperature will be checked on the next day. If it is under 20°C, the carcass is ready to be boned.

After the boning process, the carcass will be packed and stored in the chiller. The trader plays a role in getting the order from a customer, integrate it in their own system, and send it to the feedlot and abattoir as the customer demand. From the abattoir, the product is transported to the warehouse and kept until there is an order. The logistics service provider will pick up the packed chilled beef in an abattoir or in the warehouse and deliver to the retailers such as supermarkets, hotels, restaurants and caterers and end customers. The retailers are in Jakarta, West Java, Central Java, Yogyakarta, and Bali.

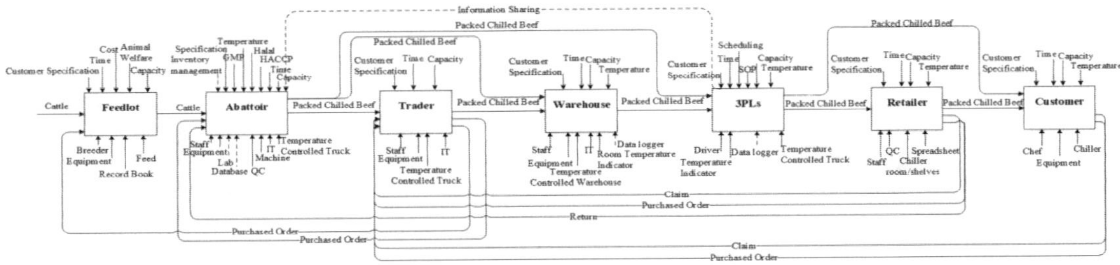

Figure 1. The business process of chilled beef product in Indonesia (IDEF0 Level 0)

3.3. *Business process analysis of an abattoir*

The business process analysis of an abattoir is visualized in Figure 2. It shows the activities including plan, source, make, deliver, and return. From the IDEF0 Level 1, the activities in the abattoir are analyzed in detail in order to identify the problem by doing the observation and in-depth interview within 1 month. The abattoir will get an order of a number of imported cattle a month in advance after an external monthly meeting with a trader. From the meeting, the slaughtering schedule is established. After the production process, the product will be stored in the warehouse. Some products are directly sent to the warehouse in Bekasi, West Java.

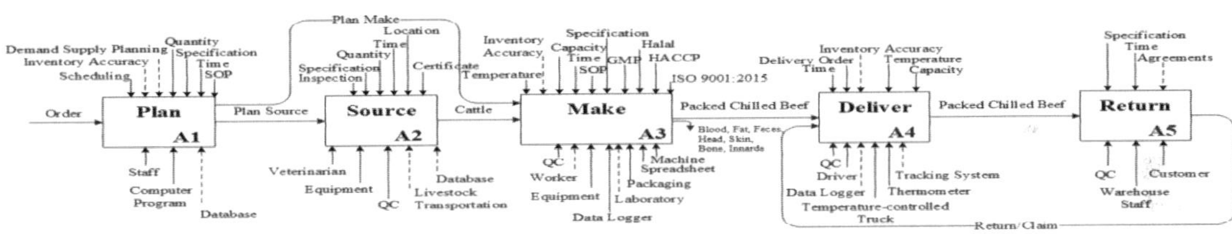

Figure 2. The business process of an abattoir (IDEF0 Level 1)

Currently, the trader will calculate customer demand and order the number of cattle to be slaughtered. Sometimes, the trader cancels the order suddenly and resulting in rescheduling the slaughtering. Then, abattoir loses the chance to serve another trader. Hence, the abattoir should establish a due date to cancel the order to protect the reschedule and implement the demand planning with the trader. In the sourcing process, inventory accuracy is quite low due to human error. It results in mismatch order and lost sales. Thus, proper inventory management should be established to reduce human error. There is no recorded information about the cattle shipment, thus they should establish the database and real-time sharing information. In make process, the low responsibility of the workforce still becomes an issue because their absenteeism affects productivity. We suggest them to do the worker performance evaluation and implement reward and punishment. Moreover, the abattoir did not have any laboratory to examine the microbial content in the meat. Due to the product is sensitive towards the contamination and the external inspection took 2 weeks, we suggest they invest in the laboratory equipment in order to obtain the result quickly and able to respond with suitable corrective action as soon as possible. In the delivery activities, there is not sufficient temperature-controlled truck to deliver products. Hence, the company outsources a third-party logistics provider who establishes a real-time temperature record during delivery. However, there is no staff who is in charge to monitor the temperature data. Only when there is an accident or rejected product during transportation, the abattoir will check the temperature data. Hence, preventive control should be implemented by real-time monitoring of the temperature. Finally, there are a lot of returned products from customers over the time frame specified by the abattoir, it is because there is no rule that controls the requirement before returning the product. Then, we suggest them to establish new agreements. The current problems and recommendations for improvement are summarized in Table 1.

Table 1. Summary of current problems and recommendations for improvement

Activities	Problems	Improvement Guideline
Plan	Order cancellation	Establish a due date for cancellation, demand planning with the trader
	Low inventory accuracy	Implement an integrated inventory management
	Lack of production plan	Establish the production planning
Source	No database	Establish database and evaluate the supplier periodically
	No real-time sharing information of cattle shipment	Establish a new system in order to track the cattle condition during shipping
	No adequate livestock transportation	Evaluate the animal welfare standard of the livestock shipping provider
Make	Low productivity	Lean/Line Balancing Implement a reward and punishment from the management policy
	Lack of worker awareness towards food safety standard	Evaluate and train the worker periodically
	Lack of laboratory inspection during production	Invest in laboratory equipment
Deliver	Insufficient truck	Outsource to qualified 3PL providers
	No monitor the temperature	Establish preventive control by monitor the temperature real-time
Return	A high amount of returned goods	Set up a rule for customers to return on the delivery date only

4. Conclusion

This research explored the chilled beef product cold chain by using IDEF0 Level 0 and 1 to analyze the business process of the abattoir in detail. Then, we identified the problems in the stakeholders activities and suggest recommendation for improvement such as implement demand and supply planning, implement an integrated inventory management, establish database and traceability system, evaluate the animal welfare standard of livestock shipping provider, evaluate the worker productivity, focus on quality inspection, hire some qualified 3PL providers,establish preventive control, and set-up a rule for customer to return the product. This research could be a guideline to help similar companies to improve their business.

5. References

[1] Agus A Budisatria IGS Ngadiyono N 2014 Road map of beef cattle industry in Indonesia (Yogyakarta: APFINDO and Faculty of Animal Science, Universitas Gadjah Mada)
[2] Agus A and Widi TSM 2018 *Asian-Austr. J. Anim. Sci.* **31** 976–83
[3] Gery J Syari S P Hidayat R D R Maulana A and Darunanto D 2018 *Proc. Advances in Transportation and Logistics Research* **1** 1233-1
[4] Nastasijevic I Lakicevic B and Petrovic Z 2017 *IOP Conf. Ser.: Earth Environ. Sci.* **85** 012022
[5] Prasertwattanakul Y and Ongkunaruk P 2018 *J. Inter. Food Research* **25** 481-6
[6] Bichou K Bell M and Evans A 2007 *Risk Management in Port Operations, Logistics and Supply Chain Security* (New York: CRC Press)
[7] Ongkunaruk P 2015 *Agriculture and Agricultural Science Procedia* **3** 35-9
[8] S P Pradita and Ongkunaruk P 2019 *IOP Conf. Ser.: Mater. Sci. and Eng.* **526** 012004

IOP Conf. Series: Materials Science and Engineering **773** (2020) 012005 doi:10.1088/1757-899X/773/1/012005

Rubberwood sawdust filled natural rubber composites: effects of filler loading and zinc oxide content

C Homkhiew[1],*, E Pianhanuruk[2], S Rawangwong[1], W Boonchouytan[1] and R Numrat[3]

[1] Materials Processing Technology Research Unit, Department of Industrial Engineering, Faculty of Engineering, Rajamangala University of Technology Srivijaya, Songkhla 90000, Thailand.
[2] Department of Rubber Technology and Polymer, Faculty of Science and Technology, Rajamangala University of Technology Srivijaya, Nakhon Si Thammarat 80110, Thailand.
[3] Plan Creations Co., Ltd., Trang 92140, Thailand.

*Corresponding author: chatree.h@rmutsv.ac.th

Abstract. The aim of this work is to investigate the effects of rubberwood sawdust (RWS) and zinc oxide contents on mechanical properties of natural rubber (NR)/RWS composites. The curing characteristics of the composites were determined and the composites were vulcanized at 150°C for 10 min and a pressure of 1000 psi using a hot press machine. The properties of the composites such as tension, hardness and morphology were characterized. The increasing additions of RWS in range of 20-100 phr increased the minimum torque (ML), maximum torque (MH) and hardness of NR/RWS composites, but tensile strength and elongation at break of the composites decreased. Further, an increment of zinc oxide contents in range of 3-5 phr into the composites slightly affected the ML, MH, hardness and tensile properties. Therefore, the addition of zinc oxide 3 phr is recommended to achieve the properties of NR/RWS composites as well as saving cost.

1. Introduction

Nowadays, children's toys were usually manufactured from wood and plastic materials because of their high strength and stiffness as well as difficult to break. However, another material that has the potential for producing the toys in Thailand is natural rubber (NR) that is harvested from rubber trees. The NR is an important unique biopolymer in many of its most significant applications, which cannot be replaced by synthetic alternatives [1]. The featured properties of the NR are high resilience (elasticity), high tensile resistance and hard tear.

The NR is a very soft and flexible material that safes for children's play. However, it is not appropriate for producing the toys that desire high shape stability, the addition of high filler content in rubber matrix is a solution. Ninyong *et al.* [2] found that the addition of wood particle 20 phr into NR composites increased the surface hardness and dimensional stability. Datta and Włoch [3] suggested that increasing fibrous filler content resulted in increasing of tensile moduli 100% of elongation and hardness, and decreasing of rebound resilience and abrasion resistance of prepared jute/natural rubber composites. Thus, rubberwood sawdust (RWS) waste from processing the wooden toy is a filler that is interesting.

The development of NR composite materials, to response the children's toy purposes, needs a better understanding in influence from the material components [4]. Hence, the effects of RWS and zinc oxide contents on mechanical properties of the NR composites are needed to be further investigated. The overall results of the current work facilitate the development of NR/RWS composites for producing the children's toys.

2. Experimental

2.1. Materials
NR of STR 5L grade (Standard Thai Rubber 5 L) was purchased from GSP Product Co. Ltd., Thailand. Zinc oxide (ZnO) was purchased from Thai-Lysaght Co. Ltd., Thailand. Stearic acid was purchased from Palm-Oleo Sdn. Bhd., Malaysia. Wingstay-L was purchased from OMNOVA Solutions Inc., USA. CBS was supplied from Chemistry Corporation Co. Ltd., Thailand. Sulfur was purchased from Miwon Chemicals Co. Ltd., South Korea. RWS was supported by Plan Creations Co., Ltd., Thailand.

2.2. Preparation of NR composites
Before blending, the RWS was sieved through a standard sieve of 40 mesh, and then was dried in an oven at 110°C for 8 h to remove moisture. NR was masticated in a two-roll mill with a rotating speed of friction ratio 1:1.11 at room temperature. To prepare pre-vulcanized NR/RWS blends, all chemicals for curing NR including zinc oxide 3/5 phr, stearic acid, Wingstay-L, RWS 0/20/40/60/80/100 phr, CBS and sulfur were added sequentially into the premixed NR/RWS blends for 15 min. All NR/RWS blends were compression-molded at 150°C for 10 min at pressure of 1,000 psi to prepare the sample specimens for testing the physical and mechanical properties.

2.3. Physical and mechanical measurements
The cure characteristics like minimum torque (ML) and maximum torque (MH) were measured on an oscillating disk rheometer (rheoTE H OD+) at 150$_o$C. Morphological analysis of the NR/RWS composites was carried out using a scanning electron microscope (Model FEI Quanta 400 from FEI Company, USA). The fractured surface of the specimen was coated with gold in order to eliminate electron charging during the imaging. They were imaged with magnifications of 1000x at an accelerating voltage of 20 kV. Hardness (shore A) of different NR/RWS composite samples was measured by a Hirosima hardness tester according to ASTM D 2240. A mechanical universal testing machine was used to investigate tensile properties according to ASTM D 412. The tensile test was conducted under ambient condition with a crosshead speed of 500 mm/min.

3. Results and discussion

3.1. Cure characteristics
The minimum and maximum torques of NR/RWS composites are showed in Figure 1. The values of minimum and maximum torques of NR composites increased clearly with an increment of RWS content. This is attributed to the increase in the stiffness of NR composites with filler content [5]. Ismail and Shaari [6] revealed that the addition of filler restricts the mobility of the NR chains, resulting the maximum torque value of NR/RWS composites increased. Furthermore, Figure 1 also displayed the effect of zinc oxide content on the minimum and maximum torque values of NR/RWS composites. It can be seen that zinc oxide content slightly affected the minimum and maximum torque values of NR/RWS composites.

3.2. Hardness property
The hardness of the NR composites with different RWS and zinc oxide contents is represented in **Figure 2.** It was found that the hardness of the NR/RWS composites linearly increased as rubberwood sawdust content increases in the NR composites. Because the rubberwood has higher hardness than that of the weak rubber matrix while the addition of wood sawdust into NR phases reduces the elasticity or flexibility of rubber chains, resulting in more rigid NR/RWS composites [7-9]. Further, an increment of zinc oxide contents in range of 3 to 5 phr insignificantly affected the hardness of the NR/RWS composites.

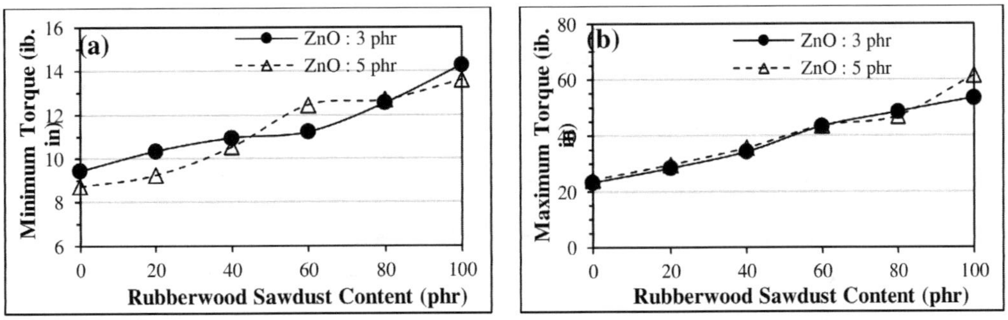

Figure 1. Effects of rubberwood sawdust and zinc oxide contents on (a) minimum torque and (b) maximum torque for NR/RWS composites.

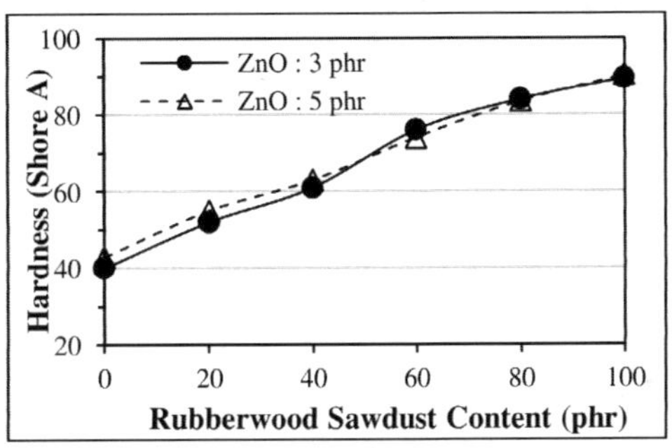

Figure 2. Effect of rubberwood sawdust and zinc oxide contents on hardness for NR/RWS composites.

3.3. Tensile properties

The effects of RWS and zinc oxide contents on tensile strength and elongation at break for NR/RWS composites are shown in Figure 3. The tensile strength and elongation at break of the NR/RWS composites were clearly decreased with the increasing content of RWS. This was mainly due to poor dispersion of the filler in the rubber matrix and weak adhesion between wood sawdust and rubber, which in turn leads to reduce the tensile strength and elongation at break of the composite materials [9]. These results could be proved with SEM images as shown in Figure 4. The NR composites reinforced with RWS 40 phr (Figure 4(a)) showed less porosities and smaller gaps between the wood sawdust and the rubber matrix than reinforced with RWS 100 phr (Figure 4(b)).

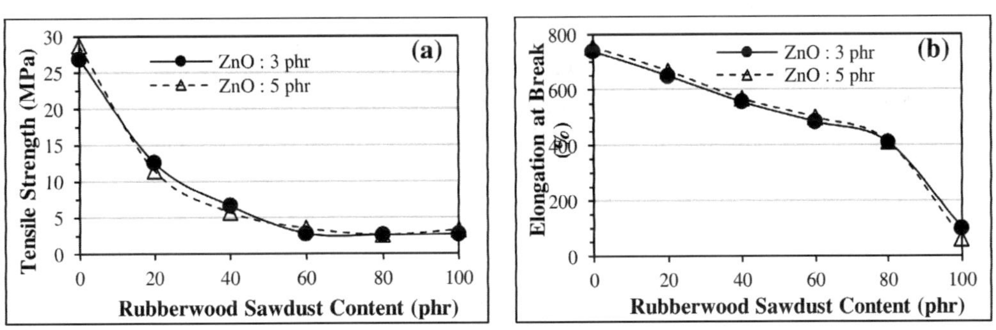

Figure 3. Effects of rubberwood sawdust and zinc oxide contents on (a) tensile strength and (b) elongation at break for NR/RWS composites.

In fact, the mechanical properties of wood-polymer composites were closely related to the efficacy of interfacial adhesion between wood and polymer matrix. The increasing addition of RWS content in the NR composites increased the dispersing difficulty of the wood sawdust and their tendency to form strong

IOP Conf. Series: Materials Science and Engineering **773** (2020) 012005 doi:10.1088/1757-899X/773/1/012005

agglomeration [8, 10]. Further, the addition of RWS in range of 60-100 phr insignificantly influenced the tensile strength of the NR/RWS composites. Likewise, the effects of zinc oxide content on the tensile strength and elongation at break for NR/RWS composites exhibited a similar behaviour to the maximum torque and hardness: the zinc oxide content lightly affected.

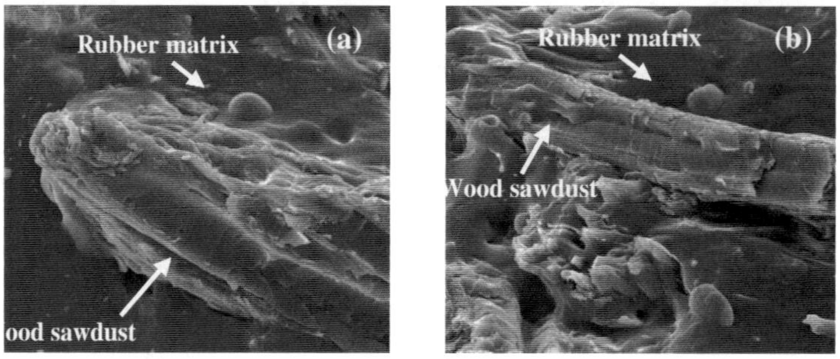

Figure 4. SEM images of NR composites reinforced with RWS (a) 40 phr and (b) 100 phr.

4. Conclusions

This research revealed the effects of RWS and zinc oxide contents on mechanical properties of NR/RWS composites. The minimum and maximum torques and hardness of the NR composites linearly increased with an increment of RWS content, but tensile strength and elongation at break of the composites decreased. Further, the addition of RWS in range of 60-100 phr insignificantly affected the tensile strength of the NR/RWS composites. Besides, the increasing addition of zinc oxide in range of 3-5 phr into the NR/RWS composites slightly affected the minimum and maximum torques, hardness, tensile strength and elongation at break. Therefore, the additions of zinc oxide 3 phr and RWS 100 phr are recommended to achieve the properties of NR/RWS composites for producing the children's toys as well as saving cost and safe to user.

5. References

[1] Beilen JB and Poirier Y 2007 *Trends Biotech.* **25** 522.
[2] Ninyong K, Wimolmala E, Sombatsompop N and Saenboonruang K 2019 *IOP Conf. Ser.: Mater. Sci. Eng.* **526** 012038.
[3] Datta J and Włoch M 2017 Polym. Bullet. **74** 763.
[4] Leu SY, Yang TH, Lo SF and Yang TH 2012 *Construc. Build. Mater.* **29** 120.
[5] Roy K, Debnath SC, Bansod ND, Pongwisuthiruchte A, Wasanapiarnpong T and Potiyaraj P 2019 *J. Mater. Cycles Waste Manag.* doi.org/10.1007/s10163-019-00939-w.
[6] Ismail H and Shaari SM 2010 *Polym. Test* **29** 872.
[7] Jamil MS, Ahmad I and Abdullah I 2006 *J. Polym. Res.* **13** 315.
[8] Homkhiew C, Rawangwong S, Boonchouytan W, Thongruang W and Ratanawilai T 2018 Inter. J. Polym. Sci. Article ID 7179527, doi.org/10.1155/2018/7179527.
[9] Homkhiew C, Boonchouytan W, Cheewawuttipong W and Ratanawilai T 2018 *J. Mater. Cycles Waste Manag.* **20** 1792.
[10] Ratanawilai T, Thanawattanasirikul N and Homkhiew C 2012 *ScienceAsia* **38** 289.

Acknowledgement

The authors would like to express their thanks to the Thailand Research Fund and National Research Council of Thailand (Grant Number RDG61T0012) for financial support throughout this work. Many thanks go to Rajamangala University of Technology Srivijaya (RMUTSV) for research facilities and other supports.

IOP Conf. Series: Materials Science and Engineering **773** (2020) 012006 doi:10.1088/1757-899X/773/1/012006

Facile preparation of magnetic carbon nanofiber composite from *nata de coco* for removal of methylene blue dye from water

P Thaveemas[1], L Chuenchom[1]*, S Techasakul[2], W Watcharin[3] and D Dechtrirat[2,4,5]*

[1] Department of Chemistry, Faculty of Science, Prince of Songkla University, Hat-Yai, Songkhla 90112, Thailand
[2] Laboratory of Organic Synthesis, Chulabhorn Research Institute, Bangkok 10210, Thailand
[3] Faculty of Biotechnology (Agro-Industry), Assumption University, Bangkok 10240, Thailand
[4] Department of Materials Science, Faculty of Science, Kasetsart University, Bangkok 10900, Thailand
[5] Specialized Center of Rubber and Polymer Materials for Agriculture and Industry (RPM), Faculty of Science, Kasetsart University, Bangkok, Thailand

E-mail: laemthong.c@psu.ac.th, fscidcd@ku.ac.th

Abstract. The facile synthesis of a magnetic 3D structured carbon nanofiber composite from *nata de coco* was demonstrated. *Nata de coco* was used as a readily available and renewable source of carbon in this study because of its high cellulose content in the form of nanofibers. Fe^{3+} was utilized as the magnetic particle source and as a pore activating agent. *Nata de coco* was freeze-dried, infiltrated with Fe^{3+}solution, and subsequently pyrolyzed at 700 °C to prepare the magnetic carbon nanofiber composite. The resulting composite was then characterized *via* scanning electron microscopy, surface area and pore size analysis, X-ray diffraction, and vibrating sample magnetometry. A magnetic carbon composite with a 3D fibrous and porous structure and a high surface area of 584 m^2/g was successfully obtained. When used as an adsorbent, the sample completely adsorbed 10 ppm of methylene blue (MB) dye within 1 min. The composite can be simply regenerated using ethanol and can thus be reused. The excellent magnetic properties of the sorbent allow it to be readily separated from aqueous solutions by applying an external magnet. Hence, the sorbent is practical to use in real systems.

1. Introduction

Water pollution is a serious global environmental threat because of its direct impact on human health and ecological systems.[1] Several methods for removing pollutants from contaminated wastewater have been developed.[2] Adsorption is considered as the most effective method due in part to its simplicity and inexpensive cost of operation with satisfactory results.[3] Porous carbon is one of the most promising candidates among existing adsorbent materials because of its high surface area and porosity with exceptional chemical stability. However, most porous carbon materials are in the form of fine powder that takes time to sediment and is difficult to separate. Hence, magnetic carbon composites have become a material of choice because they offer a fast and easy separation step using only an external magnet after a complete treatment process.[4]

In this work, a magnetic carbon nanofiber composite was straightforwardly prepared *via* freeze-drying and pyrolysis. *Nata de coco*, a bacterial cellulose, was used as a starting material to produce the carbon nanofiber because it has high cellulose nanofiber content and can be produced in a large scale. Magnetic properties and high porosity can be simply incorporated into the material using Fe_{3+} as a magnetic particle source and as an activating agent. A typically used dye in the textile industry, methylene blue (MB), was selected herein as a model for the adsorption study.

2. Methodology

Commercially available *nata de coco* (Chaokoh, Thailand) was washed in 150 mL DI water for 1 h thrice and then freeze-dried at −80 °C. The obtained sample was then soaked with 500 ppm of $FeCl_3$ solution for 3 h, freeze-dried, and pyrolyzed under N_2 atmosphere at 700 °C for 2 h at a heating rate of 2 °C/min to obtain the magnetic carbon nanofiber composite. The resulting composite was then characterized *via* scanning electron microscopy (SEM) with energy dispersive X-ray (EDX) mapping, X-ray diffraction (XRD), N_2 sorption analysis, and vibrating sample magnetometry (VSM).

Exactly 0.01 g of the magnetic carbon nanofiber composite was dispersed and shaken in 15 mL of the MB solution (100 ppm) at 30±2 °C for 0–48 h to attain the adsorption kinetics of MB. The adsorption isotherm of MB was measured by dispersing 0.01 g of the magnetic carbon nanofiber composite and shaking it in 15 mL of the MB solution in the concentration range of 1–500 ppm at 30±2 °C for 36 h. The sorbent was then separated using a magnet. The concentration of the MB solution before and after adsorption was examined using an ultraviolet–visible spectrophotometer, and the adsorbed amount was then calculated.

3. Results and Discussion

The morphology of the prepared composite was determined using a scanning electron microscope. As shown in **Figure 1A**, a nonwoven nanofibrous structure with interconnected macropores resulting from the 3D arrangement of the fibers was observed. This structure clearly confirms the successful conversion of *nata de coco* into carbon nanofibers. Furthermore, EDX mapping shown in **Figure 1B-D** also revealed that the nanofiber composite is composed of carbon, oxygen and iron species homogeneously distributed throughout the material.

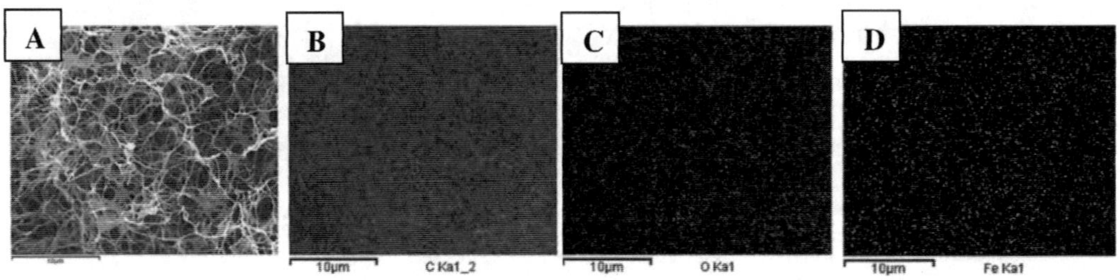

Figure 1. **(A)** SEM image of the magnetic carbon nanofiber composite **(B-D)** SEM-EDX mapping of elemental carbon, oxygen and iron in the magnetic carbon nanofiber composite.

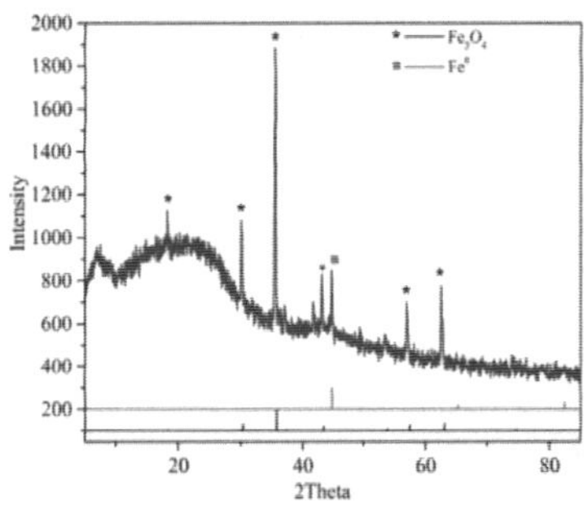

Figure 2. XRD pattern of the magnetic carbon nanofiber composite exhibiting characteristic peaks of Fe_0 and Fe_3O_4.

As shown in **Figure 2**, the XRD pattern of the magnetic carbon nanofiber composite exhibited characteristic peaks at 30.3°, 35.7°, 43.4°, 57.4°, and 63.0° corresponding to (220), (311), (400), (511), and (440) reflections, respectively, of the face-centered cubic structure of magnetite (Fe_3O_4, JCPDS No. 01-075-0449). This result confirms that Fe^{3+} can be successfully converted into the iron magnetite phase through a one-step pyrolysis. Moreover, the XRD pattern showed a sharp peak at 44.7° and a broad signal around 20°, which were indexed as the facet (110) of Fe_0 (Fe_0, JCPDS No. 06-0696) and turbostratic carbon, respectively.

N_2 sorption analysis was conducted to monitor the development of porosity of the magnetic carbon composite. As shown in **Figure 3A**, a type IV isotherm with an obvious hysteresis loop that is typically found in mesoporous materials was observed. This isotherm can be further confirmed by the narrow pore size distribution **(Figure 3B)** with an average mesopore size of 2.11 nm. Using a t-plot method, we found that the material had a mesopore fraction of 58.7%. Furthermore, the S_{BET} and the total pore volume were found to be 584 m^2/g and 0.31 cm^3/g, respectively.

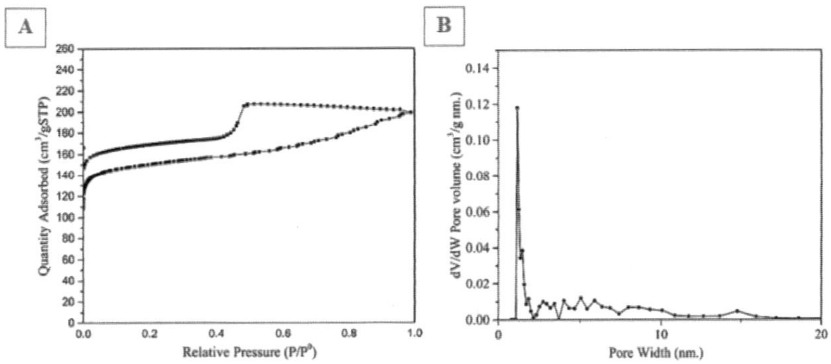

Figure 3. (A) N_2 sorption isotherm and (B) pore size distribution of the magnetic carbon nanofiber.

The magnetic properties of the resulting material are illustrated in **Figure 4**. As depicted in **Figure 4A**, the VSM result of the magnetic carbon nanofiber composite exhibited a hysteresis loop with a saturation magnetization value of 13.393 emu/g, which is more than sufficient to separate the adsorbent from the aqueous system. This separation can be clearly observed in **Figure 4B**, which shows that the magnetic carbon fiber composite was easily separated from the system using an external magnet. Furthermore, the adsorbent still exhibited excellent magnetic properties even after complete adsorption of the MB dye.

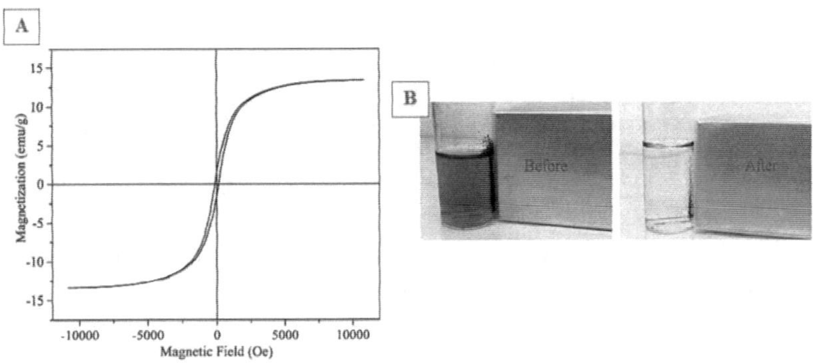

Figure 4. (A) Magnetization curve obtained using a vibrating sample magnetometer at room temperature and (B) magnetic properties of the magnetic carbon nanofiber composite before and after adsorption of MB dye (10 mg/g, 20 mL).

The adsorption kinetics plot in **Figure 5A** shows that the adsorption reached equilibrium within 36 h. The obtained data were fitted with a pseudo-second-order model, and the calculated rate constant (k_2) was found to be 2.15 g/(mg•min). Furthermore, a linear curve fitting for MB adsorption, as presented in **Figure 5B**, also exhibited a good correlation with R_2 of 0.9861. The adsorption isotherm of the magnetic carbon nanofiber composite toward MB was determined, and the result is shown in **Figure 6A**. The isotherm plot was fitted to the Langmuir adsorption model, and the correlation coefficient (R_2) was found to be 0.9955, which suggested that the data were well fitted to the Langmuir adsorption model. The calculated maximum adsorption capacity and Langmuir constant were found to be 238.10 mg/g and 0.0569 L/mg, respectively. After complete adsorption, the MB dye could be simply desorbed in ethanol. This unique property is in contrast to that of most carbon-based adsorbents that are typically regenerated under harsh conditions.

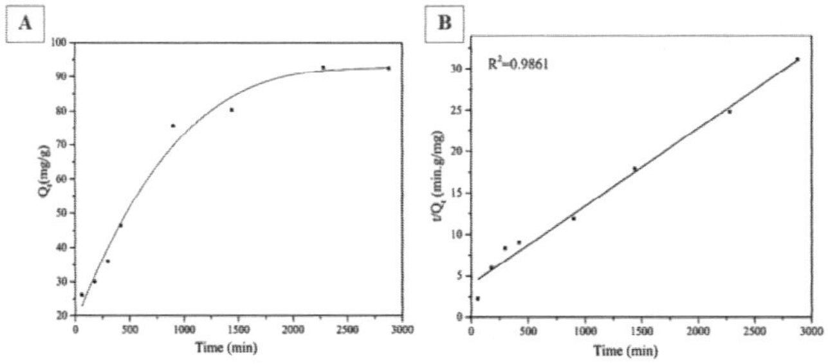

Figure 5. Adsorption kinetics toward methylene blue of the magnetic carbon nanofiber composite.

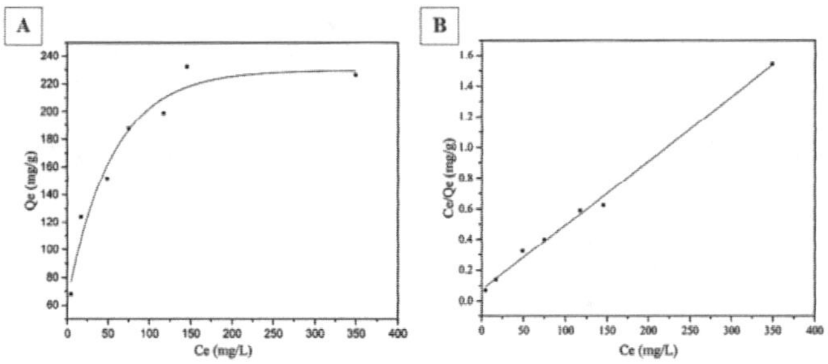

Figure 6. Adsorption isotherm: (a) nonlinear and (b) linear plots toward MB dye of the magnetic carbon nanofiber composite.

4. Conclusion

In this study, the facile preparation of a magnetic carbon nanofiber was achieved *via* freeze-drying and pyrolysis. The method developed herein does not require the use of toxic chemicals and is thus environmentally friendly and sustainable. The magnetic carbon material exhibits promising properties for environmental remediation.

5. References

[1] Peng W J, Li H Q, Liu Y Y and Song S X 2017 *Journal of Molecular Liquids* **230** 496-504
[2] Gadipelly C, Perez-Gonzalez A, Yadav G D, Ortiz I, Ibanez R, Rathod V K and Marathe K V 2014 *Industrial & Engineering Chemistry Research* **53** 11571-92
[3] Stefelova J, Slovak V, Siqueira G, Olsson R T, Tingaut P, Zimmermann T and Sehaqui H 2017 *Acs Sustainable Chemistry & Engineering* **5** 2679-92
[4] Mehta D, Mazumdar S and Singh S K 2015 *Journal of Water Process Engineering* **7** 244-65

Acknowledgements

We acknowledge the financial support provided by the Science Achievement Scholarship of Thailand (SAST).

Removal of fluoride ions from groundwater using surface-modified ultrafiltration membrane

W Yanwachiragul[1*], J Lohwacharin[1, 2, 3 *], W Sajomsang[4] and S Singto[4]

[1]Department of Environmental Engineering, Faculty of Engineering, Chulalongkorn University, Bangkok, Thailand.
[2]Reserach Network of NANOTEC-CU on Environment, Chulalongkorn University, Bangkok, Thailand.
[3]Research Unit Control of Emerging Micropollutants in Environment, Chulalongkorn University, Bangkok, Thailand.
[4]Nano-engineered Soft Materials for Green Environment Laboratory (SOF), National Nanotechnology Center, Pathumthani, Thailand.

*Corresponding author: Jenyuk.L@chula.ac.th

Abstract. Zeolitic imidazolate framework no.8 (ZIF8), a kind of metal organic frameworks (MOFs), has potential use in water treatment due to water stability. However, there are few studies of ZIF8/polyamide as a thin-film nanocomposite (TFN) membrane. Herein, a film on the surface of ultrafiltration membrane was generated with ZIF8 nanoparticles to fabricate the TFN membrane. Thus, the objective was to investigate feasibility of the ZIF8 membrane for fluoride removal from water. A scanning electron microscopy analysis demonstrates that the ZIF8 layer was successfully made on the UF surface. The ZIF8 layer rendered an increase in the fluoride rejection, probably through size sieving by small windows of the MOFs, compared to the virgin UF membrane, while sacrificed the water permeability. It is anticipated that reducing large voids between ZIF8 particles and thickness of the ZIF8 layer improve the fluoride rejection and water permeability.

1. Introduction

Metal organic frameworks (MOFs), hybrid inorganic-organic materials, have drawn great attention for various applications due to their crystalline characteristic, porous structure, and high thermal and chemical stability [1]. Among MOFs, zeolitic imidazolate framework no.8 (ZIF8), the structure of which comprises zinc(II) ion as metal nodes and 2-methylimidazole as organic linkers, offers a good stability in water; hence, it is feasibly applied for water treatment. Moreover, ZIF8 has a robust synthetic method and high purity [2]. Recently, some researchers study on the ZIF8/polyamide thin-film-nanocomposite (TFN) membranes, pressure-driven membranes for removing dye and sodium chloride from water. The studies reported a slight increase in both water permeability and solute rejection. However, the performances of the ZIF8/polyamide TFN membranes were limited since addition of large amount of ZIF8 nanoparticles into the polyamide could cause defects on the polyamide layer [3, 4].

Since groundwater in the northern part of Thailand is contaminated with high fluoride [5], pressure-driven membranes become one of alternative processes for fluoride removal from groundwater [6]. In this study, ZIF8 was synthesised as a film on the surface of ultrafiltration (UF) membrane, a structure like thin-film-composite (TFC) membrane. In addition, the synthesised ZIF8-TFC-membrane is studied on the fluoride removal performance using the synthetic water, of which the fluoride concentration and pH were referred to previous studies of groundwater contamination in the northern part of Thailand.

2. Experimental methods

ZIF8 was coated as a layer on a flat-sheet polyethersulfone (PES) UF membrane via the layer-by-layer rapid synthesis method at room temperature and atmospheric pressure [7]. Since ZIF8 was hardly coated directly on the virgin PES membrane, the surface of the PES membrane was firstly prepared by coating with thin tannin layer. The tannin solution was prepared with 10 mM of tannic acid dissolved in DI water. The solution pH was adjusted with Tris base (2-Amino-2-(hydroxymethyl)-1,3-propanediol) to7.8. The PES membrane was soaked in the tannin solution in a stirred filtration cell (Amicon, Millipore Corp., USA) and stirred at 200 rpm for 4 hours. Then, the membrane was washed by DI water. The ZIF8 solution was prepared by mixing 2 solutions: (1) zinc nitrate as a zinc(II) ion source; and (2)2-methylimidazole solution. The zinc nitrate solution was firstly added into the stirred cell and the tannin-coated PES membrane was soaked in the zinc nitrate solution for 10 minutes (stirring at 200 rpm).The 2-methylimidazole solution was then added into the cell, and the membrane was continued soaking for another 20 minutes (stirring at 200 rpm). The ratio of zinc(II) ion to 2-methylimidazole was varied to three ratios (i.e. 1:40, 1:60 and 1:80). It was found that the ratio of 1:40 was well suited because it did not destruct the PES surface.

The virgin PES, tannin-coated PES, and ZIF8/PES membranes were characterized for the surface morphology using the field-emission scanning-electron-microscopy (FE-SEM) analysis. The ZIF8/PA membrane was confirmed on formation of ZIF8 the layer via its crystalline characteristic using x-ray diffraction (XRD) analysis.

The water permeability and fluoride rejection of virgin PES and ZIF8/PES membranes were tested in batch filtration experiments. The membrane was inserted at the bottom of the membrane cell and the solution was filtrated through the membrane under a dead-end filtration without retentate recirculation. The filtration tests on the virgin PES and ZIF8/PES membranes were operated at constant pressures of 1.4 and 2.5 bar, respectively, while testing water permeability and fluoride rejection. The water permeability was calculated from an average water flux that was obtained from collected mass of DI water filtrated in 5 minutes. According to the previous study, the fluoride concentration was found in a range of 1.5 – 12.5 mg/L as F and pH was in a range of 6–8 [5]. Thus, the synthesised fluoride solution contained 1.56, 5.32 and 13.0 mg/L as F, and the solution pH was adjusted to pH 6.6 by adding Tris base. The fluoride concentration was detected using the fluoride selective probe (US EPA method 9241).

3. Results and discussion

As shown in Figure 2, the PES surface was fully covered with ZIF8 nanoparticles. The FE-SEM images of surface of the ZIF8/PES membrane illustrate the morphology of ZIF8 crystal [7]. The thick stack of ZIF8 grains and clusters were grown unconnectedly, thereby the presence of many voids between their grains and clusters on the surface.

The water permeability of the ZIF8/PES membrane was determined as ca. 1.2×10^{-12} m/s·Pa, which is considered to be in the range of ultrafiltration and nanofiltration membranes [8], in comparison to that of the virgin PES membrane of ca. 6.8×10^{-11} m/s·Pa. For the fluoride rejection, the ZIF8/PES membrane performs the rejection of 22.6%, which markedly increased from the rejection of 3.9% for the virgin PES membrane. It is anticipated that the fluoride removal is based on the size sieving mechanism by narrow windows of ZIF8 crystalline structure. The window size of ZIF8 was 0.34 nm, small enough to retain fluoride ion with a hydrated ion size of 0.70 nm, while allowing permeation of water molecules with a size of 0.28 nm. As shown in Figure 2 (c), the presence of many large voids between ZIF8 grains and clusters probably results in the observed small fluoride rejection. Jeong *et al.* [9] suggested that superhydrophilic, negatively charged nanoparticles provide preferential flow paths for water permeation and maintain solute rejection through a combination between steric and Donnan exclusion. The Donnan exclusion might be enhanced for the ZIF-8 UF membrane, compared to the virgin UF membrane, due to its more negative charge. In brief, in this study both size sieving and Donnan exclusion might be contributed to the fluoride rejection by the ZIF-8 UF membrane.

Meanwhile the pH of the permeate from the ZIF8/PES membrane was about pH 8.1, which increased compared to the feed solution pH. The increase in the permeate pH might be explained by the wash-out of remaining base on ZIF8 into the permeate, and at higher filtered volume the permeate pH gradually declined.

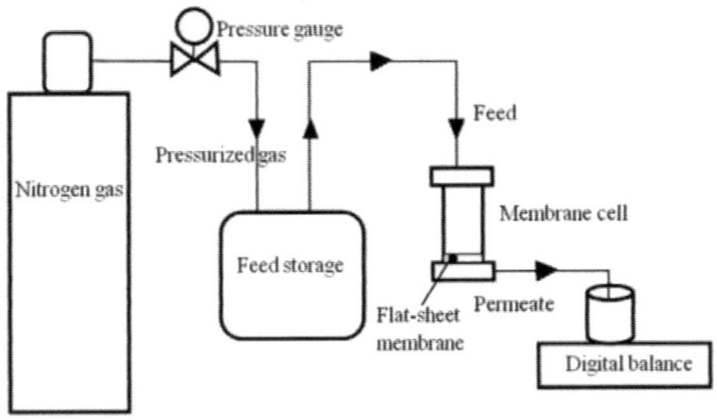

Figure 1. Set-up of membrane filtration.

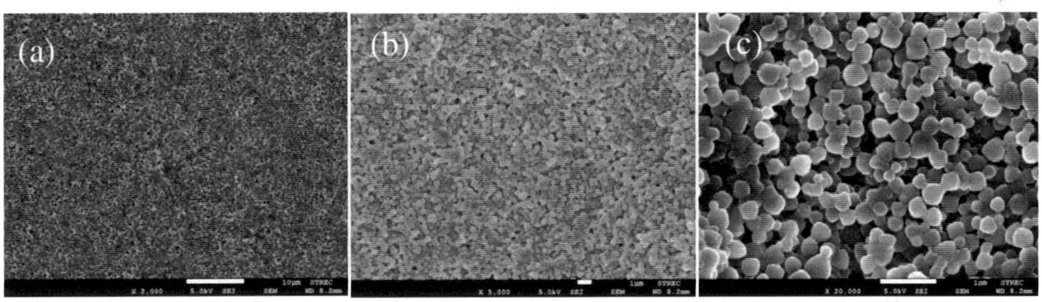

Figure 2. FE-SEM images of the surface of the ZIF8/PES membrane at different magnifications: 2k (a); 5k (b); 20k (c)

Table 1. Summary of water permeability at 25oC of the virgin PES and ZIF8/PES membranes

Membrane	Operating pressure (bar)	Water permeability ($\times 10^{-11}$ m/s·Pa)
Virgin PES	1.4	6.8
ZIF8/PES	2.5	0.12

Table 2. Summary of fluoride rejection percentage of the ZIF8/PES membrane

Initial Fluoride concentration (mg/L as F)	Fluoride rejection percentage (%)
1.56	26.7 ± 4.1
5.32	18.5 ± 0.4
13.0	23.0 ± 5.6

To confirm stability of the ZIF8 layer, the permeate has been collected and analyzed for Zn concentration that helps to determine whether the ZIF8 particles slip out of the ZIF8 layer and contaminate into the permeate.

4. Conclusions

In this research, ZIF8, a type of MOFs stable in water, was lab-synthesised and coated as a film on the surface of PES UF flat-sheet membrane, through the layer-by-layer rapid synthesis method. The experimental result shows that the water permeability of the PES membrane decreased substantially after coating with the ZIF8 layer, indicating the presence of dense layer on the UF membrane. The fluoride was rejected from the water, probably by the size sieving mechanism of the dense ZIF8 layer. A small fluoride removal might be caused by the fluoride leakage through large voids between ZIF8 particles. Further improvement requires in order to reduce large voids between the ZIF8 particles to increase the fluoride rejection while reducing the coating layer thickness is expected to increase water permeability.

5. References

[1] Kumar, P., K. Vellingiri, K.H. Kim, R.J.C. Brown, and M.J. Manos. 2017. Modern progress in metal-organic frameworks and their composites for diverse applications. Microporous and Mesoporous Materials. 253:251-256.
[2] Leus, K., T. Bogaerts, J.D. Decker, H. Depauw, K. Hendrickx, H. Vrielink, V.V. Speybroek, and P.V.D. Voort. 2016. Systematic study of the chemical and hydrothermal stability of selected "stable" Metal Organic Frameworks. Microporous and Mesoporous Materials. 226:110-116.
[3] Duan, J., Y. Pan, F. Pacheco, E. Litwiller, Z. Lai, and I. Pinnau. 2015. High-performance polyamide thin-film-nanocomposite reverse osmosis membranes containing hydrophobic zeoliticimidazolate framework-8. Journal of Membrane Science. 476:303-310.
[4] Aljundi, I.H. 2017. Desalination characteristics of TFN-RO membrane incorporated with ZIF-8 nanoparticles. Desalination. 420:12-20.
[5] Matsui, J. 2007. Fluoride removal from groundwaters using nanofiltration process. Ph.D. diss., The University of Tokyo.
[6] Shen, J., and A.I. Schäfer. 2015. Factors affecting fluoride and natural organic matter (NOM) removal from natural waters in Tanzania by nanofiltration. Reverse osmosis. Science of the Total Environment. 527-528:520-529.
[7] Kida, K., K. Fujita, T. Shimada, S. Tanaka, and Y. Miyake. 2013. Layer-by-layer aqueous rapid synthesis of ZIF-8 films on a reactive surface. Dalton Trans. 42:11128-11135.
[8] Shon, H., S. Phuntsho, D.S. Chaudhary, S. Vigneswaran, and J. Cho. 2013. Nanofiltration for water and wastewater treatment-a mini review. Drink. Water Eng. Sci. 6:47-53.
[9] Jeong, B.H., E.M.V. Hoek, Y. Yan, A. Subramani, X. Huang, G. Hurwitz, A.K. Ghosh, and A. Jawor. 2007. Interfacial polymerization of thin film nanocomposites: a new concept for reverse osmosis membranes. J. Membr. Sci. 294: 1-7.

Acknowledgement

This work has been partially supported by the National Nanotechnology Center (NANOTEC), NSTDA, Ministry of Science and Technology, Thailand, through its Research Network NANOTEC (RNN) program.

Facile preparation of monolithic magnetic porous carbon acid catalysts via surface self-assembly method and their applications in conversion of xylose into furfural

P Toumsri[1], W Auppahad[2,3], J Panpranot[2,3], and L Chuenchom[1,*]

[1] Department of Chemistry, Faculty of Science, Prince of Songkla University, Songkhla 90112, Thailand.
[2] Department of Chemical Engineering, Faculty of Engineering, Chulalongkorn University, Bangkok 10330, Thailand.
[3] Center of Excellence on Catalysis and Catalytic Reaction Engineering, Faculty of Engineering, Chulalongkorn University, Bangkok 10330, Thailand.

*Corresponding author: Email: laemthong.c@psu.ac.th

Abstract. Furfural is a versatile chemical derived from pentosane-rich agricultural and forestry residues. Considered as an important platform chemical, furfural can be converted into valuable chemicals and biofuels. The furfural production from xylose was produced through the hydrolysis and subsequent cyclodehydration reaction of xylose sugar catalyzed generally by various catalysts containing SO_3H. Nevertheless, their tedious multi-step preparation and the final catalysts usually in the form of fine powder make them difficult to be separated from the liquid systems after use, leading to time consumption for the industrial scale production. In the this work, novel magnetic porous carbonaceous acid catalyst (SO_3H-MPBG) in the form of monoliths have successfully been prepared using sugarcane bagasse as scaffold *via* a self-assembly surface coating method from phloroglucinol/glyoxylic acid precursors with soft-template F127 and Fe^{3+} as a source for magnetic iron particles. Then facile subsequent hydrothermal sulfonation with concentrated H_2SO_4 at 180°C resulted in magnetic mesoporous carbon bearing SO_3H group (SO_3H-MPBG). The SO_3H-MPBG was characterized by various techniques including XRD, FT-IR, N_2 sorption analysis, elemental analysis, and VSM. A SO_3H-MPBG showed strong magnetism due to the presence of iron magnetic particles and possessed a BET specific surface area of 252.96 m^2/g. The sample showed acceptable xylose conversion (86.13%) and furfural selectivity (55.70%) when performed at 170°C for 3 h using γ-valerolactone (GVL) as a solvent. Besides, the catalyst was easily separated after the catalytic tests using a magnet, confirming its magnetic stability.

1. Introduction

Furfural is a versatile chemical derived from pentosane-rich agricultural and forestry residues. It can be converted to valuable chemicals, for instance, furfuryl alcohol and tetrahydrofurfuryl alcohol which have broad uses in polymer, rubber, and pharmaceutical applications. The furfural production from xylose, the pentose sugar, have usually been produced through the hydrolysis and subsequent cyclodehydration reaction of xylose sugar catalyzed generally by various catalysts containing SO_3H.[1] Typically, the furfural production catalyzed by homogenous Brønsted acids catalyst such as HCl and

H_2SO_4, etc. is widely used. However, although these mineral acids are an inexpensive and effective catalyst, they pose serious disadvantages such as corrosion to the instrument, are costly and time-consuming for the separation process and poor recyclability. Thereby, the monoliths magnetically carbonaceous materials could be of interest as alternative supporting catalyst because they are stable under high temperatures and acid conditions, possess high surface activity, abundant, relatively cheap and easy to functionalize another functional group on their surface which consist of hydrophilic functional groups [2-5]. Herein, the novel magnetic porous carbonaceous acid catalyst in the form of monoliths was prepared using sugarcane bagasse as a scaffold *via* a self-assembly surface coating method and Fe^{3+} as a source for magnetic iron particles and then the facile subsequent hydrothermal sulfonation by concentrated H_2SO_4. The catalyst was used for catalyze in the selective dehydration of xylose to furfural using GVL, a biomass-derived solvent which is an environmentally friendly and has a higher boiling point than products achieved from hemicellulose as furfural (169°C) and also is the effective solvent in xylose conversion to furfural as decreasing the furfural degradation reactions causing in lower yields of monophasic systems[6, 7], as a solvent.

2. Experimental

Briefly, the sugarcane bagasse (BG) was washed with DI water and then kept at 100°C in an oven for 48 h. After that, 4 g of the dried BG was mixed in the dark solution prepared by added 1.1g of phloroglucinol, 0.8g of glyoxylic and 2.0 g of $FeCl_3.6H_2O$ in 50 mL of ethanol and was stirred at room temperature. In order to evaporate the ethanol inducing self-assembly of the carbon precursor/F127 composite on the surface of the coated BG, it was kept at room temperature for 10 h. The monoliths were heated at 80°C for 48 h, followed by calcination at 800°C for 3 h under N_2 atmosphere. The calcined material was named MPBG. Subsequently, the sulfonation of MPBG was used modified hydrothermal method[8]. The MPBG was mixed with concentrated H_2SO_4 (1 g:10 mL) in a 100 mL Teflon-lined reactor and then sealed and heated at 180°C for 24 h. The sample was washed with hot DI water until a natural pH and dried at 120°C overnight. The SO_3H-MPBG catalyst was observed. The MPBG and SO_3H-MPBG were characterized by the elemental analysis, the x-ray diffraction (XRD), N_2 sorption isotherm technique using the BET and DFT methods, vibrating-sample magnetometer (VSM), the SO_3H density was calculated based on the S content from elemental analysis, Fourier-transform infrared spectroscopy (FTIR), The catalytic activity of the SO_3H-MPBG catalyst was investigated through dehydrogenation reaction converting xylose to furfural that carried out in screw-cap Pyrex tube. In a typical experiment, a mixture of D-xylose (40 mg), catalyst (40 mg), and 2 g of GVL were mixed into the tube, sealed and heated at 170°C for specific times (0.5, 1, 2, 3, and 4 h). After that, the tube was cooled down. The catalyst was separated from the reaction residue by the external magnet. To quantify the catalytic products, the amount of xylose and furfural were analyzed by HPLC in a Waters Alliance e2695 equipment coupled to a refractive index detector (RID) kept at 50°C using a Biorad Aminex HPX-87H ion exchange column at 65 °C and a H_2SO_4 5 mmol L^{-1} solution at 0.6 mL min^{-1} as a mobile phase on isocratic mode.

3. Results and discussion

Table 1 demonstrates the specific surface areas and pore textures of the catalysts. the MPBG catalyst showed a well-developed porous structure with BET specific surface area and total pore volume are 350.63 m^2/g and 0.310 cm^3/g respectively. In the same way, the SO_3H-MPBG catalyst illustrated the BET specific surface area and total pore volume are 252.96 m^2/g and 0.295 cm^3/g respectively. The pore size distribution (PSD) of MPBG and SO_3H-MPBG catalysts are similar in the range ~ 0.5 - 15.0 nm (Figure 1 A and B (inserted)). In addition, the N_2 sorption-isotherms resemble a type-IV isotherm with evident hysteresis loops, confirming that both catalysts were consisted of mesopores (**Figure 1**). It can be seen that the texture properties of the catalyst after sulfonation (SO_3H-MPBG) were not significantly different from those of MPBG. Therefore, the sulfonation had slight effect on carbonaceous porous structure.

The elemental composition of the MPBG and SO_3H-MPBG was investigated by the elemental analysis (**Table 1**). The significant increase of S content in SO_3H-MPBG indicates that the surface of MPBG was successfully grafted with the SO_3H functional group. In addition, FT-IR spectra (Figure 1C) also confirmed the success in incorporating the sulfonated acid group onto MPBG surface with the appearance of the S=O symmetric and asymmetric stretching vibrations of SO_3H at 1023 cm^{-1} and 1125 cm^{-1}, respectively, which are the characteristic peaks of SO_3H [1, 4]. The SO_3H density was determined to be 0.531 mmol/g (**Table 1**).

Table 1 Textural properties, the elemental compositions and SO_3H density of samples

Samples	Surface area (m^2/g)			Pore volume (cm^3/g)			%mesopore	Elemental composition (%wt.)					SO_3H acidity[h]
	BET	Micro[a]	Meso[b]	Total[c]	Micro[d]	Meso[e]		C[f]	H[f]	O[g]	N[f]	S[f]	(mmol/g)
MPBG	350.63	144.63	206.00	0.310	0.061	0.249	80.291	73.603	1.031	25.276	0.090	< 0.010	< 0.003
SO_3H-MPBG	252.96	28.30	224.66	0.295	0.014	0.281	95.284	72.308	1.105	26.496	0.091	1.700	0.531

[a] t-plot micropore area, [b] t-plot external surface area, [c] Total pore volume, [d] t-plot micropore volume, [e] Difference of total pore and micropore volume.

[f] The elemental contents determined by CHNS elemental analysis. [g] O content was calculated by 100-(%C+%H+%N+%S) = %O. [h] Calculated based on elemental analysis

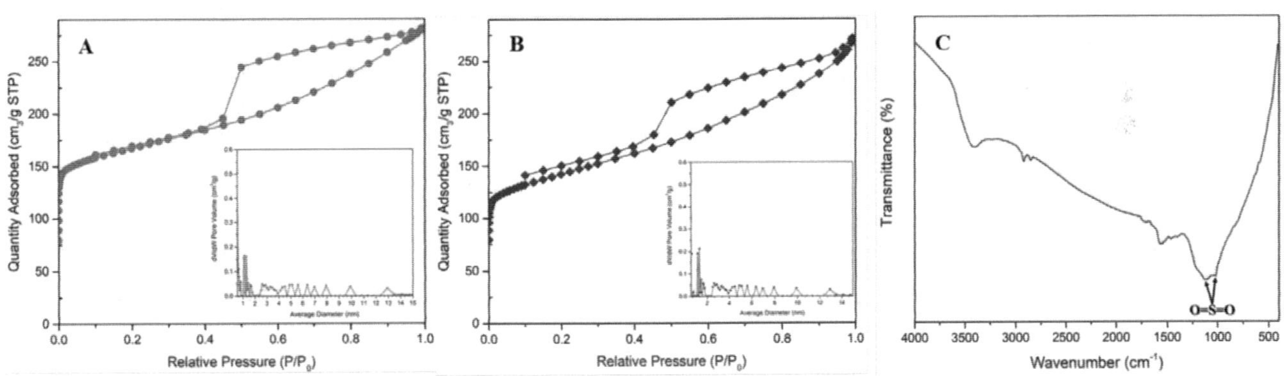

Figure 1 N_2 sorption isotherms of MPBG (A) and SO_3H-MPBG (B) and inset figure showed the pore size distribution of those calculated by DFT method. FTIR spectra of MPBG and SO_3H-MPBG (C).

The XRD patterns of the catalysts exhibit the sharp diffraction peaks at 2θ = 26.228 and 44.365°, referred to as the graphitic carbon (ICDD pdf2 number 01-075-1621) (Figure 2A). Moreover, the XRD pattern also shows the sharp Fe metal characteristic peaks of MPBG at 2θ = 44.674, 65.023, and 82.335° (ICDD pdf2 number 00-006-0696). It is noticeable that the MPBG consists of metallic iron particles because the reduction of Fe^{3+} into Fe (0) by carbon matrix. However, after sulfonation (SO_3H-MPBG), the intensity of Fe diffraction peaks remarkably decreased, due to the crease in metallic Fe by saulfonation process. The decrease in Fe content can also be reflected with the VSM results. The magnetizations of both materials are presented in **Figure 2B**, displaying that a magnetic saturation of carbonaceous catalyst significantly dropped to 0.86 emu/g (from 14.14 emu/g). The decrease of Fe peaks intensity in XRD patterns and the reduction in magnetization after sulfonation presumed that the iron particles on the catalyst surface can be leached in the concentrated sulfuric solution during sulfonation as a result in the decline of magnetic saturation [10, 11]. However, the SO_3H-MPBG catalyst still has a magnetic property, which can be separated by magnet presented in **Figure 2C**.

The catalytic activity of SO_3H-MPBG in the dehydration of xylose into furfural using GVL as a solvent at 170 °C for 0.5 – 3 h. was studied. The results indicate the perfect conversion of xylose. Furthermore, the highest furfural selectivity of 55.70 % was achieved at 170 °C for 3 h with 86.13% of xylose conversion. When the reaction time was longer than 3 h, the furfural selectivity decreased suggesting that the furfural was possibly decomposed into further products, or humin.

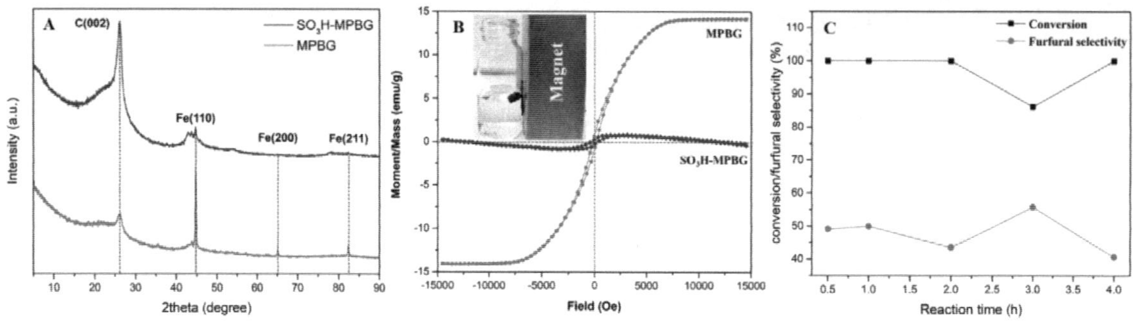

Figure 2 The XRD patterns (A), magnetization curves (B) of MPBG and SO_3H-MPBG, Magnetic separation by an external magnet (inset B) and the catalytic activity of SO_3H-MPBG (C).

4. Conclusions

The novel magnetic porous carbonaceous acid catalyst (SO_3H-MPBG) in the form of monoliths have successfully been prepared using sugarcane bagasse as a scaffold *via* a self-assembly surface coating method and subsequent sulfonation. The SO_3H-MPBG catalyst shows good catalytic activity in dehydration of xylose to furfural using GVL as a solvent when performed at 170 °C for 3 h with 86.13 % xylose conversion and 55.70 % furfural selectivity. More importantly, this catalyst was readily separated from the reaction mixture by an external magnet. In summary, the SO_3H-MPBG catalyst can serve as an efficient, recyclable and stable catalyst for the dehydration of xylose to furfural.

5. References

[1] Y. Zhu, W. Li, Y. Lu, T. Zhang, H. Jameel, H.-m. Chang, L. Ma, Production of furfural from xylose and corn stover catalyzed by a novel porous carbon solid acid in γ-valerolactone, RSC Advances, 7 (2017) 29916-29924.
[2] T. Zhang, W. Li, Z. Xu, Q. Liu, Q. Ma, H. Jameel, H.-m. Chang, L. Ma, Catalytic conversion of xylose and corn stalk into furfural over carbon solid acid catalyst in γ-valerolactone, Bioresource Technology, 209 (2016) 108-114.
[3] P.P. Upare, D.-Y. Hong, J. Kwak, M. Lee, S.K. Chitale, J.-S. Chang, D.W. Hwang, Y.K. Hwang, Direct chemical conversion of xylan into furfural over sulfonated graphene oxide, Catalysis Today, 324 (2019) 66-72.
[4] T. Chen, L. Peng, X. Yu, L. He, Magnetically recyclable cellulose-derived carbonaceous solid acid catalyzed the biofuel 5-ethoxymethylfurfural synthesis from renewable carbohydrates, Fuel, 219 (2018) 344-352.
[5] F. Zhang, X. Tian, M. Shah, W. Yang, Synthesis of magnetic carbonaceous acids derived from hydrolysates of Jatropha hulls for catalytic biodiesel production, RSC Advances, 7 (2017) 11403-11413.
[6] D.M. Alonso, S.G. Wettstein, J.A. Dumesic, Gamma-valerolactone, a sustainable platform molecule derived from lignocellulosic biomass, Green Chemistry, 15 (2013) 584-595.
[7] L. Zhang, L. Tian, R. Sun, C. Liu, Q. Kou, H. Zuo, Transformation of corncob into furfural by a bifunctional solid acid catalyst, Bioresource Technology, 276 (2019) 60-64.
[8] G. Chen, X. Wang, Y. Jiang, X. Mu, H. Liu, Insights into deactivation mechanism of sulfonated carbonaceous solid acids probed by cellulose hydrolysis, Catalysis Today, 319 (2019) 25-30.

Acknowledgement

Piyamit Toumsri was financially supported by Science Achievement Scholarship of Thailand (SAST), Prince of Songkla University. This work was financially supported by Thailand Science Research and Innovation (MRG6280106). The instrument was supported by Center of Excellence on Catalysis and Catalytic Reaction Engineering, Faculty of Engineering, Chulalongkorn University, Bangkok, Thailand.

IOP Conf. Series: Materials Science and Engineering **773** (2020) 012009 doi:10.1088/1757-899X/773/1/012009

Characterization of polylactic-epoxidized natural rubber/modified cellulosic fiber biocomposites with different silane coupling agents

S Kadea[1], J Seekaew[1], S Thongmala[1], R Hedthong[1], N Khanoonkon[2], S Witayakran[2], P Chatakanonda[2], R Chollakup[2] and T Kittikorn[1,*]

[1] Department of Materials Science and Technology, Faculty of Science, Prince of Songkla University, Hatyai, Songkhla, 90110, Thailand
[2] Kasetsart Agricultural and Agro-Industrial Product Improvement Institute (KAPI), Kasetsart University, Chatujuk, Bangkok, 10900, Thailand

*Email thorsak.k@psu.ac.th

Abstract. A purpose of this research was to investigate a potential of two differently modified cellulosic fibers for applying as reinforcing fillers in polylactic acid (PLA) and modified epoxidized rubber (ENR) base of biocomposites. Coupling agents including 3-(trimethoxysilyl) propyl methacrylate and (3-aminopropyl) trimethoxysilane were used in order to modify cellulosic fibers. These fibers were compounded with PLA-modified ENR (modified by bisphenol A diglycidyl ether) in an internal mixer and fabricated to composite sheets using a compression molding, respectively. The formulations of biocomposites were designed by weight ratio of PLA (85%): modified ENR (5%): the cellulosic fiber (10%). Accordance with the results of mechanical property analysis of their biocomposites, the modifications of cellulosic fibers with both silane types significantly improved flexural, hardness and impact properties of PLA-modified ENR based biocomposites. However, the methacrylate-silane modified cellulosic fiber provided the greatest increase in flexural properties while the amino-silane modified cellulosic fiber resulting in the highest value of hardness and impact resistance of all biocomposites. Oil absorption test of biocomposites was also investigated in this research in order to evaluate its possibility to use under oily environment. After 168 hours of experiment, the amino-silane modified composite exhibited the highest oil absorption among all materials. This result indicated the most improvement of hydrophobicity of modified cellulosic fiber by (3-aminopropyl) trimethoxysilane.

1. Introduction

A decade ago, polylactic acid (PLA) based cellulose fibre biocomposite has been considerably interested and widely developed as a new sustainable material with aiming to substitute polyolefin based cellulose biocomposite [1]. Due to full biodegradation and 100% renewable material resources, this make it expected to remedy the pollution from plastic waste as confronting nowadays [2]. Nevertheless, a problem of using cellulose fibre incorporating into PLA is due to a poor interfacial adhesion between fibre and matrix which determines thermal, mechanical properties, moisture absorption and also life time of service [3]. Hence, overcoming of this drawback by fibre surface modification is usually processed in order for improvement of the compatibility [4].

This work was aimed to solve these problems by developing a new biocomposite of cellulosic fibre/modified epoxidized rubber (ENR)/PLA. Two different types of silane: 3-(trimethoxysilyl) propyl methacrylate and (3-aminopropyl) trimethoxysilane were used to treat cellulosic fibres prior applying as reinforcing fillers in the biocomposites.

2. Experimental

2.1. Material

PLA grade of 3052D was supplied by the NatureWorks, USA. Chopped cellulous fibre and ENR 25% were kindly provided by local companies, Thailand. Bisphenol A diglycidyl ether was used to modify ENR. (3-Aminopropyl)trimethoxy silane 97% concentrations. 3-(Trimethoxysilyl) propyl methacrylate 98% concentrations were purchased from SigmaAldich. Acetic acid and ethanol of commercial grades were also used.

2.2. Methods

2.2.1. Fiber treatment. Silanization of cellulosic fibre was taken place according to the previous work [5]. Two different types of silane were supplied as reactants to improve compatibility of cellulose fibre and matrix phases.

2.2.2. Modified ENR. Bisphenol A diglycidyl ether was compounded together with ENR by a two-roll mill machine for 5 min.

2.2.3. Compounding and fabrication of the biocomposite. Biocomposites of PLA/modified ENR/cellulosic fibre were formulated by ratio of 85% :5%: 10% by weight. 3 different types of fibres as following: 1) unmodified cellulose, 2) methacrylate silanized cellulose and 3) amino silanized cellulose were compounded with PLA and modified ENR in the internal mixer. The compounding process was operated for 3 min, 60 rpm under temperature of 180$_o$C. After that, the compounded biocomposites were granulated and transferred to a compression molding unit for fabrication of biocomposite sheets. The fabrication process was carried on under a condition of 180$_o$C, pressure of 200 kgf/cm$_2$. The biocomposite sheets as obtained were kept in a desiccator for further testing in the next step.

2.2.4. Mechanical properties and oil swelling tests. In order to investigate the effect of different types of silanes onto the biocomposite performances, the mechanical testing of charpy impact test (unnotched specimen), flexure test was employed accordance with ASTM D256 and ISO 178, respectively. Hardness test was also characterized using a mode of shore A. Oil swelling test was performed accordance with ASTM D543-06. All test results were recorded and analysed followed to a statistical method.

3. Results and discussion

Figure 1 presents a comparison of hardness and impact resistance among all biocomposites. The highest value of both tests as observed in an amino silanized cellulose biocomposite could confirm a good internal interaction, probably due to a crosslink interaction between amine groups of the amino silanized cellulose and epoxy group of modified ENR. In which fibre dispersion and loading transfer from matrix to fibre was improved, these sequentially resulted to improve impact property. Meanwhile, the hardness and impact resistance of a methacrylate silanized cellulose biocomposite was also showing a better reinforcement than that of an unmodified biocomposite. This emphasises the potential of methacrylate silane treatment for improvement of fibre – matrix interaction.

Figure 2 presents and compares the flexural strength and modulus of all biocomposites. The methacrylate silanized cellulose biocomposite remarkably showed the best of all. This could be attributed by a good interfacial adhesion of the modified cellulose with PLA and ENR phases. Due to reactive groups of methacryalate silane possessing of ester and ethylene groups, these could produce the bond with carbonyl or α-carbon on backbone chain of PLA. In consequence, the reinforcement ability by loading transfer from PLA matrix to the cellulose was mostly improved.

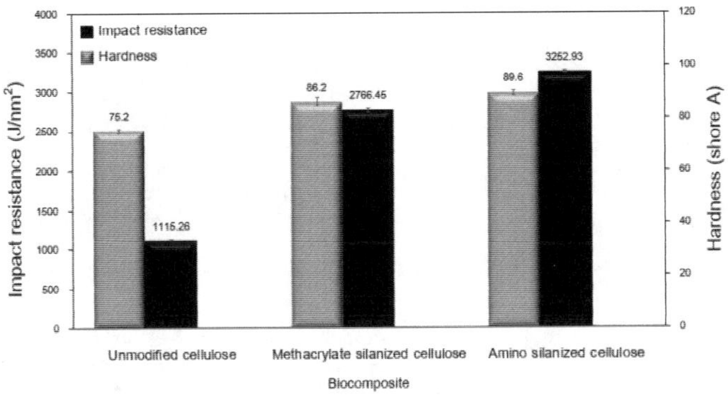

Figure 1 Impact resistance and hardness of biocomposites

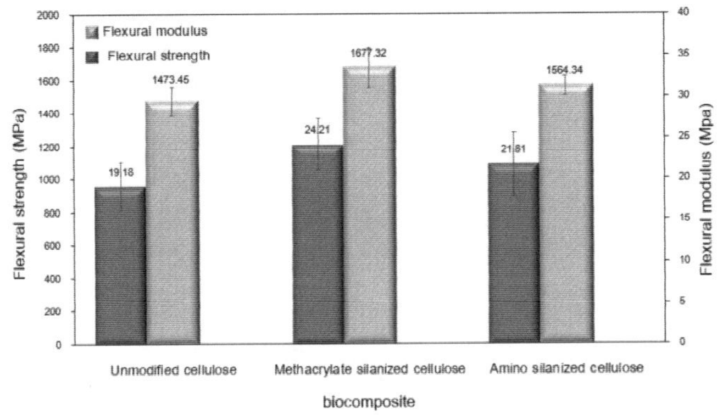

Figure 2 Flexural strength and flexural modulus of biocomposites

Figure 3 presents the oil absorption test in all biocomposites. After 168 hours of the experiment, cellulose modified with amino-silane demonstrated the most oil absorption of all materials. This is evidenced to confirm the improved hydrophobicity of this biocomposite due to a covalent bonding interaction between amine groups of the modified cellulose and epoxy groups of modified ENR. Actually, the ENR is a hydrocarbon compound which could easily introduce hydrophobic fluid like a cook oil to be absorbed while cellulose fibre performing against it because of hydroxyl groups. Hence, the most oil absorption is a crucial evidence to confirm the substantial improvement of hydrophobicity by surface alteration of (3-Aminopropyl)trimethoxy silane.

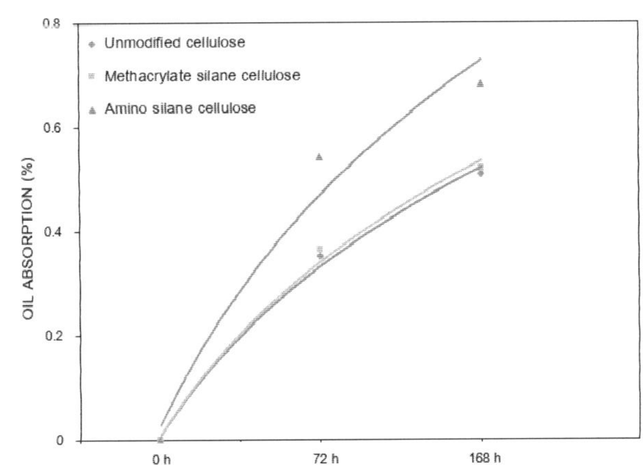

Figure 3 Oil absorption in biocomposites.

Figure 4 presents the SEM images of fracture cross section of biocomposites. It is clear that both types of silanized cellulose showed a better adhesion relative to the unmodified fiber. Sequentially, the loading transfer from matrix to fiber was improved with increasing the mechanical properties as described earlier.

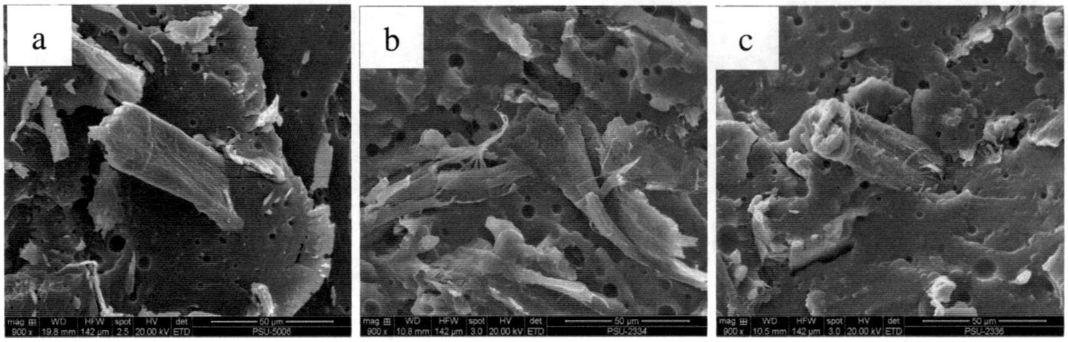

Figure 4 Cross section of fractured of biocomposite (900X): (a) unmodified cellulose, (b) methacrylate silanized cellulose, and (c) amino silanized cellulose.

4. Conclusions
The modification of cellulose with 2 different types of silane provides the mechanical properties of biocomposites in different ways. Treatment of cellulose with (3-Aminopropyl)trimethoxy silane could considerably improve the hardness and impact resistance. Meanwhile 3-(Trimethoxysilyl) propyl methacrylate results in the best of flexural strength and modulus. Furthermore, the most oil absorption of amino silanized cellulose biocomposite indicates the predominance of hydrophobicity.

5. References
[1] Silva C, Campini P, Rocha D and Rosa D 2019. The influence of treated eucalyptus microfibers on the properties of PLA biocomposites. *J. Composites Science and Technology* 179, 54-62.

[2] Janczak K, Hrynkiewicz K, Znajewska Z and Dąbrowska G 2018. Use of rhizosphere microorganisms in the biodegradation of PLA and PET polymers in compost soil. *J. International Biodeterioration & Biodegradation* 130, 65-75.

[3] Spinella S, Re G, Liu B, Dorgan J, Habibi Y, Leclère P, Raquez J, Duboisc P and Gross R 2015. Polylactide/cellulose nanocrystal nanocomposites: efficient routes for nanofiber modification and effects of nanofiber chemistry on PLA reinforcement, *J. Polymer* 65, 9-17.

[4] Paula E, Roig F, Mas A, Habas J, Mano V, Pereira F and Robin J 2016. Effect of surfacegrafted cellulose nanocrystals on the thermal and mechanical properties of PLLA based nanocomposites. *J. Eur. Polymer* 84, 173–187.

[5] Kittikorn T, Kongsuwan S and Malakul R 2017. Investigation of the durability of sisal Fiber/PLA biocomposite through evaluation of biodegradability by means of microbial growth. *J. Metals, Materials, and Minerals* 27, 23-34.

Acknowledgement
The authors would like to thank the national research council of Thailand for the grant of project number SCI6205102S. We also would like the PSU and KAPI of KU for providing facilitates and supports in this project.

IOP Conf. Series: Materials Science and Engineering **773** (2020) 012010 doi:10.1088/1757-899X/773/1/012010

Effects of post-gamma irradiation on swelling and mechanical properties of gamma vulcanized natural rubber latex (GVNRL) films

G Wichai[1], K Intharaprasit[1], E Wimolmala[2], T Markpin[2] and K Saenboonruang[1,3,*]

[1] Department of Applied Radiation and Isotopes, Faculty of Science, Kasetsart University, Bangkok, Thailand 10900

[2] Polymer Processing and Flow (P-PROF) Research Group, Division of Materials Technology, School of Energy, Environment and Materials, King Mongkut's University of Technology Thonburi, Bangkok, Thailand 10140

[3] Specialized center of Rubber and Polymer Materials for agriculture and industry (RPM), Faculty of Science, Kasetsart University, Bangkok, Thailand 10900

*Email: kiadtisak.s@ku.th, fscikssa@ku.ac.th

Abstract. Gamma vulcanized natural rubber latex (GVNRL) films have shown promising properties for various applications that require less hazardous chemicals used or released during manufacturing and/or utilization. Examples of potential applications include the uses as chemical-free latex gloves in medical or food-related facilities and as stretchable latex covers for food and agricultural products. However, due to possible degradation on swelling and mechanical properties of these films from post-gamma irradiation used for sterilization, thorough investigations are required in order to fully understand their possible changes and/or degradations in properties of interest. As a result, this work investigated on effects of post-gamma irradiation on swelling and mechanical properties of natural rubber latex (NRL) films, which were pre-vulcanized using 12-kGy and 24-kGy gamma irradiation. The properties of interest in this work included tensile modulus at 300% elongation, tensile strength, and elongation at break, as well as other related physical properties such as swelling ratios and crosslink densities. The results showed that, for samples before post-gamma irradiation, the GVNRL films with 24-kGy vulcanizing dose had higher tensile modulus, tensile strength, and crosslink density, but lower elongation at break and swelling ratio, than the films with 12-kGy vulcanizing dose. However, after post-gamma irradiation at the accumulated dose of 24 kGy, the films with 12-kGy vulcanizing dose showed significant improvements in the values of tensile modulus, tensile strength, and crosslink density, whereas the films with 24-kGy vulcanizing dose showed noticeably degradation in these properties. Hence, the overall results suggested that, while higher gamma vulcanizing doses could initially produce NRL films with higher mechanical strength, they were more negatively affected by post-gamma irradiation such that mechanical degradation could be observed.

IOP Conf. Series: Materials Science and Engineering **773** (2020) 012010 doi:10.1088/1757-899X/773/1/012010

1. Introduction

Due to possible use and/or release of hazardous chemicals and byproducts, including sulfur, -t-butyl-2-benzothiazolesulfenamide (TBBS), Wingstay-L, nitrosamines, and nitrosatable amines, from sulfur and peroxide systems used during the vulcanization of natural rubber latex (NRL), which may cause severe respiratory diseases, cancers, chemical allergies, and fatality to humans and aquatic life [1-3], gamma irradiation becomes an alternative vulcanization system that is considered safer during production, environmental-friendly, and free from accelerator-induced allergies. As a result, the interests in utilizing gamma vulcanized natural rubber latex (GVNRL) products in certain applications that require less chemicals used/released during production processes have been noticeably increasing. Examples of potential GVNRL products include NRL gloves in food and medical applications [4], of which contamination of hazardous chemicals in foods and patients is not acceptable.

In addition to using gamma irradiation to vulcanize natural rubber latex by crosslinking polymer chains, gamma rays are also used to sterilize a variety of different products such as single-use GVNRL surgical gloves, pharmaceutical packaging, and stretchable food covers [5]. The gamma sterilization process usually utilizes $_{60}$Co, which emits high-energy gamma rays (1.17 MeV and 1.33 MeV, with the average energy of 1.25 MeV) to kill microorganisms, preventing the potentials of growing or contaminating these harmful microorganisms into patients, medicines, or delicate foods. However, the possibilities of negative effects of post-gamma irradiation on mechanical properties of the irradiated products, especially in GVNRL, could not be ignored and, in fact, must be fully paid attention to as the degradation could reduce the performance and limit the full functionality of the products. As a result, this work investigated the effects of (24 kGy) post-gamma irradiation on GVNRL thin films, which was pre-vulcanized using 12 kGy and 24 kGy gamma rays and formed into testing specimens using dipping method. Properties of interest in this work include tensile properties (tensile modulus at 300% elongation, tensile strength, and elongation at break) and other related physical properties (swelling ratios and crosslink densities).

2. Experimental

2.1 Materials and chemicals

High-ammonia (HA) NRL supplied by the Rubber Authority of Thailand (RAOT) was used as the main matrix. Other chemicals that were also used in this work were 99.9% n-butyl acrylate (n-BA) and 10% potassium hydroxide (KOH) supplied by BASK (Thailand) and Gammaco (Thailand) Co., Ltd., respectively.

2.2 Preparation of GVNRL mixtures

NRL was continuously stirred using a top stirrer with a rotating speed of 300 rpm for 60 minutes. 0.25 phr of KOH was added to the NRL and stirred for another 10 minutes. Then, 5 phr of n-BA, which acted as a gamma vulcanization accelerator was added to the mixture and the stirring was continued for 40 minutes. The NRL mixture was later transferred to 5-L plastic containers and irradiated with gamma rays using a $_{60}$Co source at the accumulated doses of 12 and 24 kGy, respectively. It should be noted that the dose rate of $_{60}$Co used in this work was 2.1 kGy hour$_{-1}$ and the irradiation was carried out at the Thailand Institute of Nuclear Technology (Ongkharak, Nakhon Nayok Province). After the gamma irradiation, the pre-vulcanized NRL mixtures were kept in closed containers and stored in a freezer at the temperature of -4$_{o}$C for further uses.

2.3 Preparation of GVNRL thin films

Cylindrical glass molds with 5-cm diameter were thoroughly cleaned and dried at 60$_0$C -70$_0$C for 60 minutes using a hot-air oven. The molds were then entirely dipped into a coagulant solution for 5 seconds and dried at 60$_0$C -70$_0$C for 3 minutes. After the molds were completely dried, they were dipped into the GVNRL mixtures for 40 seconds and dried again at 100$_0$C -120$_0$C for 30 minutes. Once the GVNRL thin films were formed around the molds and completely dried, the films were rolled out of the molds and left at room temperature for 2-3 hours before being post-gamma irradiated at the accumulated dose of 24 kGy using a $_{60}$Co source.

2.4 Testing on mechanical and physical properties

The tensile properties including tensile modulus at 300% elongation, tensile strength, and elongation at break of the GVNRL thin films (both with and without post-gamma irradiation) were investigated using a universal testing machine (Autograph AG-I 5kN, Japan) following ASTM D412-06 standard testing. For the measurements of swelling ratios and crosslink densities, each specimen was cut into a circular piece with a diameter of 2 cm and soaked in toluene at room temperature, following the ASTM D471-06 standard testing method. Initial weight before immersion (w_1) and weights after 24-hour immersion (w_2) were measured and the swelling ratios were calculated using equation 1 [6]:

$$\text{Swelling ratio} = \frac{w_2 - w_1}{w_1} \times 100\% \quad (1)$$

The crosslink density was determined at the equilibrium swelling state, in which the specimens were soaked in toluene for 72 hours. The crosslink density (n) was calculated according to Flory-Rehner equation as shown in equation 2:

$$n = \frac{1}{2}\frac{\ln(1-V_r)+V_r+\chi V_r^2}{V_s(V_r^{\frac{1}{3}}-\frac{V_r}{2})} \quad (2)$$

where V_s is the molar volume of toluene (106.1 cm$_3$/mol), χ is the interaction parameter (0.393 for NR-toluene system) [7], and V_r is the volume fraction of GVNRL, which was calculated according to equation 3:

$$\frac{1}{V_r} = 1 + \left(\frac{w_2 - w_1}{w_1}\right) \quad (3)$$

3. Results and Discussion

The results of tensile properties and swelling ratios/crosslink densities of GVNRL thin films with and without 24-kGy post-gamma irradiation are shown in Table 1 and Table 2, respectively, which suggested that, for specimens without post-gamma irradiation (0 kGy), the tensile modulus, tensile strength, and crosslink density increased with increasing gamma vulcanizing doses (from 12 kGy to 24 kGy), while elongation at break and swelling ratio decreased with increasing gamma vulcanizing doses. These behaviors were observed because gamma irradiation could initiate the crosslinking of NR chains by creating free radicals inside the NRL, of which the recombination of these radicals would result in the crosslinking of NR chains into 3D network. With higher gamma vulcanizing doses (24 kGy in this work), higher crosslinking of NR chains in the specimens was observed as illustrated in Table 2, which resulted in greater restriction of the mobility of NR chains due to the increase in NR chain crosslinking structure, increasing the overall strength and stiffness of the specimens [8].

IOP Conf. Series: Materials Science and Engineering **773** (2020) 012010 doi:10.1088/1757-899X/773/1/012010

Table 1. Mechanical properties of GVNRL thin films with and without 24-kGy post-gamma irradiation. *(Numbers after ± represent the standard deviation.)*

Gamma vulcanizing dose (kGy)	Post-gamma irradiation dose (kGy)					
	Tensile modulus (MPa)		Tensile strength (MPa)		Elongation at break (%)	
	0	24	0	24	0	24
12	0.21 ± 0.03	0.23 ± 0.01	7.48 ± 1.78	8.88 ± 0.87	2387 ± 198	2393 ± 88
24	0.23 ± 0.02	0.17 ± 0.01	8.10 ± 1.80	6.80 ± 0.84	2064 ± 129	2298 ± 154

Table 2. Swelling ratios and crosslink densities of GVNRL thin films with and without 24-kGy post-gamma irradiation. *(Numbers after ± represent the standard deviation.)*

Gamma vulcanizing dose (kGy)	Post-gamma irradiation dose (kGy)			
	Swelling ratio (%)		Crosslink density (mol cm-3)	
	0	24	0	24
12	977±47	875±34	1.25×10^{-5}	1.53×10^{-5}
24	835±31	1111±19	1.66×10^{-5}	9.9×10^{-6}

However, as the specimens were post-gamma irradiated with the accumulated dose of 24 kGy, specimens, which were pre-vulcanized with 12 kGy and 24 kGy gamma rays, behaved in different ways. For instance, the specimens with 12-kGy gamma vulcanizing dose showed increasing tensile modulus, tensile strength, and crosslink density (lower elongation at break and swelling ratio) after post-gamma irradiation, while the specimens with 24-kGy gamma vulcanizing dose showed decreasing tensile modulus, tensile strength, and crosslink density (higher elongation at break and swelling ratio) after post-gamma irradiation. These behaviors could be explained by the fact that, since post-gamma irradiation could initiate both crosslinking and degradation, especially chain scission mechanism, to GVNRL thin films, the effects of crosslinking from post-gamma irradiation still dominated effects of chain scission as shown by the decrease in swelling ratio and the increase in crosslink density (Table 2), resulting in the overall increases of strength and stiffness of the specimens (Table 1). On the other hand, for specimens with already high gamma vulcanizing dose (24 kGy), the crosslink density created during the vulcanization process was already high. As a result, once the post-gamma irrradiation was carried out on the specimens, the addition of crosslinking mechanism was lower and suppressed by the chain scission mechanisms, resulting in the overall reduction of strength and stiffness of the specimens [9]. From the overall results, it could be concluded that, although higher gamma vulcanizing doses could initially produce GVNRL thin films with higher crossling density, tensile strength, and stiffness than lower gamma vulcanizing doses, the post-gamma irradiation used during product's sterilization could negatively affected the products by lowering overall crosslink density and strength of specimens, implying thorough optimization of gamma doses used for NRL vulcanization and sterilization must be investigated and carried out.

4. Conclusions

GVNRL thin films have great potentials to be utilized in various applications that require chemical-free and environmental-friendly products. However, the effects of post-gamma irradiation used during the sterilization could have negative effects on mechanical properties of the GVNRL thin films. As a result, this work investigated mechanical properties of GVNRL thin films after being post-gamma irradiated with the accumulated dose of 24 kGy. The overall results revealed that higher gamma vulcanizing doses could initially produce GVNRL thin films with higher crosslink density, mechanical strength, and stiffness than the films with lower gamma vulcanizing doses. However, after being post-gamma irradiated at 24 kGy, films with initial higher gamma vulcanizing doses (24 kGy) showed more pronounced degradation than the ones with lower gamma vulcanizing doses (12 kGy), resulting in lower overall crosslink density and mechanical strength.

5. References

[1] Intharapat P, Kongnoo A and Kateungngan K 2013 *J. Polym. Environ.* **21** 245-58
[2] Ibrahim S, Badri K, Ratnam C T and Ali N H M 2018 *Radiat. Eff. Defects Solids.* **173** 427-34
[3] Abad L V, Relleve L S, Aranilla C T, Aliganga A K, San Diego C M and de la Rosa A M 2002 *Polym. Degrad. Stab.* **76** 275-9
[4] Tanpantree S, Opaprakasit P, Loykulnant S, Kangwansupamonkon W and Tangboriboonrat P 2010 *J. Appl. Polym. Sci.* **117** 1798-803
[5] Oliver S and Tomlinson A H 1960 *J. Hygiene.* **58** 465-72
[6] Ge X, Li M C, Li X X and Cho U R 2015 *Appl. Clay Sci.* **118** 265-75
[7] Chenal J M, Chazeau L, Guy L, Bomal Y and Gauthier C 2007 *Polym.* **48** 1042-6
[8] Moonlek B and Saenboonruang K 2019 *Radiat. Eff. Defects Solids.* **174** 452-66
[9] Markovic G, Marinovic-Cincovic M S, Jovanovic V, Samarzjia-Jovanovic S and Budinski-Simendic J 2012 *Compos. Part B-Eng.* **43** 609-15

Acknowledgements
Authors would like to acknowledge financial supports from the Faculty of Science, Kasetsart University under the Undergraduate Research Matching Fund (URMF). Authors also would like to thank the Thailand Institute of Nuclear Technology (TINT) for the equipment and technical support.

IOP Conf. Series: Materials Science and Engineering **773** (2020) 012011 doi:10.1088/1757-899X/773/1/012011

Determination of radioactivities in gamma vulcanized natural rubber latex (GVNRL) for the assessment of radiological safety

D Toyen[1,2] and K Saenboonruang[1,3,*]

[1] Department of Applied Radiation and Isotopes, Faculty of Science, Kasetsart University, Bangkok, Thailand 10900
[2] Scientific Equipment and Research Division, Kasetsart University Research and Development Institute (KURDI), Kasetsart University, Bangkok, Thailand 10900
[3] Specialized center of Rubber and Polymer Materials for agriculture and industry (RPM), Faculty of Science, Kasetsart University, Bangkok, Thailand 10900

*Corresponding author: kiadtisak.s@ku.th, fscikssa@ku.ac.th

Abstract. Due to great potentials of utilizing products made from gamma vulcanized natural rubber latex (GVNRL) in applications that require less use or less release of hazardous chemicals during rubber vulcanization, the radiological health risks in rubber producers/users caused by radioactivities from natural radioisotopes containing in GVNRL must be thoroughly determined and evaluated. As a result, this work determined the activity concentrations of ^{226}Ra, ^{232}Th, and ^{40}K in GVNRL samples that were pre-vulcanized with 12-kGy and 24-kGy gamma irradiation using gamma spectroscopy. The results showed that the average radioactivity concentrations (± standard deviation) of 226Ra, 232Th, and 40K in 12-kGy (24-kGy) GVNRL were 30.8±1.5 (31.2±2.7) Bq kg^{-1}, 89.1±0.6 (89.7±0.4) Bq kg^{-1}, and 95.4±1.6 (93.2±2.2) Bq kg^{-1}, respectively. These results implied that GVNRL vulcanized with 12-kGy and 24-kGy gamma irradiation had statistically similar radioactivity concentrations and higher gamma doses did not result in increasing radioactivies. Furthermore, when compared these results with the values from non-vulcanized natural rubber latex (NRL), of which the activity concentrations of ^{226}Ra, ^{232}Th, and ^{40}K were 29.9±1.2 Bq kg^{-1}, 95.2±1.1 Bq kg^{-1}, and 96.2±1.0 Bq kg^{-1}, respectively, it could be concluded that the use of gamma irradiation for rubber vulcanization did not statistically increase or change activity concentrations of the mentioned radioisotopes, thus, it is radiologically safe for related personnel to work or use GVNRL products.

1. Introduction

Gamma vulcanized natural rubber latex (GVNRL) has gained much attention in recent years, mainly due to its preferable characteristics compared with typical sulfur-vulcanized or peroxide-vulcanized natural rubber latex, including being safer during production and use, less hazardous chemicals used and released, more environmental-friendly, and free from accelerator-induced allergies [1-3]. Examples of the uses of GVNRL include surgical latex gloves, stretchable food covers, and pharmaceutical packaging.

Despite being safer in terms of chemical-free products, issues related to the elevation of natural radionuclides such as ^{226}Ra, ^{232}Th, and ^{40}K in environment and consumer's products, which also include non-vulcanized natural rubber latex and GVNRL, have become serious concerns for the health safety of human and living organisms. The elevation of these radionuclides in the present circumstances could be due to several factors such as the increase in use of fertilizer in agriculture, more residues left from mineral and coal mining, and possible byproducts from gamma irradiation [4-6]. Examples of reports, which showed the activity concentrations of these natural radionuclides in plants and consumer's products include

- the average activity concentrations of ^{226}Ra, and ^{40}K in medicinal plants in Nigeria were 25.0 ± 3.2 Bq kg^{-1} and 172 ± 6 Bq kg^{-1}, respectively, [7] and
- the average activity concentrations of ^{226}Ra, ^{232}Th, and ^{40}K in commercial infant formulas in Malaysia were 1.87±0.17 Bq kg^{-1}, 0.89±0.13 Bq kg^{-1}, and 213.00±2.09 Bq kg^{-1}, respectively [8].

Although there already is widely available data on activity concentrations of several plants and products, the report on radioactivities in natural rubber latex (NRL) and GVNRL is still unavailable as no official reports have yet been released to the public, posing uncertainty in safety to workers and users who might have direct contact or use these GVNRL products in daily basis. As a result, this work aimed to determine and to compare the activity concentrations of ^{226}Ra, ^{232}Th, and ^{40}K in both non-vulcanized NRL and GVNRL using gamma spectroscopy. The results obtained from the measurement would be thoroughly discussed and reported in this work.

2. Experimental

2.1. Materials and chemicals
High-ammonia (HA) NRL supplied by the Rubber Authority of Thailand (RAOT) was used as the main matrix. Other chemicals that were also used in this work were 99.9% n-butyl acrylate (n-BA) supplied by BASK (Thailand) and 10% potassium hydroxide (KOH) supplied by Gammaco (Thailand) Co., Ltd.

2.2. Preparation of GVNRL mixtures
NRL was continuously stirred using a top stirrer with a rotating speed of 300 rpm for 60 minutes. 0.25 phr of KOH was added to the NRL and stirred for another 10 minutes. Then, 5 phr of n-BA, which acted as a gamma vulcanization accelerator was added to the mixture and the stirring was continued for 40 minutes. The NRL mixture was later transferred to 5-L plastic containers and irradiated with gamma rays using a ^{60}Co source at the accumulated doses of 12 and 24 kGy, respectively. It should be noted that the gamma irradiation was carried out at the Thailand Institute of Nuclear Technology (Nakhon Nayok, Thailand) and the dose rate of ^{60}Co source used in this work was 2.1 kGy hour^{-1}. After the gamma irradiation finished, the GVNRL mixtures as well as the non-vulcanized NRL were kept in separate 500-mL Marinelli-type beakers and stored in a freezer at the temperature of -4°C for at least 30 days to let ^{226}Ra, ^{232}Th, and their progenies reach secular equilibriums prior to gamma spectroscopy.

2.3. Determination of activity concentrations
The activity concentrations of natural radionuclides of interest were determined using a reverse electrode, closed-end, coaxial HPGe gamma spectrometer (Canberra; Model GR2519; Serial number 12946022; 51.7 mm crystal diameter; 58.5 mm length), which have a relative efficiency of 20% and an energy resolution of 1.9 keV (FWHM) at the 1.33 MeV peak of ^{60}Co.. The measurement was carried out inside a 12-cm thick cylindrical lead shield with a movable cover to reduce external and background radiation. The detector was connected to a 16k channel integrated multichannel analyzer (Canberra; Model DSA1000). Spectra of gamma rays emitted from the samples were analyzed using the Genie 2000 (3.2) software, in which the calibration of the energy and efficiency was performed using a mixed gamma standard-type EG-ML developed by Eckert & Ziegler (California, USA). All measurements were performed with a 60,000 s counting time period.

The activity concentrations of ^{226}Ra, ^{232}Th, ^{40}K, and ^{137}Cs in the samples were calculated using Eq.1 [4]:

$$A = \frac{N_{net}}{m \times t \times P_{\gamma} \times \epsilon_{\gamma}} \quad (1)$$

where A, N_{net}, m, t, P_{γ}, and ε_{γ} are the activity concentration (Bq kg^{-1}), the net count of gamma rays at the respective energy per second (cps), the mass of the sample (kg), the collection time (s), the probability of the transition of the interested radionuclide at the respective gamma energy, and the efficiency of the gamma spectrometer at the respective gamma energy, respectively. The values of ε_{γ} for a respective gamma energy are shown in Figure 1.

3. Results and discussion

The activity concentrations of ^{226}Ra, ^{232}Th, and ^{40}K in both non-vulcanized NRL and GVNRL are shown in Table 1 and the contributions of each radionuclides to the total activity concentration are shown in Figure 2.

Table 1. Activity concentrations of ^{226}Ra, ^{232}Th, and ^{40}K and total activity concentrations in non-vulcanized NRL and GVNRL.

Sample (gamma vulcanizing dose)	Activity concentration (Bq kg^{-1})			Total activity concentration (Bq kg^{-1})
	^{226}Ra	^{232}Th	^{40}K	
NRL (0 kGy)	29.9±1.1	95.2±1.2	96.2±1.0	221.3±3.3
GVNRL (12 kGy)	30.8±1.5	89.1±0.6	95.4±1.6	215.3±3.7
GVNRL (24 kGy)	31.2±2.7	89.7±0.4	93.2±0.4	214.1±3.5
Average	30.6±1.8	91.3±0.7	94.9±1.0	216.9±3.5

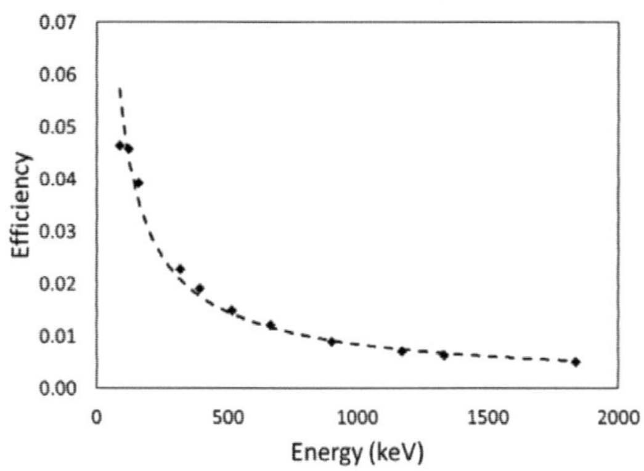

Figure 1. Efficiencies of the gamma spectrometer at respective gamma energies

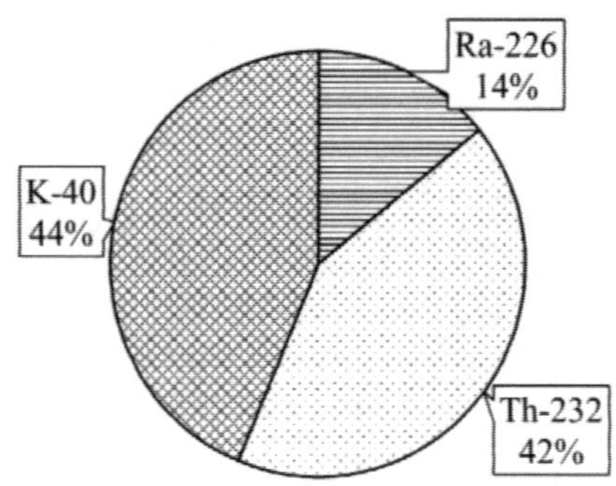

Figure 2. Contributions of ^{226}Ra, ^{232}Th, and ^{40}K in non-vulcanized NRL and GVNRL to the average total activity concentration.

As shown in Table 1, the activity concentrations of ^{40}K were higher than the values of ^{226}Ra and ^{232}Th, ranging from 93.2±0.4 Bq kg^{-1} to 96.2±1.0 Bq kg^{-1} with the average value of 94.9±1.0 Bq kg^{-1}. Interestingly, the activity concentrations of ^{232}Th was only slightly less than the values of ^{40}K, with the average activity concentrations of 91.3±0.7 Bq kg^{-1}, while ^{226}Ra had the least activity concentrations, with the average value of 30.6±1.8 Bq kg^{-1}. The total activity concentrations contributed from natural radionuclides of interest ranged from 214.1±3.5 Bq kg^{-1} to 221.3±3.3 Bq kg^{-1}, with the average total activity concentration of 216.9±3.5 Bq kg^{-1}, in which ^{40}K contributed the highest percentage of activity concentrations to the total value (44%). The contributions from each radionuclide to the total activity concentration are shown in Figure 2.

The higher values of ^{40}K were expected as K is an essential element and the most abundant inorganic mineral nutrient in plant cells due to its functions as a stabilizer in metabolism and an osmoticum contributing to cellular hydrostatic pressure, growth, and environmental changes, leading to high accumulation of K in all parts of the plants, which could be transferred and accumulated in NRL [9]. On the other hand, for ^{226}Ra and ^{232}Th, the results showed that they also significantly accumulated in both non-vulcanized NRL and GVNRL, as seen by their contributions of 14% and 42% to the average total activity concentrations, respectively. Their accumulations could be explained by the fact that although ^{226}Ra and ^{232}Th are not essential elements for plants, they are chemically similar to some essential elements such as calcium, resulting in possible transportation of these radionuclides along with other elements. Another interesting result shown in Table 1 was that when the activity concentrations of each radionuclide in non-vulcanized NRL and GVNRL (both 12-kGy and 24-kGy gamma doses) were compared, the values were not statistically different. This behaviour was observed because both 12-kGy and 24-kGy gamma irradiation used during the vulcanization could not activate or produce radioactive isotopes inside the NRL, resulting in relatively unchanged total activity concentrations in all samples. As a result, it could be concluded from the overall results that, in terms of radiological safety, GVNRL was safe for production and usage as they had similar total activity concentrations as the non-vulcanized NRL.

4. Conclusions

In this work, the activity concentrations of ^{226}Ra, ^{232}Th, and ^{40}K in non-vulcanized NRL and GVNRL that were gamma irradiated with the accumulated doses of 12 kGy and 24 kGy, respectively, were determined using a gamma spectroscopy. The results showed that ^{40}K had the highest activity concentration and contributed the most percentage to the total values. The results also indicated that using gamma irradiation for rubber vulcanization did not statistically changed both individual and total activity concentrations from the natural radionuclides of interest. Hence, based on the overall results, it could be concluded that GVNRL was safe for both production and usage.

5. References

[1] Intharapat P, Kongnoo A and Kateungngan K 2013 *J. Polym. Environ.* **21** 245-58.
[2] Ibrahim S, Badri K, Ratnam C T and Ali N H M 2018 *Radiat. Eff. Defects Solids.* **173** 427-34.
[3] Moonlek B and Saenboonruang K 2019 *Radiat. Eff. Defects Solids.* **174** 452-66.
[4] Saenboonruang K, Phonchanthuek E and Prasandee K 2018 *J. Environ. Radioact.* **184-185** 1-5.
[5] Saenboonruang K, Phonchanthuek E and Prasandee K 2018 *Chiang Mai J. Sci.* **45** 821-31.
[6] Poltabtim W and Saenboonruang K 2019 *Chiang Mai J. Sci.* **46** 778-86.
[7] Njinga R L, Jonah S A and Gomina M 2015 *J. Radiat. Res. Appl. Sci.* **8** 208-15.
[8] Priharti W, Samat S B, Yasir M S and Garba N N 2016 *J. Radioanal. Nucl. Chem.* **307** 297-303.
[9] Dreyer I and Uozumi N 2011 *FEBS. J.* **278** 4293-303.

Acknowledgements

Authors would like to acknowledge financial supports from the Faculty of Science, Kasetsart University under the Undergraduate Research Matching Fund (URMF). Authors also would like to thank the Thailand Institute of Nuclear Technology (TINT) for the technical support and the Specialized Center of Rubber and Polymer Materials for Agriculture and Industry (RPM), Faculty of Science, Kasetsart University, for publication support.

IOP Conf. Series: Materials Science and Engineering **773** (2020) 012012 doi:10.1088/1757-899X/773/1/012012

The effect of hydroxylamine sulfate on the storage hardening of natural rubber

Khwanchat Promhuad[1], and Wirasak Smitthipong [1,2,*]

[1] Specialized center of Rubber and Polymer Materials in agriculture and industry (RPM), Department of Materials Science, Faculty of Science, Kasetsart University, Bangkok, Thailand.
[2] Office of natural rubber research program, Thailand Science Research and Innovation, Bangkok, Thailand.

*Corresponding author (W. Smitthipong): fsciwssm@ku.ac.th

Abstract. Natural rubber (NR) consists of rubber and non-rubber components, which is mostly *cis*-1,4-polyisoprene. The main non-rubber components are proteins and phospholipids which are attached at each end of the NR chains. If we store NR for prolong period, the proteins and phospholipids promote formation of a NR network. This causes the gel formation with increasing the Mooney viscosity of NR, and this is called storage hardening. The objective of this work is to study the effect of hydroxylamine sulfate on the storage hardening of natural rubber. Three types of natural rubber sample were prepared: pure NR, NR with 0.2 phr of hydroxylamine sulfate, and NR with 2.0 phr of hydroxylamine sulfate. Samples were characterized immediately (time 0) and after 12 weeks of storage at room temperature. We found that gel content and Mooney viscosity of NR without hydroxylamine sulfate are increased with the storage hardening for 12 weeks. However, both NR samples with hydroxylamine sulfate represent constant viscosity values because hydroxylamine sulfate inhibits the network and gel formation in NR. This is important as the stabilization of NR properties in the function of time is required by rubber industry.

1. Introduction

Natural rubber (NR) can be obtained from *Hevea brasiliensis* plant. NR latex has been used in industry for surgical glove, pillow and mattress, elastic band, medical products, and tire industry, plus many more. NR consists of rubber and non-rubber components, the main component is the *cis*-1,4-polyisoprene. The non-rubber fractions consist of proteins (ω-terminal) and phospholipids (α-terminal) which are not only attached at each end chain of polyisoprene but also included in the serum [1]. When we store the NR for prolonged period of time, proteins and phospholipids at the end chain could promote formation of NR network through protein-protein interactions at the ω-terminal or phospholipid-phospholipid interaction at the α-terminal. The increase of branch chain or network of NR as a function of time can be determined by the Mooney viscosity *via* the storage hardening [2]. The storage hardening of NR can be inhibited by a stabilizer. Nimpaiboon *et al.*, [3] studied the effect of polar chemicals on the storage hardening of NR. Three types of polar chemicals were used: phenol, diethylene glycol, and hydroxylamine hydrochloride. They found that gel content, Mooney viscosity, and Wallace plasticity values of NR samples with phenol and diethylene glycol are increased with storage hardening. However, NR sample with hydroxylamine hydrochloride represents constant values of those three parameters.

IOP Conf. Series: Materials Science and Engineering **773** (2020) 012012 doi:10.1088/1757-899X/773/1/012012

2. Experimental

Four hundred grams of rubber samples from natural latex (supplied by Num rubber and latex. Co. Ltd, Trang, Thailand) were masticated using a two-roll mill (Yong Fong). Three types of NR sample were prepared: pure NR, NR with 0.2 phr of hydroxylamine sulfate (HS from Sigma-Aldrich), and NR with 2.0 phr of hydroxylamine sulfate. Then, samples were characterized part same day after preparation (time 0 or 0 week) and after 12 weeks of storage at room temperature. The gel content was measured by dissolving NR sample in toluene (RCI Labscan). The solution was kept in the dark at room temperature, for one week. Finally, the solution was filtered, weighed, and calculated with respect to the original NR sample [4]. Mooney viscosity analysis was performed using a Mooney viscometer (TECHPRO, viscTECH+) within a sealed, pressurized and heated cavity. The rubber sample was heated at 100°C for 1 min before analysis. After that, the rubber samples were continuously measured for 4 min by the torque required to keep the rotor rotating at a constant rate as a function of time for reading the Mooney viscosity which was recorded as torque in Nm. The functional group in rubber samples was determined by using attenuated total reflection Fourier transform infrared (ATR-FTIR, VERTEX 70 BRUKER). Rubber samples were cut into small pieces and put on Ge crystal probe at 600-4000 cm^{-1}. The processability under strain sweep modes of the rubber samples was investigated by a rubber processing analyser (RPA 2000, Monsanto). The strain sweep mode was in the range of 0.5–100%, the process was carried out at 100°C and 1 Hz.

3. Results and discussion

Mooney viscosity relates to the flow of elastomers like natural rubber, the value depends on the macrostructure of rubber. The results of the Mooney viscosity of samples at various storage times are shown in Table 1. The Mooney viscosity of pure NR increases after 12 weeks, in contrast, Mooney viscosity of NR with hydroxylamine sulfate is stable whatever the concentration of HS.

Table 1. Mooney viscosity of NR samples for 0 and 12 weeks

Samples	**Mooney viscosity (± 5 a.u.)**	
	0 week	**12 weeks**
Pure NR	40	52
NR with HS 0.2 phr	45	46
NR with HS 2.0 phr	40	38

The results of the gel content are shown in Table 2. At time 0, the pure NR has gel content larger than that of NR with hydroxylamine sulfate. After 12 weeks, pure NR has a higher gel content resulting from gel formation during prolonged storage. The gel content depends on proteins and lipids that are responsible for the gel formation. In contrast, the gel content is stable for NR with hydroxylamine sulfate. The effect of proteins and lipids causes a gel network that calls storage hardening in NR. This result is in good agreement with the result of Mooney viscosity.

Table 2. Gel content of NR samples for 0 and 12 weeks

Samples	**Gel content (wt.%)**	
	0 week	**12 weeks**
Pure NR	57.0 ± 0.4	64.9 ± 1.9
NR with HS 0.2 phr	54.7 ± 3.2	54.6 ± 5.6
NR with HS 2.0 phr	53.3 ± 2.2	53.7 ± 0.5

The FTIR spectra have been used to determine the composition of samples, especially with respect of the N-H bonds in NR with hydroxylamine sulfate samples at 3600 – 2600 cm^{-1} (Figure 1), indicating that at time 0, samples showed peaks of the amine group at 3280 cm^{-1}, the fatty acid ester group at 1740 cm^{-1}, the aldehyde group at 1710 cm^{-1} and the amide I group at 1660 –1630 cm^{-1} (data not shown here). After 12 weeks, NR with hydroxylamine sulfate samples represented rather no change in the FTIR spectra.

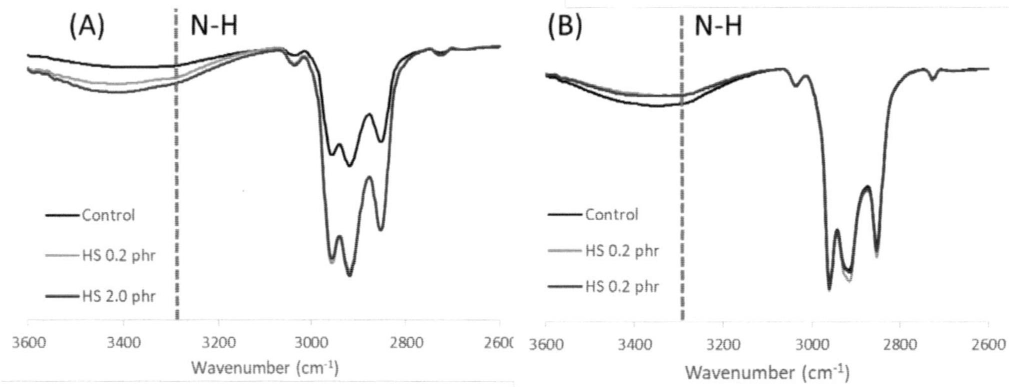

Figure 1. FTIR spectra of NR with/without hydroxylamine sulfate samples: (A) 0 week and (B) 12 weeks

In the case of pure NR, the peak of the amine group at 3280 cm^{-1} and the peak of the amide I group at 1660 – 1630 cm^{-1} were increased respectively after storage time for 12 weeks. The results of the storage modulus (G') and the strain of NR with hydroxylamine sulfate samples for 0 and 12 weeks were shown in Figure 2. At time 0, the results showed that all the samples represent the same level of rigidity, this result is in good agreement with the result of Mooney viscosity. After 12 weeks, the results of NR with hydroxylamine sulphate rather decreased compared to those at time 0, except for pure NR where its G' increased.

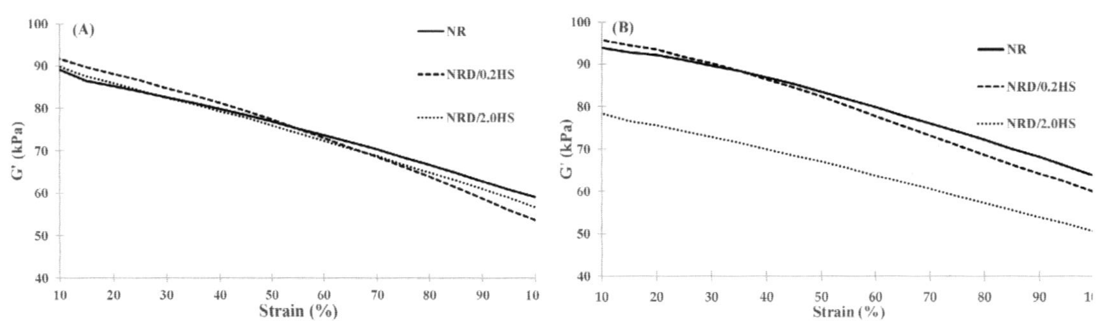

Figure 2. Relationship between storage modulus (G') and the strain of NR with/without hydroxylamine sulfate samples for 0 week (A) and 12 weeks (B)

4. Conclusions

Hydroxylamine sulfate can be used to inhibit storage hardening in NR. The dynamic gel networks and Mooney viscosity of NR with hydroxylamine sulfate are stable compared to the NR without hydroxylamine sulphate. This means that the viscosity stabilizer restrained the intermolecular links between non-rubber components of the rubber chains, resulting in an almost constant Mooney viscosity and gel content. The result of processability under strain sweep is in good agreement with that of Mooney viscosity.

Acknowledgement
Authors would like to thank Num rubber and latex Co. Ltd., Trang, Thailand for the support of natural latex and Thai Eastern Rubbers Co. Ltd. for the support of hydroxylamine sulfate.

5. References

1. Chollakup, R., et al., *New approach on structure-property relationships of stabilized natural rubbers.* Journal of Polymer Research, 2019. **26**(2): p. 37.
2. Tarachiwin, L., et al., *Structural characterization of α-terminal group of natural rubber. 2. Decomposition of branch-points by phospholipase and chemical treatments.* Biomacromolecules, 2005. **6**(4): p. 1858-1863.
3. Nimpaiboon, A., M. Sriring, and J.T. Sakdapipanich, *Molecular structure and storage hardening of natural rubber: Insight into the reactions between hydroxylamine and phospholipids linked to natural rubber molecule.* Journal of Applied Polymer Science, 2016. **133**(31).
4. Smitthipong, W., et al., *Study of tack properties of uncrosslinked natural rubber.* Journal of adhesion science and technology, 2004. **18**(12): p. 1449-1463.

IOP Conf. Series: Materials Science and Engineering **773** (2020) 012013 doi:10.1088/1757-899X/773/1/012013

Study of rubber/calcium carbonate composites

N Phuhiangpa[1], S Phongphanphanee[1], and W Smitthipong[1,2,*]

[1] Specialized center of Rubber and Polymer Materials in agriculture and industry (RPM), Department of Materials Science, Faculty of Science, Kasetsart University, Bangkok, Thailand.
[2] Office of natural rubber research program, Thailand Science Research and Innovation, Bangkok, Thailand.

*Corresponding author (W. Smitthipong): fsciwssm@ku.ac.th

Abstract. Fillers have been widely used in the rubber industry for many applications such as tile floor, vehicle tire, etc. Calcium carbonate ($CaCO_3$) is one of the important inorganic powders and it is widely used as filler in order to reduce the cost in rubber industry. The objective of this study is to investigate physical and mechanical properties of the rubber composites with $CaCO_3$ powder. We studied the $CaCO_3$ additive in natural rubber (NR) with 25, 50, 75 and 100 parts per hundred rubber (phr). The effect of $CaCO_3$ on the properties of rubber composite, such as Mooney viscosity, bound rubber, Mullins effect and Payne effect, was investigated. The result of Mullins effect of rubber composite filled with $CaCO_3$ is in good agreement with the result of bound rubber: higher bound rubber leads to higher stress to pull the rubber composites. The Payne effect shows that the value of different storage moduli ($\Delta G'$) of rubber compound filled with 25 phr $CaCO_3$ is the lowest due to weaker filler network. While, the NR supplemented with 100 phr $CaCO_3$ represents more significant $\Delta G'$ with the strain increasing. This type of material could be applied for tailoring the properties of rubber products.

1. Introduction

Natural rubber (NR) is a major agricultural product, which is widely used in industry, for example, tile floor, tire, glove, pillow and mattress, medical products, etc. NR obtained from *Hevea brasiliensis* consists of *cis*-1,4-polyisoprene and nonrubber components (proteins, phospholipids, sugar, salt, etc.). Until now, there are no synthetic materials that can replace natural rubber obtained from plants [1]. NR is often reinforced by incorporation of filler to improve its mechanical properties: modulus, hardness, tensile strength, abrasion resistance and tear resistance [2].

Recently, fillers have been widely used in the rubber industry for many purposes such as improvement of mechanical properties, efficient production and reducing cost of rubber products. Generally, there are three main groups of filler, including reinforcing fillers, semi-reinforcing fillers, and non-reinforcing fillers. Efficiency of the reinforcing filler depends on several factors such as particle size, surface area and shape of filler [3]. Among commercial fillers, carbon black and silica (SiO_2) are the most important reinforcing fillers. They are added to improve the mechanical properties of rubber compounds. However, there are some fillers that can be used as none or semi-reinforcing fillers (such as clay, talc, and calcium carbonate, etc.) for either reducing cost or improving mechanical properties [4]. Calcium carbonate ($CaCO_3$) is one of the important inorganic powders and widely used as filler in paints, plastics, and rubber industry in order to reduce the material cost [5]. The main aim of this work is to study the effect of $CaCO_3$ in rubber composites on the properties, in particular physical and mechanical properties.

2. Experimental

For the sample preparation, the natural rubber (NR) grade STR 5L was first masticated with a two-roll mill at 70°C for 5 minutes. Composite samples were prepared by mixing NR with $CaCO_3$ at 25, 50, 75 and 100 phr for 10-15 minutes (totally four samples). All samples were then cut into form of testing sheets. For scanning electron microscopy analysis (SEM: FEI, Quanta 450 FEI), rubber samples were cut into small pieces, and coated with gold for the morphology analysis. Mooney viscosity was determined by using Mooney viscometer of the rubber compounds. Bound rubber was considered a quantitative measure of the filler surface activity and the rubber-filler interaction. Bound rubber was determined by immersing 1 gram of rubber composite in 100 ml of toluene solvent at room temperature for 7 days. After dissolution, a piece of rubber that is insoluble in toluene was filtered and weighed [6]. Mullins effect of the rubber composite was measured using Universal Testing Machine with strain range: 1–100% in cycle mode. Payne effects of $CaCO_3$ supplemented rubber composite was analyzed using rubber process analyzer under temperature 60 °C, frequency 1 Hz, strain range: 1–100%. The value of different storage moduli ($\Delta G'$) means $G'_{max} - G'_{min}$.

3. Results and discussion

Before using as rubber filler, the $CaCO_3$ material was analyzed by scanning electron microscopy (SEM). This proves the narrow size distribution around 1 micron of particles (data not shown). Analysing the Mooney viscosity of NR with $CaCO_3$ additives the results show steady increase of viscosity of the rubber composites between 58-62 a.u. for the composition of 25-100 phr of filler (Figure 1). The $CaCO_3$ are solid particle fillers, as expected, when mixed with NR, the product has increased viscosity. This relates to the hydrodynamic effect [7], which causes the increase in the rubber viscosity when supplemented with none-reactive solid additive. This also suggests good blending of used compounds.

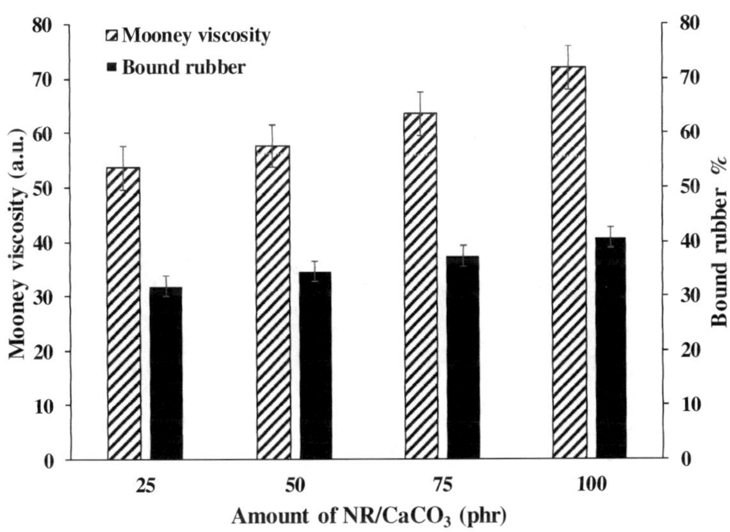

Figure 1. Linear increase of the Mooney viscosity and bound rubber content with increased amount of the $CaCO_3$ (25-100 phr) in NR.

The bound rubber refers to the fraction of polymer which cannot be extracted with a good solvent from a rubber-filler mix. To evaluate the interaction between filler particles and NR, the bound rubber of uncrosslinked composites was measured by extraction with toluene at room temperature. The percentages of bound rubber in all samples (25-100 phr of $CaCO_3$) are also summarized in Figure 1. The highest bound rubber is for natural rubber samples with 100 phr $CaCO_3$ added. These results are in good agreement with the Mooney viscosity analysis.

The Mullins effect is a phenomenon observed in rubber material where the stress-strain equilibrium changes with the softening. This is regarded as a strain-induced softening character and is observed in elastomeric polymers when forces are repeatedly applied to a material which leads to weaken of rubber material. The Mullins effect was analyzed by stress strain curve (3 loops of go-return curves) for natural rubber with addition of $CaCO_3$ at 25, 50, 75 and 100 phr respectively (data not shown). When increasing $CaCO_3$ in NR, the stress of NR composites tends to be increased, especially at high loading of $CaCO_3$ (75 and 100 phr).

The Payne effect is a feature of the stress-strain behaviour, which is related to the shape change by causing strain in the rubber with fillers. This phenomenon is regarded to the storage modulus (G') and loss modulus (G") in shear deformation conditions. The reason for this phenomenon is the formation of a network formed by filler-filler interaction in the composites at low strain. In real life, this means the energy loss (Tan delta) and the efforts are to maximize this effect in filled rubber [8]. For instance, the result of shear modulus of the NR composites corresponding to the amount of $CaCO_3$ filler shows differences. This phenomenon is caused by the filler–filler networks and adsorption–desorption of polymeric chains at the filler interface of the rubber composites. Looking at the Payne effect results (Figure 2), the value of different storage moduli (ΔG') of rubber compound filled with 25 phr $CaCO_3$ is the lowest due to weaker filler network. While, the NR supplemented with 100 phr $CaCO_3$ represents more significant ΔG' with the increasing of strain. The value of damping factor (Tan delta) corresponds to energy loss showed that $CaCO_3$ particle at 100 phr has more pronounced values. Energy loss from the straining also produced heat caused increasing of temperature, higher ΔG' higher Tan delta.

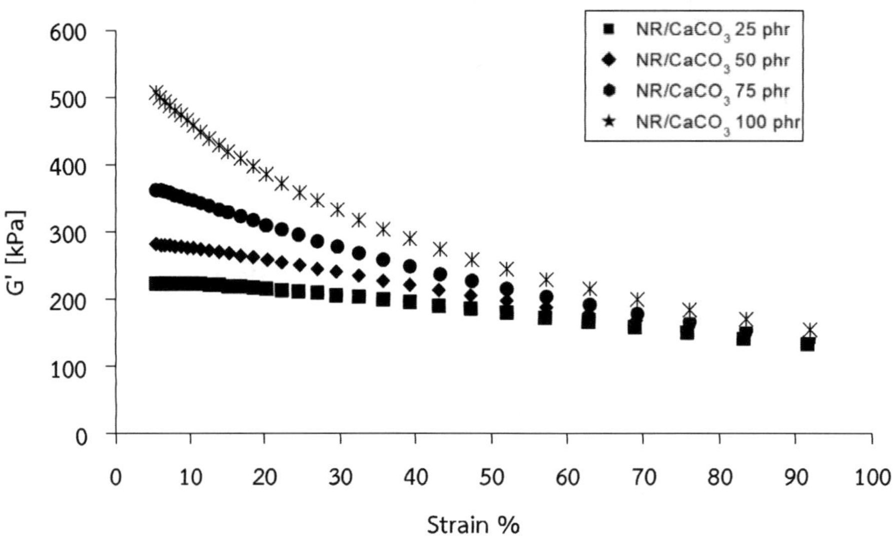

Figure 2. Payne effect as a function of strain for NR with $CaCO_3$ particles at 25-100 phr.

4. Conclusions

In this research, we found that the Mullins effect of rubber composite filled $CaCO_3$ is in good agreement with the result of bound rubber, higher bound rubber of rubber composite higher stress to pull the rubber composites. From the Payne effect point of view, the value of different storage moduli (ΔG') of rubber compound filled with 25 phr $CaCO_3$ is the lowest due to weaker filler network, while the NR supplemented with 100 phr $CaCO_3$ represents more significant ΔG', higher ΔG' higher Tan delta.

5. References

[1] Smitthipong, W., Suethao, S., Shah, D., & Vollrath, F. (2016). Interesting green elastomeric composites: Silk textile reinforced natural rubber. *Polymer Testing, 55*, 17-24.

[2] Findik, F., Yilmaz, R., & Köksal, T. (2004). Investigation of mechanical and physical properties of several industrial rubbers. *Materials & Design, 25*, 269-276.

[3] Khan, I., & Bhat, A. H. (2014). Micro and Nano Calcium Carbonate Filled Natural Rubber Composites and Nanocomposites. In (pp. 467-487).

[4] Prasertsri, S., Vudjung, C., Inthisaeng, W., Srichan, S., Sapprasert, K., & Kongon, J. (2018). Comparison of Reinforcing Efficiency between Calcium Carbonate/Carbon Black and Calcium Carbonate/Silica Hybrid Filled Natural Rubber Composites. *Defect and Diffusion Forum, 382*, 94-98.

[5] Gopinath, C. S., Hegde, S. G., Ramaswamy, A. V., & Mahapatra, S. (2002). Photoemission studies of polymorphic CaCO3 materials. *Materials Research Bulletin, 37*(7), 1323-1332.

[6] Mohapatra, S., & Nando, G. (2015). Analysis of Carbon Black–Reinforced Cardanol-Modified Natural Rubber Compounds. *Rubber Chemistry and Technology, 88*, 289-309.

[7] Fröhlich, J., W. Niedermeier, and H.D. Luginsland, *The effect of filler–filler and filler–elastomer interaction on rubber reinforcement.* Composites Part A: Applied Science and Manufacturing, 2005. **36**(4): p. 449-460.

[8] Fang, Q., Song, B., Tee, T.-T., Sin, L., Hui, D., & Bee, S.-T. (2014). Investigation of dynamic characteristics of nano-size calcium carbonate added in natural rubber vulcanizate. *Composites Part B: Engineering, 60*, 561-567.

Acknowledgement

The authors gratefully acknowledge the financial support from the Siam United Rubber Co., Ltd.

IOP Conf. Series: Materials Science and Engineering **773** (2020) 012014 doi:10.1088/1757-899X/773/1/012014

The study of a small-sized Thai chili paste supply chain

J Srilarp and P Ongkunaruk*

Department of Agro-Industrial Technology, Faculty of Agro-Industry, Kasetsart University, 50 Ngam Wong Wan Rd. Lad Yao, Chatuchak, Bangkok 10900, Thailand.

*Corresponding author: pornthipa.o@ku.ac.th

Abstract. The highly competitive lifestyle results in the increasing demand for ready to eat food. In Thailand, chili paste is a traditional food that has been passed down from generation to generation. It plays an important role in Thai meals and becomes more popular in convenience stores and hypermarkets. The objectives of this study are to explore the current supply chain of a small-sized Thai chili paste enterprise, to analyze the business processes and identify the problem and provide the recommendation for improvement. First, we observe and perform an in-depth interview of the owner and staff for realizing the business process. Then, the result is analyzed by using the Integration Definition for Function Modeling (IDEF0) to show the existing stakeholders and current key activities in Thai chili paste. The major problems are lack of data collection, low forecast accuracy, lack of inventory management, and low productivity. The proper controls are proposed. Finally, our guidelines could enhance the productivity of other small-sized enterprises which can increase the potential capability for domestic and international markets.

1. Introduction

Chili Paste or Namprik is a traditional food and nationwide favorite side dish in Thai cuisine. It reflects Thai ancient wisdom since eating spicy food encourages the body to perspire profusely and, provide a body suitable for the tropical climate [1]. Thai chili paste is made of various components i.e., a fresh or dried chili for a spicy taste, salty from the fish sauce or fish paste, and herbs or spices such as garlic, shallots, and kaffir lime rind. However, other components can be included in the chili paste based on preferences and regions. It is consumed with rice, fish or meat, and fresh or boiled vegetables to reduce the resulting tongue-burning from hot and spicy. It encourages consumers to eat more vegetables and appreciate the taste of rice and fish. Nowadays, urbanization lifestyle and advanced food processing technology drive a need for an instant or packed chili paste, especially for a migrated people who stay in Bangkok and other cities. It satisfies rural people to continue their local chili paste in the meal [2]. Hence, instant chili paste becomes more popular in convenience stores and hypermarkets. Most consumers decided to buy the instant chili paste based on taste satisfaction at convenience stores and hypermarkets. Thai SME's contributed to the 2018 Gross Domestic Product (GDP) 43 percent of the country's overall production. The food manufacturing sector contributed to 258.8 Billion Bath [3]. At present, there are more than 1.2 hundred thousand Thai SMEs in the food industry. The increment of potential capability and productivity of SMEs can raise the competitiveness for the food industry sector [4]. Our objective was to study a chili paste supply chain of the small-sized enterprise that produces the instant chili paste. Next, the activities and problems of manufacturer were identified, and the improvement methods were proposed so as to increase the productivity of the manufacturer. This research would help the SMEs to strengthen the ability for competitive advantage throughout the supply chain.

2. Methodology

In this study, we conducted an in-depth interview with the owner and staff by a questionnaire to figure out the activities of the chili paste supply chain. Then, we identified the major activities and analyse the business processes based on Integration Definition for Function Modeling (IDEF0), which represents the flow of activities in the form square such as planning, sourcing, making delivering and returning. There are four types of arrows that identify the input, output, control, and mechanism of activities [5]. First, the input consists of the factors that drive the activities, represented by a left arrow directed into an activity box. Second, the output is the result of doing the activities, represented by a right arrow directed out of the activity box. Third, the control is the standard, regulation or requirement of the activity, represented by an arrow above an activity box. Finally, the mechanism includes resources such as staff, equipment, and machines used to accomplish the activities represented by an arrow above an activity box [6]. The code located in the right corner of each activity box designates the rank of the activity such as A1, A2, …, A5 [7]. In addition, a solid line represents current activities [8] and we proposed implementation best practice to increase their efficiency as shown by the dashed line [9] in Figure 1.

3. Results and discussion

3.1. Background of the Thai chili paste factory

A case study company is a Thai chili paste factory established in 2006 with over 30 years of food experience from small community shops in Samut Prakan. Throughout the years, they continuously improve product quality and receiving many awards such as Top of the best OTOP and Thai Star Packaging Awards 2014. At present, the factory is certified GMP, HACCP, and Halal. A unique recipe has been passed down from generation to generation of the owner family that makes their product quite popular in convenience stores and hypermarkets. They have two types of products. First, chili paste in a small, medium and large plastic container. Second, chili paste in a small plastic box for one meal. There are various flavors that provided a variety of choices to customers where the best seller is Narok chili paste. The channels of distribution are modern trade, convenience stores, middlemen, provincial souvenir stores and online. In the future, the products will be exported to CLMV countries through the agent so as to increase sales. Hence, the factory needs to improve the efficiency of the supply chain for the future export markets.

3.2. Business process analysis of the Thai chili paste supply chain

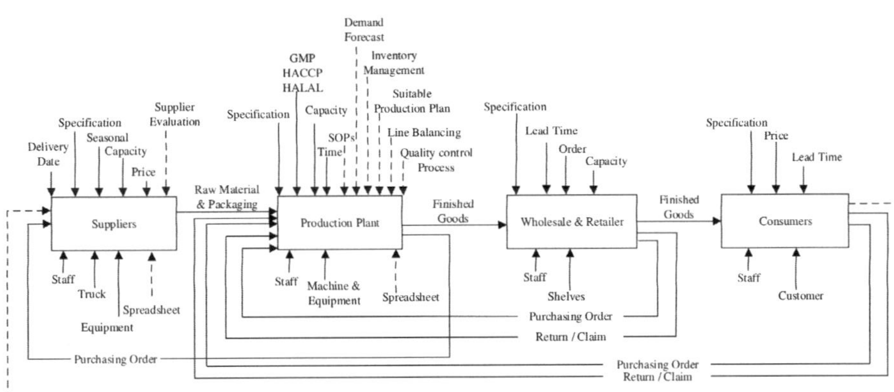

Figure 1. The business process of Thai chili paste supply chain (IDEF0 Level 0).

The business process analysis in the organization level from upstream to downstream was analysed by IDFE0 level 0. The solid line represents current activities with input, output, control, and mechanism from upstream to downstream of a supply chain and the dashed lines are improvements that we suggested to increase their efficiency. The major stakeholders of the Thai chili paste supply chain are suppliers, the factory, wholesalers and retailers, and consumers. The suppliers provide raw material such as chili, fish, other ingredients and packaging. For vegetables, the factory outsources the logistics service providers to deliver from the wholesale market. The manufacturer processes chili paste and sells through wholesalers and retailers. Finally, consumers buy chili paste through various channels such as modern trades, convenience stores or online channels.

3.3. Business analysis of the Thai chili paste factory

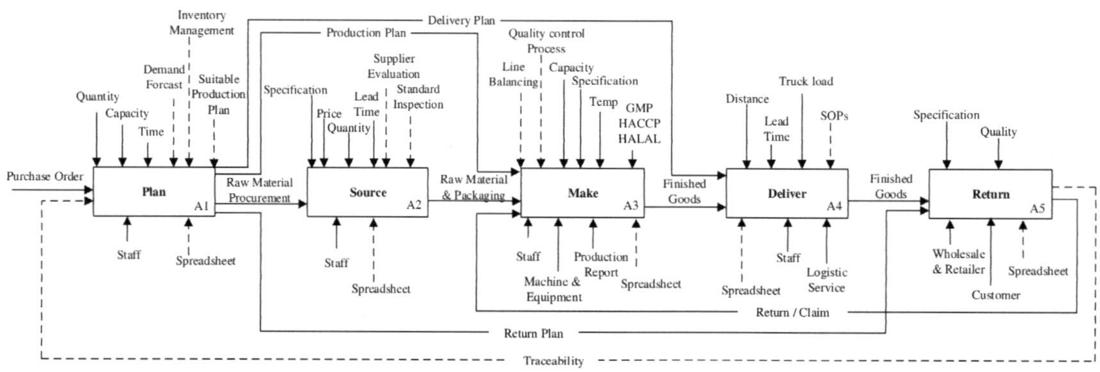

Figure 2. The business process of Thai chili paste factory (IDEF0 Level 1).

The business process analysis in activities level was analyzed by IDFE0 level 1 which showed all activities in the factory consisting of planning, sourcing, making, delivering and returning in Figure 2. The production plan was estimated based on the demand for products from the major customers accounted for 82 percent of total demand. Most of the production was made-to-order, but the safety stocks for online orders were also maintained. For the major customer, the orders were placed every Tuesday and Friday and the products were delivered on Friday and Tuesday, respectively. The delivery lead time was 3 days. The products that are delivered to convenience stores must be manufactured less than 7 days before with first in first out basis. Due to a lack of precise plan, we suggested that the sales data should be collected to perform demand forecasting and create a suitable production plan and manage inventory. The sourcing activities were to order raw materials and packaging. For vegetables, their quality was fluctuated due to seasonality. The factory procured each raw material from a single supplier which is quite risky for shortage. In addition, there was no standard measurement for inspection raw material from suppliers. Then, the inspection standard should be established including the maximum level of acceptable defects. The production processes were receiving and preparing raw materials by removing foreign matters, weighing, mixing ingredients by a mixer which controls time and temperature due to a critical point to control process, drying by hot air oven to control water activities for enhancing shelf life, filling and packing, respectively. At present, products could not be finished on time hence there occurred the over-time production while there were some delayed and waiting times in some processes. Hence, a line balancing technique should be studied for productivity improvement. For wholesale and retail customers, the deliver was outsourced to logistics service providers who delivered products to the distribution center. On the other hand, for the other customers, the products were delivered via multiple channels i.e., postal service, logistics service providers. We suggest creating a standard operation procedure (SOPs) for a logistic service provider to deliver a good quality product to customer. Finally, the return activity which the customer returns a non-conformity product back to the factory. We suggested to build up a traceability system to identify the root causes of the non-conformity product and prevent not to produce and pass the unsafe products in the future by setting the preventive plan. The summary of problems and recommendations for improvement was explained in Table 1.

Table 1. The problems and recommendations for improvement for main activities

Activies	Problems	Recommendation for improvement
Plan	lack of precise plan and strategic forecast	Collect demand data, demand forecasting, create a suitable production plan and inventory policy
Source	No standard measurement for receiving the raw materials from suppliers	Create the standard for raw material inspection and keep a record in the spreadsheet
	The quality of vegetables has fluctuated	Supplier Evaluation
	Buying raw materials from a single supplier	Find alternative suppliers to prevent raw materials shortage and negotiate the best price
Make	The delay or waiting time in the process	Using the line balancing technique, quality control, process improvement for productivity improvement
	Lack of systematic data collection	Create a spreadsheet to record data
Deliver	No standard for control delivery system	Create the standard operation procedure for delivery
Return	No management system for returned goods	building up a traceability system, identify root-cause of the nonconformity product and the preventive plan

4. Conclusions

The business process analysis of the Thai chili paste supply chain was analysed by IDFE0. There were five major stakeholders i.e. supplier, the production plant, wholesalers, retailers, and consumers. We identified the problems and provided the recommendation for improvements such as collect historical data to perform demand forecasting, creating a suitable production plan, evaluating suppliers and creating a standard inspection for raw materials, using suitable techniques for productivity improvements such as line balancing, quality control tools and traceability system. These guidelines could assist other small-sized enterprises in the food industry to understand the problems and solutions so as to escalate the potential capability for domestic and international markets.

5. References

[1] Seubsman S -A, Suttinan P, Dixon J, Banwell C 2009 *Meals in Sci. and Practice* 413-451.
[2] Chaveesuk R and Jitareetep C 2017 *KnE Life Sci.* 136-145.
[3] https://www.sme.go.th/en/download.php?modulekey=94; November 2019.
[4] https://www.sme.go.th/upload/mod_download/download-20190909103733.pdf; November 2019.
[5] Prasertwattanakul Y and Ongkunaruk P 2016 *KnE Life Sci.* 156-160.
[6] Ongkunaruk P 2015 *Agric. Agric. Sci. Proc.* **3** 35-9.
[7] Prasertwattanakul Y and Ongkunaruk P 2018 *Int. Food Res. J.* 25(2) 481-6.
[8] Yupensuk W and Ongkunaruk P 2017 *KnE Life Sci.* 162-169.
[9] Pradita S P and Ongkunaruk P 2019 *IOP Conf. Ser.: Mater. Sci. Eng.* **526** 012004

Acknowledgement
The authors would like to thank the owner and staff of the chili paste enterprise who kindly provided information for this research.

Synthesis of Graphene Foam Filter by Chemical Vapor Deposition Using Palm Oil as a Carbon Source for $PM_{2.5}$ Capture

T Naksakul[1], P Sunpong[2] and A Ruammaitree[1,*]

[1] Department of Physics, Faculty of Science and Technology, Thammasat University, Pathum Thani 12120, Thailand.
[2] Department of Gems and jewelry, Faculty of Gems, Burapha University, Chanthaburi 22170, Thailand.

*Corresponding author: u4605070@hotmail.com

Abstract. Particulate matter which has aerodynamic diameter less than 2.5 μm ($PM_{2.5}$) has seriously impacted on human health such as cardio-respiratory diseases. Graphene foam is a 3-dimensional graphene, in which carbon atoms are arranged into a honeycomb lattice. In this study, $PM_{2.5}$ capture using graphene foam filter is demonstrated. The graphene foam is synthesized by chemical vapor deposition using nickel foam and palm oil as a template and carbon source, respectively. After that the sample is soaked with $FeCl_3$ to dissolve the nickel foam template. The presence of graphene on nickel foam is confirmed by Raman spectroscopy. The morphologies of graphene and $PM_{2.5}$ particles are measured by field emission scanning electron microscope. The element composition of $PM_{2.5}$ is investigated by energy-dispersive X-ray spectroscopy (EDX). The results show that graphene foam can capture various types of $PM_{2.5}$ particles (such as aluminium, silicon, phosphorus and so on) without additional energy.

1. Introduction

Particulate matter which has aerodynamic diameter less than 2.5 μm ($PM_{2.5}$) [1] has seriously impacted on human health especially cardio-respiratory diseases [2]. Nowadays, people in many cities are still suffering from $PM_{2.5}$ air pollutants although air purifier which is expensive and high energy consumption can remove contaminants (include $PM_{2.5}$) from the air. In order to make the air purifier available to everyone, the cost and energy consumption of the air purifier must reduce.

Graphene is a 2-dimensional material in which carbon atoms are arranged into a honeycomb lattice. Graphene foam is a 3-dimensional connected network structure of graphene. Therefore, the surface area of graphene foam is much more than that of general graphene resulting in the graphene foam can capture a large quantity of $PM_{2.5}$. Palm oil is a cheap and available vegetable oil but contains a large amount of carbon. Currently, there are many methods can synthesis graphene such as mechanical exfoliation of graphite [3], improved Hummer's method [4], annealing SiC [5-8], Chemical vapor deposition (CVD) [9-11] and so on. CVD is a simple and low-cost method for graphene foam synthesis. Therefore, the synthesis of graphene foam filter by CVD using palm oil as a carbon source for $PM_{2.5}$ capture is demonstrated in this study. SEM and EDX are utilized to investigate sizes and element composition of $PM_{2.5}$ particles which are captured on the graphene foam.

2. Experimental

A nickel foam plate with a size of 3×20×0.16 cm^3 was cleaned in acetone by ultrasonic cleaner for 5 min. After the acetone was evaporated, a tray of palm oil with quantity of 25 ml and the cleaned nickel foam substrate were put into the quartz tube in the area of furnace 1 and furnace 2, respectively, as shown in Figure 1. The air in the quartz tube was evacuated until the pressure in the quartz tube is lower than 3×10^{-2} torr. Then, the nickel foam substrate was annealed to the growth temperature of graphene at 800°C by furnace 2. At the graphene growth temperature, the vapor of palm oil was introduced to the area of furnace 2 for 30 min by annealing the palm oil at 400°C using furnace 1. After that the nickel foam substrate was fast cool down to room temperature. Thereafter, nickel metal was dissolved by sinking the sample in $FeCl_3$ to obtain the graphene foam. Finally, the graphene foam was soaked in water to rinse off remaining $FeCl_3$.

For the test of $PM_{2.5}$ capture, the graphene on nickel foam and graphene foam with electric potential of 0V, 5V and 10V were put in a box. The $PM_{2.5}$ particles were generated from incense smoke and introduced into the box for 60 min. A density of $PM_{2.5}$ particles in the box is 2000 μg/m^3. The presence of graphene on substrate was confirmed by Raman spectroscopy with a laser excitation wavelength of 532 nm. The size of the laser spot was 1 μm. The morphology and element analysis of $PM_{2.5}$ particles were measured by SEM and EDX, respectively.

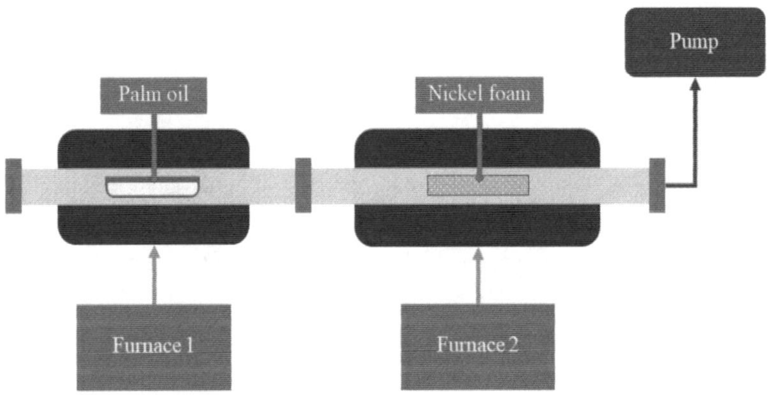

Figure 1. Schematic of experimental setup of graphene growth on nickel foam by CVD using palm oil as a carbon source.

3. Results and discussion

Figure 2 shows Raman spectrum of graphene on nickel foam underwent the CVD process using palm oil as a carbon source. D, G and 2D bands appear at about 1350 cm^{-1}, 1580 cm^{-1} and 2700 cm^{-1}, respectively. Generally, graphene has 2 main bands in the Raman spectrum at about 1580 cm^{-1} (G band) and 2700 cm^{-1} (2D band) which are originated from E_{2g} vibrational mode and second-order two-phonon mode, respectively [12]. Nevertheless, The D band at ~1350 cm^{-1} is also another specific Raman peak of graphene because it will appear where the symmetry of graphene structure is broken by edge or defect [13]. Therefore, the appearance of G and 2D bands in Figure 2 confirms the presence of graphene on the nickel foam substrate.

Figure 3 shows SEM images of graphene on nickel foam (Figure 3 (a)-(c)) and graphene foam (Figure 3 (d)-(f)) after testing $PM_{2.5}$ capture with electric potential of 0V (Figure 3 (a) and (d)), 5V (Figure 3 (b) and (e)) and 10V (Figure 3 (c) and (f)). The EDX results (Table 1) show that there is no $PM_{2.5}$ particle on the graphene on nickel foam samples (Figure 3 (a)-(c)). Conversely, Figure 3 (d)-(f) show many tiny particles with the size of less than 2 μm on the surfaces of graphene foams. The EDX results confirm that the $PM_{2.5}$ particles contain various elements such as aluminium, silicon, phosphorus and so on. Moreover, they also show that there are Fe and Cl elements remain on the graphene foam surfaces. The $PM_{2.5}$-capture test show that the graphene foam can capture $PM_{2.5}$ particles without applying electric potential (Figure 3 (d)). It is different from the reduced graphene-oxide filter which needs electric potential of 5V for capturing $PM_{2.5}$ particles [14]. The $PM_{2.5}$-capture ability of graphene foam may be attributed to the electrostatics which originate from the remaining Fe and Cl elements on the sample surface.

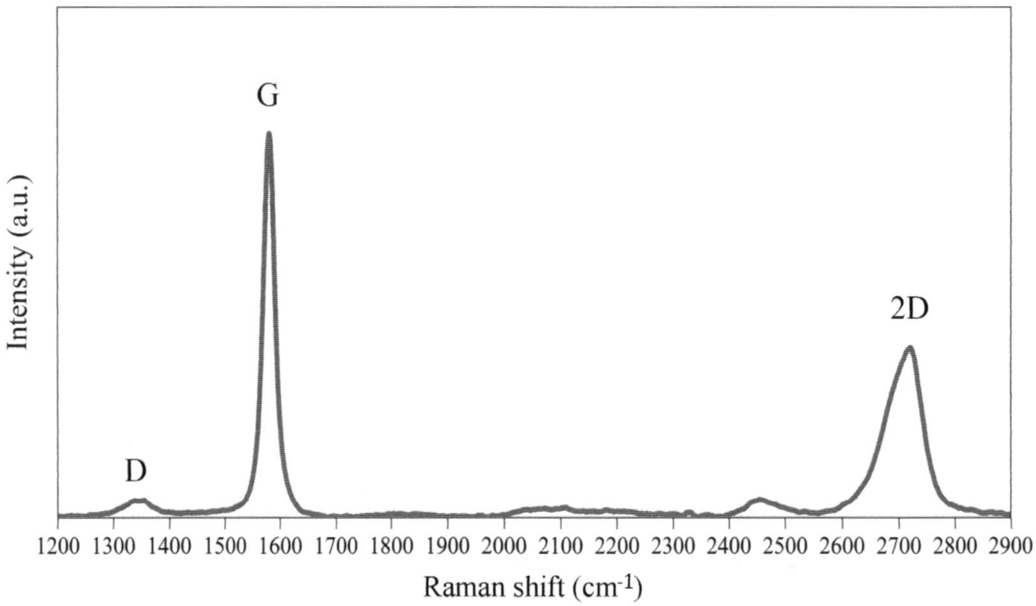

Figure 2. Raman spectrum of the graphene on nickel foam substrate.

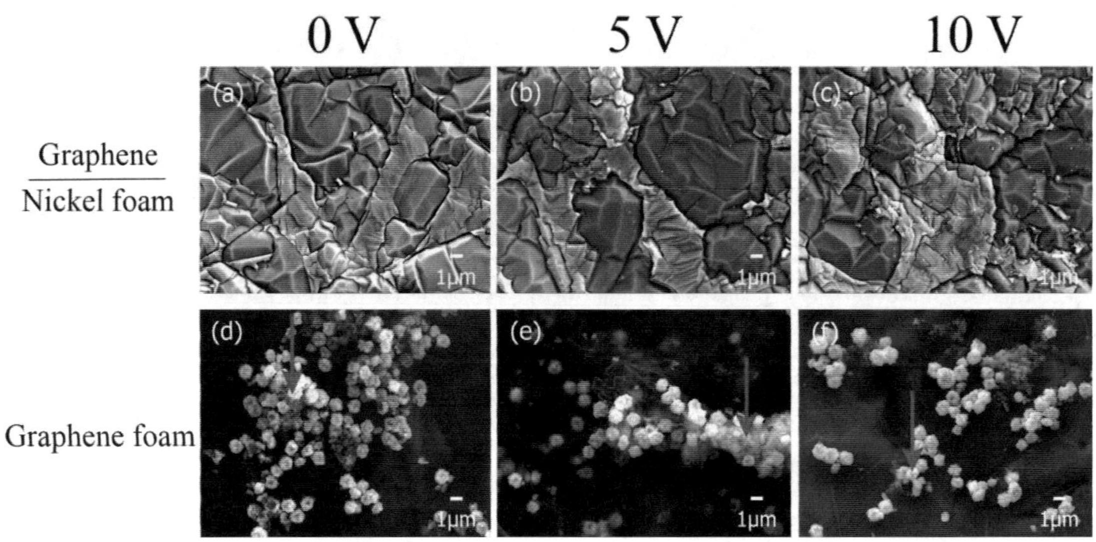

Figure 3. SEM images measured on (a)-(c) graphene on nickel foam and (d)-(f) graphene foam after testing $PM_{2.5}$ capture. (b) and (e) the samples were applied electric potential of 5V during the test of $PM_{2.5}$ capture. (c) and (f) the samples were applied electric potential of 10V during the test of $PM_{2.5}$ capture.

Table 1. the EDX results show the element and its weight% on the surface in Figure 3 (a)-(c) and at the red arrows position in Figure 3 (d)-(f)

Element	Graphene on Nickel Foam Weight %			Graphene Foam Weight %		
	0V	5V	10V	0V	5V	10V
C	45.59	51.64	44.77	46.17	44.80	52.34
O	-	-	0.83	37.40	36.33	36.36
Al	-	-	-	0.21	0.27	0.19
Si	-	-	-	2.90	3.32	-
P	-	-	-	4.00	3.16	3.27
S	-	-	-	-	0.22	-
Cl	-	-	-	0.42	0.95	0.23
Fe	-	-	-	8.91	10.29	7.46
Cu	-	-	-	-	-	0.15
Ni	54.41	48.36	54.41	-	0.67	-

4. Conclusions

In this research, the synthesis of graphene foam by CVD using palm oil as a carbon source for $PM_{2.5}$ capture is demonstrated. the graphene foam can capture $PM_{2.5}$ particles on its surface without additional energy. The graphene foam filter can capture various types of $PM_{2.5}$ particle such as aluminium, silicon, phosphorus and so on.

5. References

[1] Bell Michelle L, Dominici F, Ebisu K, Zeger S L and Samet J M 2007 *Environmental Health Perspectives* **115** 989.
[2] Pui D Y H, Chen S C and Zuo Z 2014 *Particuology* **13** 1.
[3] Novoselov K S, Geim A K, Morozov S V, Jiang D, Dubonos S V, Zhang Y, Grigorieva I V and Firsov A A 2004 *SCIENCE* **306** 5.
[4] Chen J, Yao B, Li C and Shi G 2013 *Carbon* **64** 225.
[5] Ruammaitree A, Nakahara H and Saito Y 2014 *Surf. Interface Anal.* **46** 1156.
[6] Ruammaitree A, Nakahara H, Akimoto K, Soda K and Saito Y 2013 *Applied Surface Science* **282** 297.
[7] Ruammaitree A, Nakahara H and Saito Y 2014 *Applied Surface Science* **307** 136.
[8] Hu H, Ruammaitree A, Nakahara H, Asaka K and Saito Y 2012 *Surf. Interface Anal.* **44** 793.
[9] Ruammaitree A, Phokharatkul D, Nuntawong N and Wisitsoraat A 2018 *Surface Review and Letters* 1840003 (6 pages).
[10] Ruammaitree A, Phokharatkul D and Wisitsoraat A 2018 *Solid State Phenomena* **283** 173.
[11] Kokmat P, Donnuea N, Nuntawong N, Wisitsoraat A and Ruammaitree A 2019 *IOP Conf. Series: Materials Science and Engineering* **526** 012008.
[12] Allen M-J, Tung V-C and Kaner R-B 2010 *Chem. Rev.* **110** 132.
[13] Machac P and Hrebicek T 2017 *J Mater Sci: Mater Electron* **28** 12425.
[14] Jung W, Lee J S, Han S, Ko S H, Kim T and Kim Y H 2018 *J. Mater. Chem. A* **6** 16975.

Acknowledgement

This work was supported by Agricultural Research Development Agency (Public Organization) (contract number CRP6205032080). The authors thank Center of Scientific Equipment for Advanced Research, Thammasat University for SEM and EDX measurements.

The improvement of the high value-added supply chain: a cordyceps beverage case study

P Poochinya and P Ongkunaruk*

Department of Agro-Industrial Technology, Faculty of Agro-Industry, Kasetsart University, 50 Ngam Wong Wan Rd. Lad Yao, Chatuchak, Bangkok 10900, Thailand

*Corresponding author: pornthipa.o@ku.ac.th

Abstract. A healthy food trend is becoming popular which makes people become more aware of taking care of themselves. This causes the growth of healthy business phenomena such as supplements and herb in Thailand. Cordyceps which is known as "winter worm summer grass" is a famous Chinese herbal medicine and very expensive because it is rare. People believe that it provides benefits for enhancing health and boosting energy. Today Thai farmers and entrepreneurs can cultivate cordyceps by themselves resulting in widespread use and lower the price. The objective of this research is to analyze a cordyceps beverage supply chain of a small-sized entrepreneur using Integration Definition for Function Modeling (IDEF0). It explains the overall flow of business and related activity of cordyceps beverage. Then we conduct an analysis of the current problems and recommend how to improve the efficiency of the supply chain. Currently, there is a lack of systematic data collection, no proper KPI and inventory policy and low forecast accuracy. We recommend the entrepreneur to use the spreadsheet for forecasting and record data, create KPI to evaluate performance, set standard and target. Establish an inventory policy to manage material and evaluate supplier's performance. We hope our research will promote the entrepreneur to stay competitive in the dynamic market.

1. Introduction

A healthy food trend is becoming popular which makes people become more aware of taking care of themselves. This results in the growth of healthy business phenomena such as supplements and herb in Thailand [1]. Cordyceps which is known as "winter worm summer grass" is a famous Chinese herbal medicine and very expensive because it is rare. People believe that it enhances health and boosts energy. Cordyceps includes many species. The popular one for cultivating is Cordyceps militaris which contains a high dose of nucleosides which is the substances providing benefits mentioned previously. In general, the quality of cordyceps is identified by quantity and quality of substance "adenosines" which is part of nucleosides. The study found that cordyceps from cultivation has more nucleosides than cordyceps from natural. This happens because the cordyceps from cultivation was raised in controlled conditions from several trials to get the optimal result. In the past, cordyceps cultivating happens in the laboratory or for study only. Until recently, the research and development institute of Chiang Mai Rajabhat University arrange workshops for the general public [2]. Nowadays, Thai farmers and entrepreneurs can cultivate cordyceps by themselves resulting in widespread use and lower the price. This is a challenge to any party who involves in this business. Even though farmers and entrepreneurs have expertise in cultivating and developing but they lack proper supply chain management which causes raw material and packaging out of stock issues [3]. They also do not use the right technology to manage the system [4]. The objective of this research is to analyze a cordyceps beverage supply chain of a small-sized entrepreneur. Then, we conduct an analysis of the current problems and recommend how to improve the efficiency of the supply chain.

2. Methodology

We conducted an in-depth interview with a small-sized entrepreneur [5]. The reason we selected this company because they are open-mind and willing to share information with us for study purpose. They also a well-known company in the mushroom business which not only makes a profit from selling products but also a learning center to teach people how to create a job which contributes to the country. This includes an interview by phone call, meeting in person and on-site observation in the farm and factory. Next, we identify the overall activity of the business by using Integration Definition for Function Modeling (IDEF0) which is the series of rectangular diagrams connected by arrows that explain the overall flow of business [6]. The related activities of cordyceps beverage are mentioned inside the diagrams. Arrows that point to the left side of the diagram are inputs, arrows point to the top are controls such as specification, regulation, and procedure, arrows point to the bottom are resources that require to operate activities and arrows point out of the diagram to the right side are outputs from activities. The outputs of the previous activities are the inputs of the successor activities. The solid lines of arrows are the current activities and the dash ones are the improvement that were suggested for implementing. We conduct an analysis of IDEF0 for 2 level which is level 0 and level 1. Level 0 is the big picture of the business including supplier, farm, factory, and customer [7]. Level 1 is about analysis more detail of the activity in the factory [8-9]. Finally, we use the Key Performance Indicator (KPI) principle to measure the outcome of each activity [10].

3. Results and Discussion

3.1. Company background

The entrepreneur is a mushroom farmer in Saraburi province. In the past, he cultivated fruits and ornamental plants, but the business did not go well. Then he started to do research about straw mushroom cultivation and then extended his research to various types of mushrooms such as pearl oyster mushroom, log white fungi and Yanagi mushroom. Theses mushroom can make good profits to him at last. This mushroom farm does not only sell products but also being a complete mushroom cultivation learning center, that hosted many farmers, researchers, entrepreneurs and students who would like to learn from him. With his progressive nature and eager for knowledge, he proceeded to learn about Reishi mushroom and Cordyceps which are considered as medicinal mushrooms and have high profit due to it requires more complex processes for cultivation where not many people can do at that time. He successfully implemented these 2 new mushrooms in his farm later and became a well-known learning center of Reishi mushroom and Cordyceps.

3.2. Business process analysis of cordyceps beverage supply chain

The business process was analyzed by using IDEF0 level 0 or an organizational level as shown in Figure 1. The upstream suppliers are various suppliers who provide silkworms, rice, other ingredients, packaging, and label. The main raw material for preparing cordyceps cultivation is frozen silkworm for creating culture medium. According to their research, worms are selected by specific species and their age because it will provide the best nutrition and the optimal thickness to reduce the damage to the blender. The lead time of ordering frozen silkworm from local farmers is 30 days and the order quantity is 100 kg. Next, Sang Yod rice is obtained from Phatthalung province at a lot of 200 kg and an order lead times of 5 days. For the other ingredients, they were ordered from local vendors and markets which lead time of 3 days. All the materials will be stored in the central warehouse in the farm. At the mushroom farm, which is the main activity to cultivate cordyceps cultivation by the entrepreneur and his family using the machine, measurement equipment, and controlled temperature room which takes 75 days. The control of the cultivation is mushroom cultivation formula, work procedure, measurement standard, humidity, and recording. The Cordyceps will be dried and kept in the storage room. The

Cordyceps beverage manufacturer located near the farm produces the Cordyceps beverage and delivery to the customer. Then, customers are consumers or retailers who order the product via social media such as LINE, Facebook. Some customers order from an off-line channel such as phone calls and directly buy from the company. If there is an issue about the product, the customer will send details to the entrepreneur. After the entrepreneur review and decide that its defect product, the replacement product will deliver to the customer.

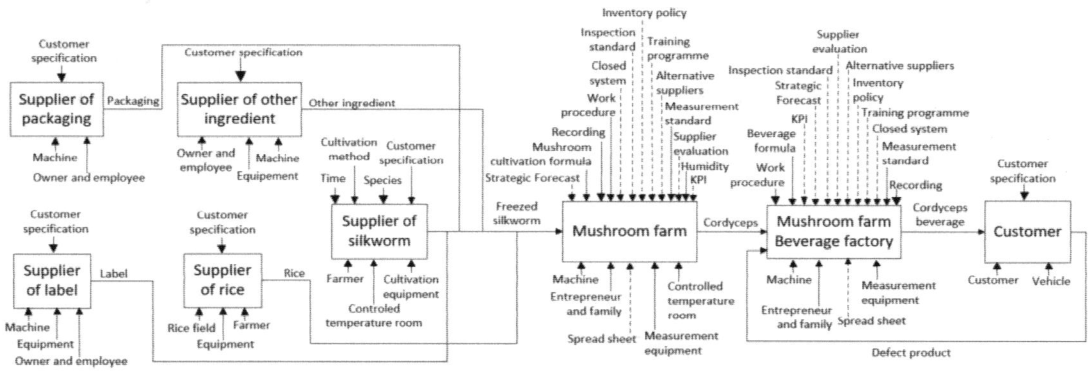

Figure 1. The business process of the Cordyceps beverage supply chain (IDEF0 Level 0).

3.3. Business process analysis of cordyceps beverage factory

The business process of the cordyceps beverage factory was analyzed using the IDEF0 level 1 or activity level which composed of five major activities as explained in Figure 2. The first activity is planning which usually managed by the entrepreneur alone. The information on inventory, capacity and lead time were gathered before creating a production plan using a notebook. Normally, production starts when the finished goods inventory is lower than 500 bottles. Second, the source or procurement process is to order raw materials from the suppliers. The third is the production process or make, composed of mixing beverage, filling, sterilizing by retort and packing which will be finished within a day. The finished goods are kept in the warehouse. Fourth, the delivery will start when the customers place an order where the staff pack the product in the afternoon and outsourced the shipment to the logistic service provider on the same day. If the order arrives after 2 p.m., it will be delivered the next day. Finally, the return occurred if customers find defect products, they will contact the entrepreneur directly to get the new one.

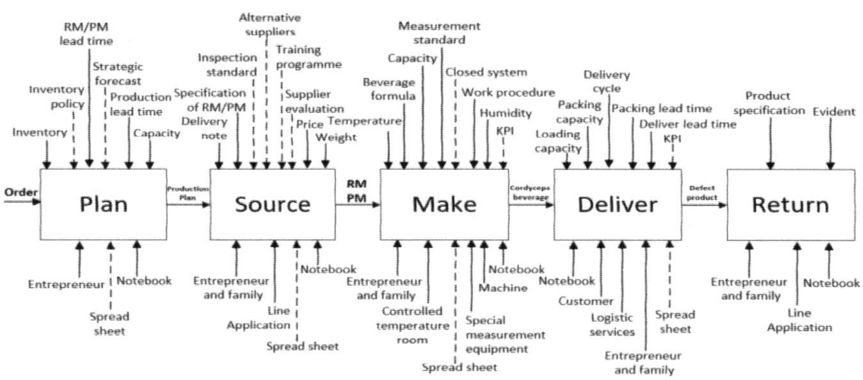

Figure 2. The business process of the Cordyceps beverage factory (IDEF0 Level 1).

3.4. Problem analysis and proposed strategies for an improvement

Currently, the entrepreneur records all data using a notebook and does not share information with employees and does not verify the calculation. Sometimes, packaging materials are a shortage and could not start production. There is only a single supplier for silkworm and some time, the worms are too old and their silks are too thick which can damage the blender. In addition, in the mold spreading period starting from October to November, the cordyceps cultivation activity cannot be done. At present, a key performance indicator (KPI) for the cordyceps cultivation is accepted within 10% of defects. However, the current defect rate is more than 10%. The defective Cordyceps become waste and no value-added. There is no KPI for packing on the delivery process. We summarized the problem and propose the improvement plan as shown in Table 1.

Table 1. Summary of current problems and the improvement plan.

Activities	Problems	Improvement guidelines
Plan	Packaging material shortage	Record data of demand, production, and inventory. Create a strategic forecast and inventory policy
	Lack of systematic data collection	Create a template to record data and use Google sheet to share information in the organization
Source	A single supplier for silkworm	Find an alternative supplier, perform supplier evaluation
	Silkworms are out of standard	Create an inspection standard, train employees to implement and evaluate the supplier
	Lack of systematic data collection	Create a template to record data and use Google sheet to share information in the organization
Make	Cannot run the production during Oct-Nov due to the mold spreading	Investigate the production site, establish a closed system such as a greenhouse to eliminate the contamination. Increase production output before the mold spreading
	A high number of defects	Identify the root causes and implement improved processes for KPI
	Lack of systematic data collection	Create a template to record data and use Google sheet to share information in the organization
Deliver	No KPI for the packing process	Create KPIs for the packing process and do continuous assessment
	Lack of systematic data collection	Create a template to record data and use Google sheet to share information in the organization

4. Conclusion

The business process of the Cordyceps beverage supply chain of the small-sized company has been analyzed. The major problems were caused by lack of data, proper forecast and suitable inventory management and standard process. In addition, most activities were managed solely by the entrepreneur himself. The suggestion was to improve the planning process by starting systematic data collection, creating a strategic forecast and inventory policy, finding an alternative supplier performing supplier evaluation, establishing proper KPIs and inspection standards. Other agri-businesses that have similar products may use this research as a guideline to improve their efficiency.

5. References

[1] FiBL & IFOAM Organics International 2019 *the World of Org. Agr. Stat. & Emerg. Trends 2019* 187-205
[2] Tapingkae T 2018 *Cordyceps Mushroom Cultivation* ed Na Songkhla P (Nonthaburi: Kehakaset)
[3] Ongkunaruk P and Prawat R 2011 *21st Int. Conf. Flex. Autom. Int. Manuf. (FAIM2011)* 1-6
[4] Teerada K and Ongkunaruk P 2015 *KnE Life Sci.* **3**(3) 1-5
[5] Ongkunaruk P 2015 *Agric. Agric. Sci. Proc.* **3** 35-9.
[6] Colquhoun G J., Baines R W. and Crossley R 1993 *Int. J. Comp. Integ. M.* **6**(4) 252-64
[7] Pradita S P and Ongkunaruk P 2019 *IOP Conf. Ser.: Mater. Sci. Eng.* **526** 012004
[8] Prasertwattanakul Y and Ongkunaruk P 2018 *Int. Food Res. J.* **25**(2) 481-6
[9] Tangkham K and Ongkunaruk P 2019 *Int. Conf. Eng. Sci. and Ind. App. (ICESI)*, 1-5
[10] Cai J, Liu X, Xiao Z and Liu J 2008 *Decis. Support Syst.* **46** 512–21

IOP Conf. Series: Materials Science and Engineering **773** (2020) 012017 doi:10.1088/1757-899X/773/1/012017

Y_2BaCuO_5 Particle Size Control via Ultrasonication and its Effect on Bulk $YBa_2Cu_3O_y$ Superconductor

S Pinmangkorn, M Miryala*, S S Arvapalli and M Murakami

Graduates School of Science and Engineering, Regional Environment System, Superconducting Material Laboratory, Shibaura Institute of Technology, 3-7-5 Toyosu, Koto-Ku, Tokyo 135-8548, Japan

*Corresponding author: miryala1@shibaura-it.ac.jp

Abstract. Recently, bulk $LRBa_2Cu_3O_y$ (LRE: Y, Gd, Sm, NEG) superconducting materials became a popular choice for engineering applications such as high-field magnet systems for medical devices, rotating machine, magnetic bearings, etc. $YBa_2Cu_3O_y$ (Y-123) is a trending high temperature superconducting material with a high critical temperature around 91 K. The superconducting performance of bulk $YBa_2Cu_3O_y$ (Y-123) can be dramatically improved by controlling initially added Y_2BaCuO_5 (Y-211) secondary phase particles and optimizing its processing conditions in the top seeded melt-growth (TSMG). In this work, we employed a low-cost ultrasonication technique to control size of initially added secondary phase Y-211 particles. The ultrasonication time is varied from 20 to 80 minutes along with a 20 minutes interval while maintaining constant power of 300 W and 20 kHz frequency. We had successfully prepared nanometre sized secondary phase Y-211 particles. Top seeded melt textured bulk Y-123 (20 mm diameter) were prepared with utilizing refined Y-211. The magnetization measurements exhibited a sharp superconducting transition temperature with $T_{c,onset}$ around 91 K. Critical current density (J_c) and trapped field measured at 77 K exhibited higher value in bulk prepared with Y-211 secondary phase particles ultrasonicated for 80 minutes. The highest trapped field of 0.42 T at 77 K was recorded 0.3 mm above centre of the Y-123 bulk's surface. The self-field critical current density was around 47 kA/cm^2 at 77 K (H//*c*-axis). The present results demonstrate that the performance of bulk Y-123 can be improved by controlling Y-211 secondary phase particle size via a low cost ultrasonication process.

1. Introduction

The melt textured process is the most eminent process for synthesizing bulk $YBa_2Cu_3O_y$ (Y-123) superconductors due to its high critical temperature (T_c), high critical current density (J_c) and superior trapped field value around liquid nitrogen temperature (77 K) [1-3] in end product. These parameters are well suited for most of practical applications such as superconducting magnets, magnetic motors, flywheel, etc [4, 5]. In order to achieve high superconducting performance, the smaller Y-211 secondary phase particles can be introduced in Y-123 matrix. So many techniques such as ball milling, microemulsion, carbon nanotube and solution blow spinning technique were employed to refine and control Y-211 secondary phase particles size. The limitations of those techniques follows as introducing contamination into Y-211 powder and possess to be difficult in nature for optimization process [6-9]. We tried to find an alternative technique to save processing time as its easier to work with while producing high purity Y-211 secondary phase particles that are effectively refined.

In this study, we employed ultrasonication technique to refine Y-211 particles. Several bulk Y-123 superconductors were processed using top-seeded melt growth technique, with and without chemical doping condition alongside the refined Y-211 secondary phase precursor. The superconducting performance as a function of ultrasonic time were discussed.

2. Experimental

Firstly, we fabricated the $YBa_2Cu_3O_{7-\delta}$ (Y-123) and Y_2BaCuO_5 (Y-211) powders from commercial Y_2O_3, BaO_2 and CuO powders. The initial powders were weighed stoichiometrically to make the Y-123 and Y-211 by conventional solid-state reaction. Y-123 powder was repeatedly calcined for 4 times at 870, 900, 920 and 940°C each for 15 hours to ensure complete formation of Y-123. Similarly, the Y-211 powder was also calcined for 4 times at 840, 870, 890 and 900°C for 15 hours each. More importantly, we employed Mitsui ultrasonic homogenizer model UX-300 while using constant power (300 w) and frequency (20 kHz) to control and refine particles of Y-211. The ultrasonication was performed for various durations ranging from 20-100 minutes and followed by detailed characterization. The Y-211 powder after ultrasonicated were analysed by scanning electron microscope (SEM). For synthesized Y-123 sample via TSMG, Y-123 and Y-211 powders were mixed in 10:5 molar ratio with 0.5 wt% of PtO_2 and pulverized for 2 hours. The mixed powder was pressed into pellets under pressure of 20 kN. The Nd-123 generic seed was placed on the top surface. Samples were fabricated by TSMG process with optimized temperature profile and heating time. The sample was heated 100 °C/h to 880°C and dwell for 15 minutes, heat to 1050°C in 50 minutes and dwell for 50 minutes. The temperature was cooled down to 1005°C in 50 minutes and slowly cooled to 985°C in 200 hours. Then, bulk Y-123 samples were cool down followed by furnace cooling to room temperature. For oxygen annealing process, the Y-123 samples were annealed at 450°C for 400 hours under O_2 flow control at rate 300 ml/minute.

For trapped field measurement, the bulk Y-123 samples were cooled to 77 K in an applied magnetic field 1 T for 15 minutes. The profile of the trapped magnetic flux density was measured with hall probe sensors at a distance of 1.3 mm above the top sample surface. Superconducting transition temperature (T_c) and magnetization hysteresis loops in field from 0-5 T were measure at 77 K using a SQUID magnetometer (model MPMS5). For the measurement of magnetic properties, small samples were cut below seed along *c*-axis around 1.5x1.5x0.5 mm^3 in dimensions. Critical current density (J_c) was estimated from M-H loops by the extended Bean's critical state model formula given as $J_c = 20\Delta m/(a^2c(b - a/3))$, where a,b are cross section dimensions with $b \geq a$, c is thickness of sample and Δm is the difference of magnetic moment between decreasing and increasing field in the *M-H* loop [10].

3. Results and discussion

The fine Y-211 secondary phase particles in bulk Y-123 superconductors were expected to improve superconducting properties. The ultrasonication technique was employed to refine Y-211 particle. The particle size of Y-211 as a function of ultrasonic time were displayed on Figure 1. Majority of Y-211 particles were prepared as Y-211 powder were bigger than 5 μm (around 48%) and also a few particles less than 2 μm (around 7.5%). After we employed ultrasonication, the Y-211 particles became smaller than before. With 40 minutes and 80 minutes of ultrasonication, particles were less than 1 μm had been estimated to be 18.5% and 96.25%, respectively. For 80 minutes, the Y-211 particles were less than 500 nm estimated to be around 1.25%. The Y-211 particles were toning down with increasing ultrasonication time as it can be seen in FE-SEM image on Figure 1. The high energy wave from ultrasonic probe will disrupt segregated Y-211 particles into division of individual and small particles [11]. In this system, PtO_2 was considered as grain refiner and additive that improves superconducting properties of Y-123 bulk. For comparing the individual effects of ultrasonication, we had synthesized bulk Y-123 samples with and without PtO_2 addition. After oxygenation of Y-123 bulks, trapped field distribution under an applied field of 1 T at 77 K at 0.3 mm above the top surface was measured with respect to ultrasonic time as shown in Figure 2. The highest value was observed in sample with PtO_2 addition along with 80 minutes of ultrasonicated Y-211 powder that had reached 0.42 T. Figure 3 shows superconducting transition temperature (T_c) of all samples exhibited $T_{c.onset}$ around 91 to 92 K with ΔT_c less than 1.5 K similar to nature of bulk Y-123 superconductors [12]. Critical current density (J_c) at 77 K (H//*c*-axis) shows higher value in sample with PtO_2 condition and Y-211 particles refined at 80 minutes of ultrasonic time. Figure 4 shows J_c values as a function of ultrasonic time at zero field and 2 T at 77 K. High J_c was observed in the same sample similar to that of trapped fields results which is around 47.5 kA/cm^2 at zero field. In case of high field (2 T) J_c values, the samples exhibited very close performances both with and without PtO_2 added samples. Especially, the values were around 10 kA/cm^2 in 80 minutes based samples. This indicates that for superior superconducting properties in bulk Y-123 superconductors, both PtO_2 addition as well as smaller initial Y-211 particles were crucial. The PtO_2 was observed for helping in containing/restricting sub micro size of Y-211 secondary phase in Y-123 matrix [13, 14]. From the present comparisons of superconducting performance, the highest value was exhibited in the sample with PtO_2 addition and Y-211 refined 80 minutes of ultrasonication.

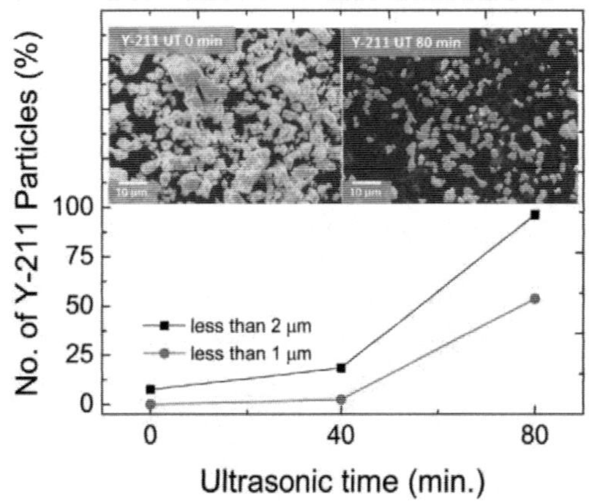

Figure 1. Average particles size of Y-211 powders as a function of ultrasonic along with inset FE-SEM images of Y-211 ultrasonicated at 0 and 80 min.

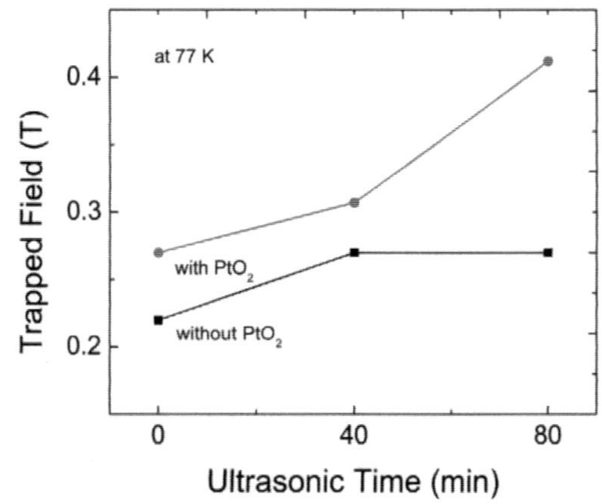

Figure 2. Trapped field value measured under a magnetic field of 1 T at 77 K as a function of ultrasonic time.

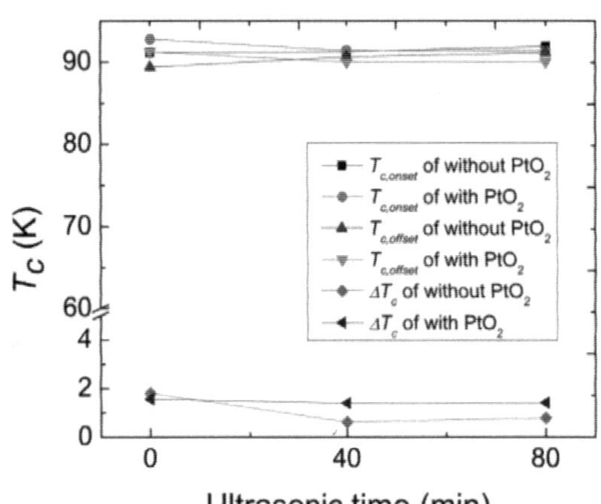

Figure 3. Critical temperature (T_c) as a function of ultrasonic time

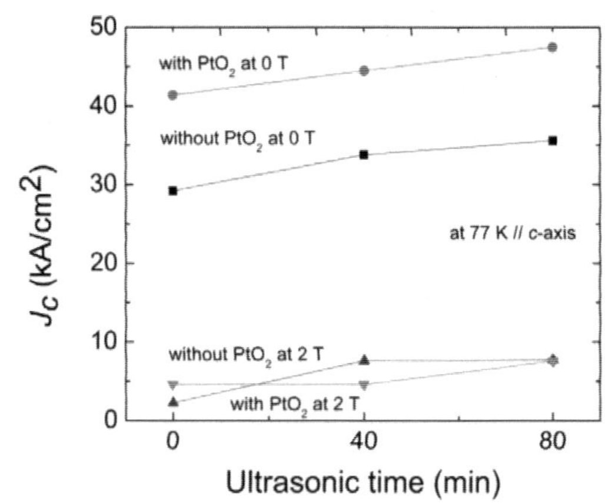

Figure 4. Critical current density (J_c) at 0 T and 2 T as a function of ultrasonic time

4. Conclusions

We employed the novel ultrasonication technique, to refine Y-211 secondary phase particle and the particle size analysis indicated that the size reduction is proportional to the ultrasonication time. Good quality bulk Y-123 samples (T_c~91 K) prepared from these powders via TSMG processing show an improvement in trapped field performance and critical current density. Among all, sample with 80 minutes ultrasonicated Y-211 along with PtO_2 addition showed highest performance such as 0.42 T trapped field (magnetized under 1 T,77 K) and 47.5 kA/cm_2 critical current density at 0 T, 77 K. Microstructural studies on bulk revealed that tiny Y-211 particles are dispersed throughout the Y-123 matrix which acting as pinning centers. In addition, the Y-211 size can be made finer with increase in ultrasonication time, which enhanced the pinning performance resulting in higher trapped field and critical current density. The ultrasonication technique is a reliable applicable method to refine and control Y-211 particle size which is also quite a cost effective and a less contaminating in synthesizing bulk Y-123 superconductors.

5. References

[1] Cloots R, Koutzarova T, Mathieu J P and Ausloos M 2004 *Superconductor Science and Technology* **18** R9-R23.
[2] Muralidhar M and Murakami M 2000 *Physical Review B* **62** 13911.
[3] Murakami M 1992 *Melt Processed High-Temperature Superconductors.*
[4] Coombs T A, Hong Z, Zhu X and Krabbes G 2008 *Superconductor Science and Technology* **21** 034001.
[5] Durrell J H, Ainslie M D, Zhou D, Vanderbemden P, Bradshaw T, Speller S, Filipenko M and Cardwell D A 2018 *Superconductor Science and Technology* **31** 103501.
[6] Rotta M, Namburi D K, Shi Y, Pessoa A L, Carvalho C L, Durrell J H, Cardwell D A and Zadorosny R 2019 *Ceramics International* **45** 3948-53.
[7] Shi Y, Hasan T, Babu N H, Torrisi F, Milana S, Ferrari A C and Cardwell D A 2012 *ACS Nano* **6** 5395-403.
[8] Muralidhar M, Kenta N, Zeng X, Koblischka M R, Diko P and Murakami M 2016 *physica status solidi (a)* **213** 443-9.
[9] Li F, Vipulanandan C, Zhou Y X and Salama K 2006 *Superconductor Science and Technology* **19** 589-95.
[10] Chen D X and Goldfarb R B 1989 *Journal of Applied Physics* **66** 2489-500.
[11] Qian M, Ramirez A and Das A 2009 *Journal of Crystal Growth* **311** 3708-15.
[12] Mahmood A, Chu Y S and Sung T H 2012 *Superconductor Science and Technology* **25** 045008.
[13] Kim C-J, Kim K-B, Kuk I-H and Hong G-W 1997 *Physica C: Superconductivity* **281** 244-52.
[14] Chow J C L, Lo W, Dewhurst C D, Leung H T, Cardwell D A and Shi Y H 1997 *Superconductor Science and Technology* **10** 435-43.

Acknowledgement

This work was partly supported by Shibaura Institute of Technology (SIT) Research Center for Green Innovation and Grant-in-Aid FD research budget code: 112282. Two of the authors (Sunsanee Pinmangkorn and Sai Srikanth Arvapalli) acknowledges support from SIT for providing the financial support for the doctoral program.

The use of physical simulation to evaluate thermal properties of food containers in cold chain logistics

N Chaitanoo[1] P Ongkunaruk[2,*] D Leingpibul[3]

[1]Department of Mechanical Engineering, Faculty of Engineering, Rajamangala University of Technology Lanna, Chiang Mai 50300, Thailand
[2]Department of Agro-Industrial Technology, Faculty of Agro-Industry, Kasetsart University, 50 Ngam Wong Wan Rd. Lad Yao, Chatuchak, Bangkok 10900, Thailand
[3]Haworth College of Business, Western Michigan University, Kalamazoo, MI 49008, USA

*Corresponding author: pornthipa.o@ku.ac.th

Abstract. Cold chain logistics has gained significant attention among all stakeholders in the food and pharmaceutical delivery industries due to the stricter food safety enforcement and tighter quality assurance [1-3]. Most cold chain logistics providers seek to determine the right packaging that ensures temperature control delivery in order to gain and/or retain their competitive advantage in the market place. Nowadays, materials such as Polystyrene Foam (PS), Polyethylene-Nylon (PE-Nylon), and Vacuum Insulation Panels (VIP) are widely used in the construction of delivery boxes. The objective of this study is to use a physical simulation approach to determine which materials are most suitable for delivery boxes. First, three delivery boxes with different materials (PS, PE-Nylon and VIP) were used as test subjects in this study. Second, the cold chain delivery process was simulated for each box and the temperature at each critical point had been measured throughout the delivery process. Third, data analysis was performed by comparing the temperature pattern among all three subjects. The thermal conductivity data for each box was calculated by mathematic equations to help interpret the results. The result showed that the box made from VIP material showed the best thermal resistance. The application of this thermal conductivity estimation procedure might help guide food delivery related practitioners to determine the best materials for their delivery box equipment and the findings might help effectively and efficiently improve their delivery services.

1. Introduction

The current global trend of urbanization accelerates the growth of last-mile food delivery in many large cities [1]. With the exponential growth in temperature-controlled food delivery, the proper delivery box plays a critical role to help ensure food safety and minimize food loss [2,3]. Any temperature control issue in cold chain logistics will increase food loss in the food industry sector and elevate the food safety risk in the consumer sector [3]. The delivery box is considered the primary protection of food safety. Practitioners need to understand which box/material should fit their delivery conditions. Therefore, research in this area is urgently needed to support the fast-growing numbers of new business operations that will critically impact such a large number of stakeholders worldwide. For the delivery providers, the box selection process involves several factors such as cost, dimension, strength, design, and insulation properties. There are many materials used as insulation for food delivery such as Expanded

Polystyrene Foam (EPS) [5]. Non-woven feather fiber packaging liners from the poultry industry is a sustainable material with a comparable thermal conductivity as the EPS [6]. The insulation property is traditionally determined by the thermal conductivity resistance factor (R) or thermal conductivity constant (k) by practitioners. This research proposes a more holistic approach by conducting a physical delivery simulation along with the use of traditional R and k for thermal properties. This particular research aims to answer, "Which material (VIP, PE-Nylon, and PS) best insulates chilled food?" according to the results of a physical delivery simulation approach.

2. Methodology

The three different commonly used delivery boxes made of VIP, PE-Nylon, and PS materials were selected as the subjects of study to replicate industry usage. Based on focus group interviews, the sizes of three boxes are comparable and fall into the same small category delivery box. For each box, weight, dimension, thickness, and volume were carefully measured and recorded in Table 1. Then, the thermal conductivity experiment was conducted using the following approaches.

Table 1. Weight, dimension, thickness, and volume of three delivery boxes

Delivery Box	Weight (kg)	Dimension (m)						Thickness (cm)	Volume (m^3)
		Internal			External				
		Width	Length	Depth	Width	Length	Depth		
VIP	5.70	0.46	0.25	0.35	0.53	0.32	0.39	4.00	0.04
PE-Nylon	0.33	0.24	0.32	0.28	0.26	0.34	0.30	0.80	0.02
PS	0.29	0.26	0.30	0.30	0.30	0.38	0.34	2.00	0.02

2.1 Experimentation

To minimize the impact from the ambient temperature and replicate the delivery condition in Thailand, the temperature in the lab was controlled at 25±1.5°C during the whole physical simulation experiment. Next, each box was filled with 1000 ml of cold water (8.0±1.5°C) as a cold food item. The water is used to minimize the interference of thermal conductivity from uneven food physical properties. Four data loggers were placed throughout the box and the temperatures were recorded every 5 min interval over the simulation period of 3 h for each box. There are recommended by the practitioner focus groups. The positions of the data logger were evenly distributed as shown in Figure 1.

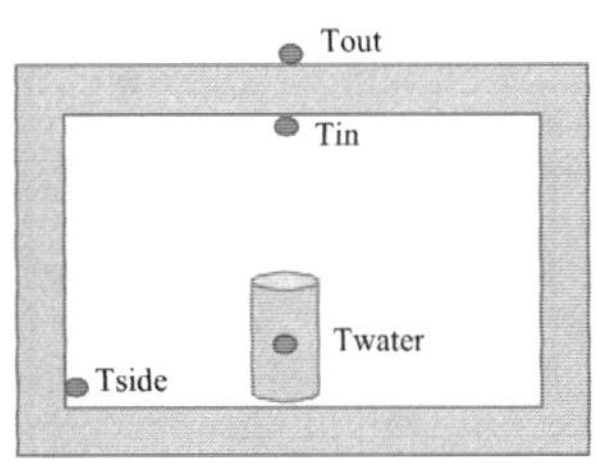

Figure 1. The position of data logger in the box.

2.2 Heat transfer model

According to the Zeroth law of thermodynamics [7], when temperatures of two points in the nearby location are different, the heat will be transferred from the higher point to the lower heat point until the transfer process reaches thermal equilibrium. The changes in the water temperature indicated that heat transfer exists inside the box. This transfer can be explained by the heat capacity [7] as in Equation (1).

$$Q = mC_p \frac{dT}{dt} \tag{1}$$

The water temperature rises due to heat from the external air traveling through the box walls with some of that heat being absorbed by the box wall itself. However, these factors are still insufficient to identify the overall thermal property. This heat transfer can be calculated further using Fourier's law as in Equation (2). Then, the thermal conductivity (k) can be derived from Equation (3). Next, the thermal resistance (R) can be derived by Equation (4) [8]. These thermal properties can be used to compare box performance. The thermal conductivity implies how much heat can be absorbed, hence in this case, lower absorption is better. Finally, the thermal resistance identifies how much the heat will be resisted (or insulated) in the box; hence, higher resistance is better:

$$Q = kA\frac{dT}{dx} \tag{2}$$

$$mC_p\frac{dT}{dt} = kA\frac{dT}{dx} \tag{3}$$

$$R = \frac{\Delta x}{k} \tag{4}$$

where Q = The heat transfer (W), k = Thermal conductivity (W/m-k), R = thermal resistance, A = Surface of box (m_2), x = Box thickness (m), dT/dx = Temperature gradient (K/m), m = Mass of water (kg), Cp = Specific heat of water (kJ/kg-K), dT/dt = The change in temperature (°C) during a time interval (dt) or 5 min in our study. Then, the time-temperature profiles of each box are analysed and calculated for the values of k, R and temperature increment. Finally, the statistical analysis is performed on all three boxes using a one-way analysis of variance (ANOVA) followed by a Tukey's test at a 95% confidence level.

3. Results and Discussion

3.1 Time-temperature profile of three materials

During the first 20 min, thermal equilibrium occurred since the inside box temperature was slightly decreased as shown in Figure 2(a) and 2(b). The heat transfer to the chilled water results in temperature increasing as shown in Figure 2(d). When the temperature at two sides of the boxes is significantly different, then heat transfer occurs. In Figure 2(c), the upper outer box temperature is average at 25±1.5°C while the upper inner average temperature of VIP, PE-Nylon, and PS are 20.07±0.76°C, 23.81±0.24°C and 22.04±0.37°C respectively. For the side inner box, the average temperatures are 19.08±0.92°C, 22.87±0.24°C and 21.10±0.37°C for VIP, PE-Nylon, and PS, respectively. The box with the best performance is of VIP construction, which has the smallest incremental temperature in 3 h for 7.53±0.61°C. The next best performance is PS at 8.41±0.17°C and finally PE-Nylon at 8.91±0.22°C

3.2 Thermal properties of the delivery box

The thermal conductivity property of delivery boxes plays an important role in raising the temperature of chilled water inside three boxes. The highest temperature differences between the inside and outside temperatures occurred at the VIP box (~5°C) while the PE-Nylon and PS show a lower temperature differences at ~2°C for PS and ~1°C for PE-Nylon, the inner wall of VIP box also maintains the same low temperature for over an hour (Figure 2(a) and 2(b)). The temperature of the chilled water inside the VIP box raised only 1.5°C (Figure 2(d)). These temperature values perform according to R (and k in an inverse relationship to R) of each material. As shown in Table 2, VIP has the lowest k and the highest R the same R/k inverse relationship also applies in PE-Nylon and PS cases.

Table 2. Comparison the thermal conductivity, thermal resistance and temperature increment of 3 types of box

Delivery Box	*k* (W/m-K)	*R* (m2-K/W)	Temperature Increment (°C)
VIP	0.029±0.011$_b$	1.614±0.740$_a$	7.53±0.61$_a$
PE-Nylon	0.054±0.021$_a$	0.179±0.097$_c$	8.91±0.22$_a$
PS	0.038±0.011$_b$	0.583±0.238$_b$	8.41±0.17$_a$

*the same character indicate no significant difference at 95% confidence interval

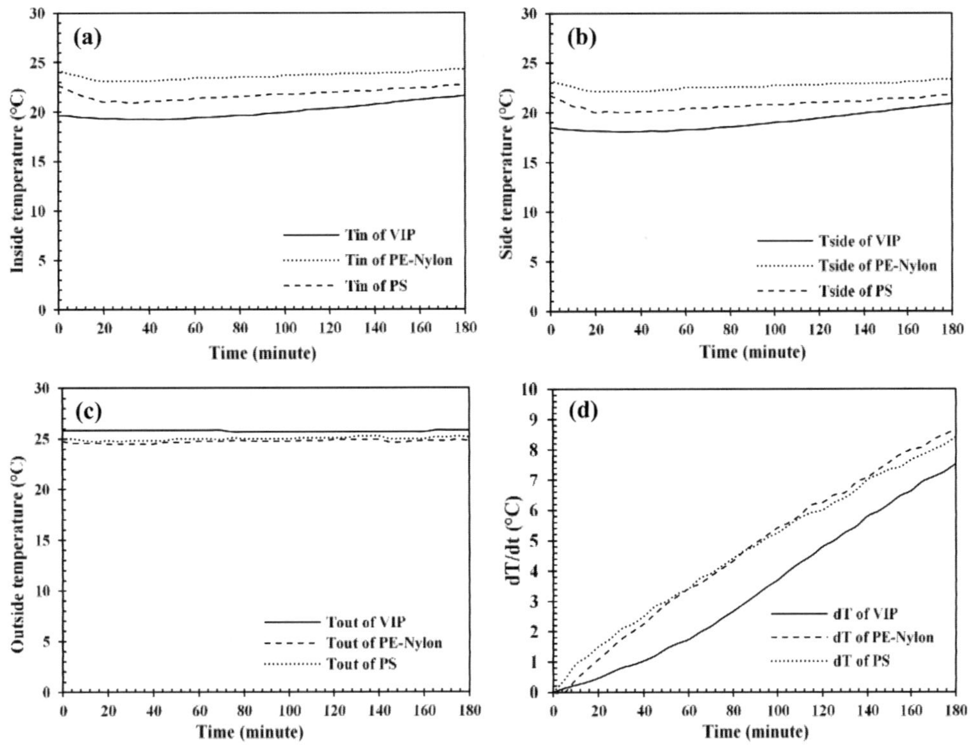

Figure 2. Comparison of changing the temperature by time at each location (a) upper inner box (b) side inner box (c) upper outer box and (d) difference of water temperature.

4. Conclusion

Based on the *R* and *k* concept, three boxes of VIP, PE Nylon, and PS materials should perform heat insulation differently. The temperature increment of chilled water should be significantly different among three boxes with the lowest temperature with the high *R*. However, the results show that the temperature chilled water in three boxes is not significantly different at the 95% confidence interval, among three boxes, based upon the physical simulation approach. While the *R* and *k* thermal properties concept is used to differentiate materials scientifically, the usage condition also plays very important role in helping define which materials are the best performer. Practically, the high *R* material also comes with a higher price; therefore, the premium price of a VIP box might yield similar results to less expensive boxes such as NE-Nylon or PS. The finding recommends the effective selection of a delivery box should involve a holistic approach by using *R* and *k* along with a physical delivery simulation since the *R* and *k* concept alone might be misleading and prompt the selection of a non-economically viable option.

5. References

[1] Ndraha N, Sung W C and Hsiao H I 2019 *J. Food Eng.* **242** 21.
[2] Chung K H, Park, M S, Kim H Y and Bahk G J 2019 *Food Control.* **98** 187.
[3] Ndraha N, Hsiao H I, Vlajic J, Yang M F, and Lin H T 2018 *Food Control.* **89** 12.
[4] Rattanawong, A and Ongkunaruk, P 2016. *KnE Life Sciences* 253.
[5] Hsiao, H I and K L Huang. 2016. *Food Control* **64** 181.
[6] Dieckmann E, Nagy B, Yiakoumetti K, Sheldrick L, and Cheeseman C 2019 *Food Packag. Shelf Life.* **21** 100360.
[7] Cengel Y A and Boles M A 2007 (Eds.) In Thermodynamics, *McGraw-Hill*, 976 pages.
[8] Holman J P 2002 (Eds.) In Heat Transfer, *McGraw-Hill*, 665 pages.

IOP Conf. Series: Materials Science and Engineering **773** (2020) 012019 doi:10.1088/1757-899X/773/1/012019

Effect of acrylonitrile butadiene rubber on the properties of silica-reinforced natural rubber

K Sengloyluan[1,*], O Sangampai[2], N Uthaipan[1] and V Tanrattanakul[1,2]

[1] Sino-Thai International Rubber College, Prince of Songkla University, Songkhla, Thailand 90110.
[2] Department of Materials Science and Technology, Faculty of Science, Prince of Songkla University, Songkhla, Thailand 90110.

*Corresponding author: karnda.s@psu.ac.th

Abstract. The effect of acrylonitrile butadiene rubber (NBR) as a compatibilizer in silica-reinforced natural rubber (NR) is studied. NBR, Emulcril 3380 with 32-34%, was various at 5, 10, 15, 20 and 25 part per hundred parts of rubber (phr) to be used as a compatibilizer for silica-filled NR. The use of NBR as a compatiblizer can improve cure time (T_{90}), cure rate index (CRI), and Mooney viscosity of silica-filled NR compounds, but the re-agglomeration of filler still observes. The use of NBR does not show significantly effect on bound rubber contents. However, tensile strength, reinforcement index and tear strength are increased with increasing amount of NBR contents due to a presence of some rubber-filler interaction through acrylonitrile groups of NBR and silanol groups on the silica surface. It leads to observe more surface roughness on tensile fractured surfaces.

1. Introduction

Silica is widely used as a reinforcing filler for white applications. In tire industry, silica has been used as a filler to provide saving energy tires because the use of silica can improve a rolling resistance and wet grip, while wear resistance is still remained. However, silica basically shows a poor dispersion and a low filler-rubber interaction because of a high surface polarity. Silane coupling agent is required to use in the silica-filled rubber compound to improve silica-rubber interaction such as bis-(triethoxysilylpropyl) tetrasulfide (TESPT)[1] and bis-(triethoxysilylpropyl) disulfide (TESPD).[2] Mixing condition of silica and silane coupling agent has been considered to obtain a good silanization reaction during mixing.

Surface polarity of silica tends to from silica agglomerates and shows low interaction with non-polar rubbers. Silica-filled NR shows very poor silica dispersion and silica-rubber interaction, leading to low physical and dynamic mechanical properties. On the other hand, the mixing of silica with a polar rubber can provided some interaction through their functional groups. Chloroprene rubber (CR) can generate a silica-rubber interaction via polarity of chlorine atom and silanol groups on the silica surface.[3] Epoxidized natural rubber (ENR) shows a good improvement of dynamic mechanical properties because epoxy groups in ENR structure can generate some interaction with silanol groups which become a strong bonding at high temperature.[4] For silica-filled NBR compounds, the interaction between silica and NBR is observed though hydrogen bond and the interaction increases with amount of acrylonitrile in NBR.[5]

In this work, NBR was used as a compatibilizer to improve the properties of silica-filled NR compounds. Cure characteristics, bound rubber content, Payne effect and mechanical properties were studied.

2. Experimental methods

Ribbed smoke sheet No.3 (RSS#3) was used as a matrix in this experiment. Emulcril 3380 (INSA GPRO Synthetic rubber, China) is an acrylonitrile butadiene rubber (NBR) with acrylonitrile contents of 32-34%, was studied on various amounts of 5, 10, 15, 20, and 25 phr. Chemicals in this experiment were a precipitated silica 185-Micro Pearl (Ecopower, China) for 50 phr, bis-(triethoxysilylpropyl)-tetrasulfide (TESPT) (Evinik, Germany) 5 phr for a reference compound, Treated distillated aromatic extract oil (TDAE oil) (Hansen and Rosenthal, Germany) 5 phr, 2,2,4-Trimethyl-1,2-dihydroquinoline (TMQ) 1 phr, zinc oxide (ZnO) 3 phr, stearic acid 1 phr, diphenylguanidine (DPG) 1 phr, N-cyclo-hexyl-2-benzothiazyl sulphenamide (CBS) 1.5 phr and sulphur (all from Behnmayer, USA) 1.5 phr.

Silica-reinforced NR compounds were prepared by using an internal mixer (Haake PolyLab, Germany) with initial mixing temperature at 110$_o$C and rotor speed at 60 rpm. Mixing procedures were provided in two steps; the first step was prepared in the internal mixer and the second step was mixed on a two-roll mill as details in Table 1.

Table 1. Mixing procedures.

Mixing steps	Time (min)
Step 1: Internal mixer	
- Mastication of NR	2
- Mixing of a first half of silica (TESPT) and oil	5
- Mixing of a second half of silica (TESPT) and oil	5
- Mixing of TMQ, ZnO, stearic acid and DPG	3
Step2: Two roll mill	
- Mixing of CBS and Sulphur	5

All of compounds were tested a Mooney viscosity at 100$_o$C with a large rotor (ML1+4, 100$_o$C) (Montech, Germany). Cure characteristics of all compounds were studied using a moving die rheometer (MDR) (Montech, Germany) at 150$_o$C and compressed to reach their maximum cure times.

Bound rubber contents of silica-reinforced NR compounds were investigated at room temperature by immersing 0.25 g of the compounds in 30 ml of toluene for three days and dried it at 80$_o$C for 24 hr, then immersed it again in 30 ml of toluene for another three days, dried it at 80$_o$C for 24 hr and measured the residual weights. Total bound rubber content was calculated by using an equation (a);

$$Total\ bound\ rubber\ (\%) = \frac{m - m_s}{m_r} \times 100 \qquad (a)$$

where m is the weight of the sample after extraction, m_s is the weight of silica in the sample and m_r is the original weight of rubber in the sample.

Tensile properties and tear resistance were analyzed according to ASTM D412 and ASTM D624, respectively. Tensile fracture surface was tested by using a stereomicroscope (Zeiss, Germany) with 50x magnification.

3. Results and discussion

Cure characteristics of silica-reinforced NR compounds with various types and concentration of NBR are shown in Figures 1. The use of NBR-3380 as a compatibilizer in the silica-filled NR compounds is shown the same trend of cure characteristics. Cure time (T_{90}) and cure rate index (CRI) are improved with increasing NBR contents. Hydrophilic surface of silica can create some interaction with acrylonitrile groups in NBR leading to a lower curatives absorbed on its surfaces as seen in the improvement of cure properties. However, the presence of NBR in the silica-filled NR compounds is still clearly seen the effect of re-agglomeration of silica aggregates under shear and high temperature condition as an increase of cure torques at the beginning of cure curves, as known as a flocculation phenomenon, but does not observe in the TESPT compound.

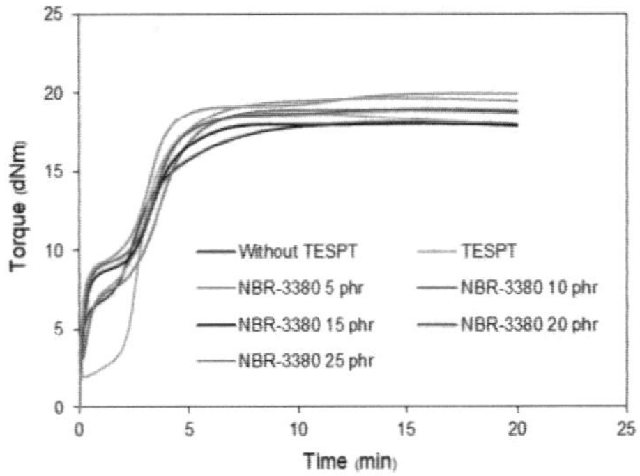

Figure 1. Cure characteristics of silica-reinforced NR with NBR-3380 as a compatibilizer.

Payne effect is used to explain filler-filler interaction in the compounds, as shown in Figure 2a. It is clearly seen that the silica-filled NR compound without any compatibilizer shows the highest filler-filler interaction, in contrast the addition of TESPT shows the lowest one. While the use of NBR as a compatiblizer in silica-filled NR compounds tends to decrease filler-filler interaction. However, there is not significantly different when amount of NBR is increased.

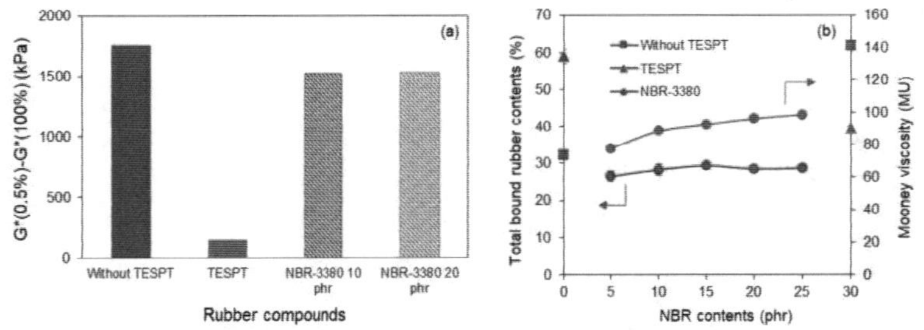

Figure 2. Payne effect (a), total bound rubber contents and Mooney viscosity (b) of silica-reinforced NR with different types of NBR and concentrations.

Figure 2b shows total bound rubber contents and Mooney viscosity of silica-filled NR compounds. The presence of TESPT gives the highest bound rubber content due to a good interaction after a silanization reaction during mixing at a certain temperature. While the use of NBR as a compatibilizer shows a slightly improvement of silica-rubber interaction with increasing NBR concentration. The silica-filled NR compound without compatibilizer basically shows a high amount of trapped rubbers which will be acted as hard particles in the compound and cannot be extracted. Then, it tends to high bound rubber contents in the compound compared with the compounds with NBR. Mooney viscosity of the silica-filled compounds evidently improves by presence of NBR in the compounds as seen in a lower Mooney viscosity.

Tensile strength and reinforcement index of the silica-filled NR vulcaniztes are shown in Figure 3a. The use of NBR as a compatibilizer can improve tensile strength with increasing amount of NBR. Maximum tensile strength is observed when NBR-3380 at 20 phr has been used, however, it is still lower than those of the use of TESPT. Reinforcement index is identified by 300% modulus divided by 100% modulus as shown in Figure 3a. It shows the same trend as well as tensile strength and total bound rubber contents (Figure 2b). The reinforcement index is observed when the NBR has been used up to 20 phr, then it tends to decrease in a further addition of NBR in the silica-filled NR compound. Tear strength of silica-filled NR with NBR as a compatibilizer is shown in Figure 3b. The use of NBR-3380 shows an improvement of tear strength of silica-filled NR. It is found that the increasing of NBR-3380 contents leads to increase of tear strength.

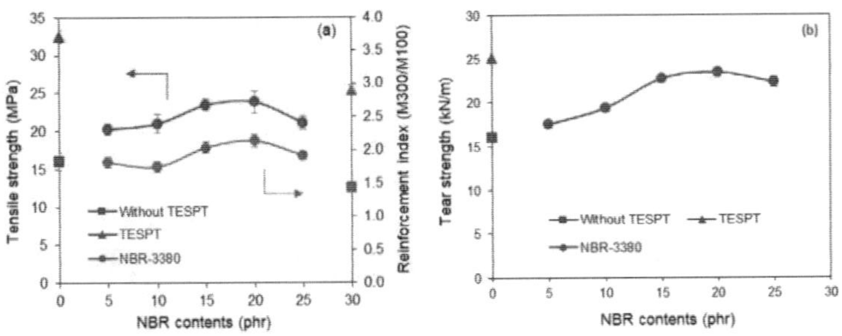

Figure 3. Tensile strength, reinforcement index (a) and tear strength (b) of silica-reinforced NR with NBR as a compatibilizer.

The tear strength is improved to the same level as a use of TESPT when NBR-3380 at 20 phr was added into the silica-filled NR rubber. It proves that acrylonitrile groups in NBR structure can generate some interaction between silica and rubber as well as an improvement of silica dispersion. These are affected on the improvement of cure characteristics, tensile and tear properties.

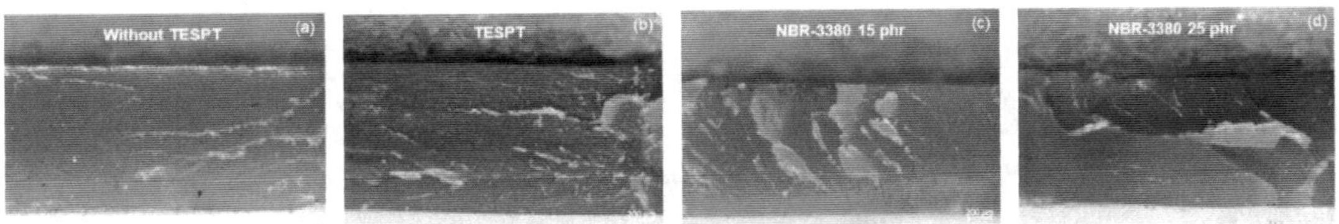

Figure 4. Tensile fractured surfaces of silica-filled NR without TESPT (a), with TESPT (b) and various amount of NBR contents (c-d).

Comparison of tensile fractured surfaces of silica-filled NR vulcanizates is shown in Figures 4a-d. The silica-filled NR without TESPT has a smoother fractured surface than the silica-filled NR with TESPT due to a poor silica dispersion and a low silica-rubber interaction, leading to the lowest tensile strength as seen in Figure 3. Tensile fractured surfaces of silica-filled NR with NBR as a compatibilizer are clearly seen the surface roughness increased by increasing NBR concentrations because of enhancement of silica-rubber interaction as observed on the improvement of overall properties.

4. Conclusion

The properties of silica-filled NR can improved by using NBR as a compatibilizer. The increase of NBR contents leads to decrease of Mooney viscosity, increase of cure rate index, tensile and tear strengths. From the overall properties, it is found that the use of NBR-3380 at 20 phr as a compatibilizer shows a good improvement to reach optimum. Tensile fracture surface roughness is increased by increasing amount of NBR content due to a better silica-filler interaction and silica dispersion.

5. References

[1] Lin CJ, Hergenrother WL, Hilson AS 2002 *Rubber Chem. Technol.* **75** 215.
[2] Kim K-J, Vanderkooi J 2002 *Kautsch. Gummi Kunstst.* **55** 518.
[3] Kapgate BP, Das C, Das A, Basu D, Wiessner S, Reuter U, Heinrich G 2016 *J. Appl. Polym. Sci.* **133(30)** 43717.
[4] Sengloyluan K, Sahakaro K, Dierkes WK, Noordermeer JWM 2014 *Eur. Polym. J.* **51** 69.
[5] Suzuki N, Ito M, Ono S 2005 *J. Appl. Polym. Sci.* **95** 74.
[6] Kantala C, Wimolmala E, Sirisinha C, Sombatsompop N 2009 *Polym. Adv. Technol.* **20** 448.

Acknowledgement

Sino-Thai International Rubber College and research fund No. RDO6202091S, Prince of Songkla University, Songkhla, Thailand.

IOP Conf. Series: Materials Science and Engineering **773** (2020) 012020 doi:10.1088/1757-899X/773/1/012020

Enhanced interfacial dielectric polarization in PVDF-HFP copolymer with treating PPy by using silane coupling agent

A Salea[1], R Saputra[1,2] and C Putson[1,*]

[1] Materials physics laboratory, Physics department, Faculty of Science, Prince of Songkla University, Songkhla, 90112, Thailand
[2] Physics Department, Faculty of Science and Technology, UIN Sunan Kalijaga Marsda Adisucipto street, No 1, Yogyakarta, 55281, Indonesia

*Corresponding author: email: chatchai.p@psu.ac.th

Abstract. Dielectric materials are polar materials for energy storage applications such as capacitors, transformer, and other electrical devices. The great dielectric properties generally depend on easily switchable polarization and higher-order structure in a material. Filler composite in the flexible dielectric polymer is then considered to rearrange polymer chain. However, the filler becomes agglomeration easily at high loading content in polymer, resulting in high energy loss and low electrical breakdown. This work presents the treated Polypyrrole (PPy) filler by 3-Aminopropyltriethoxysilane for avoiding agglomeration in PVDF-HFP thin film. These 30 μm PVDF-HFP film thickness is fabricated by tape casting method with *N,N*-dimethylformamide (DMF) solvent. The distributions of PPy filler on PVDF-HFP are observed by SEM image. Dielectric constant, dielectric loss, and conductivity are analyzed. As a result, the maximum silane content was found on 1 wt% for 1 wt% PPy/PVDF-HFP to maximized dielectric constant and reduce dielectric loss and conductivity. The conductive of PPy filler was lowered by covering with electrical insulating silane, resulting in decreased dielectric loss and conductivity. Then, polymer chain with silane bonding easily polarized under the electric field, resulting in an intensification of dielectric constant around 2.5 times compared with non-silane. Afterward, this dielectric constant clearly decreased when it reached to exceeded silane content as 5-20 wt%. Treated PPy with the suitable silane content in PVDF-HFP performs good dielectric properties for advanced energy storage in this work.

1. Introduction

A dielectric polymer is a flexible polar material that produces a large net dipole moment under an electric field. High dielectric properties, high chemical resistance, and low cost are key features to consider for dielectric polymer matrix in this field such as Polyvinylidene fluoride-hexafluoropropylene (P(VDF-HFP)). Conductive filler in dielectric polymer has been one of the other techniques to improve these properties. However, the large amount of this filler easily agglomerates, resulting in failure from energy loss heat. For example, Putson and colleagues found that dielectric loss of PVDF-HFP was increased with conductive filler [1]. This problem can be solved by using a Silane coupling agent to connect between filler and polymer on PVDF composite [2, 3]. An electrical insulator of Silane covered conductive fillers, bonding with polymer chain. The specific polymer has a specific silane along with filler content [4]. One of the high conductive fillers as Polypyrrole (PPy) is interesting to improve the dielectric properties of PVDF-HFP. There are no prior works to develop dielectric properties on PVDF-

HFP by using a silane coupling agent on PPy filler. In this work, the 3-Aminopropyltriethoxysilane coupling agent is considered for preventing agglomeration and aiming to achieve better dielectric performance.

2. Experimental

Composite film preparation, the polypyrrole (PPy) as 0.25, 0.50, 0.75, and 1.00 wt% is blended and stirred with Polyvinylidene fluoride-hexafluoropropylene (P(VDF-HFP)) (purchased from Solvay Solexis, Belgium) in 99% purity *N,N*-dimethylformamide (DMF) (purchased from RCI Labscan Limited, Thailand) before cast on the smooth glass by tape casting method. The suitable PPy content in PVDF-HFP is observed.

For the silane coupling agent process, the PPy is treated by Hydrogen peroxide 30% (H_2O_2) to get OH group on the surface by sonicating 30 min (250W) and drying 12 h (100°C). Afterward, the 3-Aminopropyltriethoxysilane (silane) is hydrolyzed by DI water and Ethanol (50 ml: 50 ml) for 20 min to open the functional group. The silane contents are considered as 1, 5, 10, and 20 % of PPy filler. These solutions are completely stirred with the suitable PPy content for 24 h. Resulting, treated PPy is obtained by removing the exceeding silane by DI water (5 times centrifuges). The treated PPy is completely dried 100°C (12 h). The treated PPy powder with several silane content is finally prepared to fabricate composite films as mentioned in the previous step.

For characterization, morphological PPy/PVDF-HFP image is observed by using the SEM (FEI Quanta 400, USA). Dielectric loss (ε_r'') and capacitance (C, in F) are recorded by applying 1 V ac voltage to the sample by two electrodes of LCR meter (IM 3533, HIOKI, Japan) in frequency from 1 to 10^5 Hz. Moreover, dielectric constant (ε_r) and conductivity (σ_{ac}, S/m) are calculated by using the equation (1) and (2), respectively. Here d is sample thickness (m), A is the area of the electrode (m^2), and ε_0 is free space permittivity (8.853 x 10^{-12} F m^{-1}), f is frequency (Hz).

$$\varepsilon_r = Cd/\varepsilon_0 A \quad (1)$$

$$\sigma_{ac} = 2\pi f \varepsilon_0 \varepsilon_r \varepsilon_r'' \quad (2)$$

3. Results and discussion

3.1. The PPy filler contents on PVDF-HFP transformation

Figure 1 shows the relationship between dielectric properties and frequency. The 0, 0.25, 0.5, 0.75, and 1 %PPy in PVDF-HFP is measured (Figure 1 (a)). The low frequency (10-10^3 Hz) provides the largest dielectric constant (ε_r) because interfacial polarization is mainly active and easily switchable explaining by the Maxwell-Wagner-Sillars (MWS) theory [5]. In PPy content, ε_r is increased with PPy content. For example, 0.75%PPy immediately increase ε_r from 5 (pure PVDF-HFP) to 76 at 10-10^3 Hz. The ε_r of 1 wt% PPy at low frequency is clearly disturbed by conductive filler loading to be unstable value, resulting in larger conductivity (σ_{ac}) and dielectric loss (ε_r'') as in Figure 1 (b) and (c), respectively. For an example of inthe crease σ_{ac} includes pure PVDF-HFP (10^{-10} S/m) to 1 wt% PPy (10^{-4} S/m) around 10^6 times. In addition, ε_r'' are clearly increased with PPy content. The fixed dipole with conductive filler is easily polarized under electric field, resulting in larger ε_r'' and σ_{ac} when adding conductive filler, as prior theory [5].

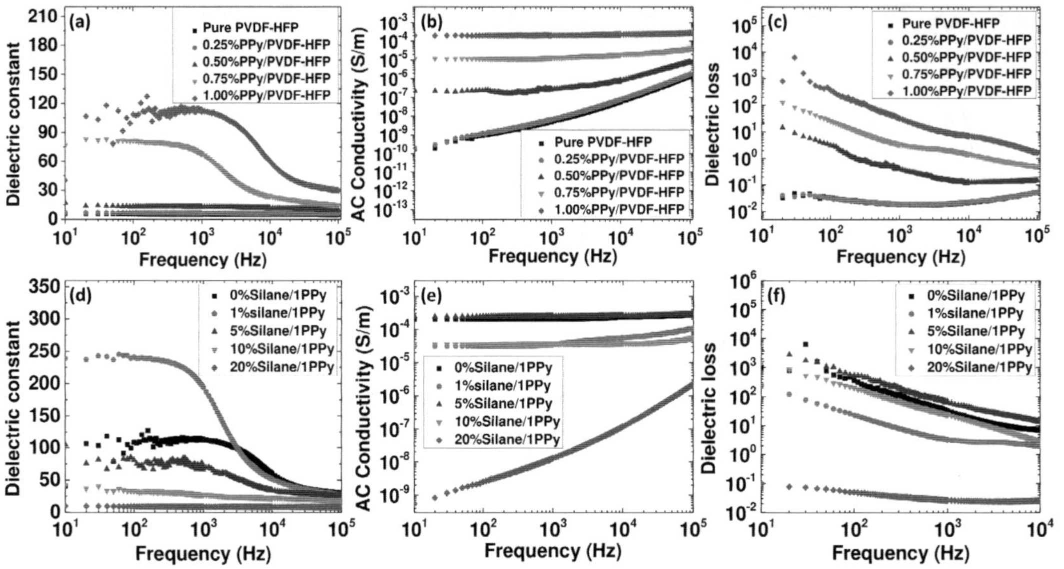

Figure 1 Dielectric constant (a), conductivity (b), and dielectric loss (c), for PPy content in PVDF-HFP. Dielectric constant (d), conductivity (e), and dielectric loss (f), for and for silane content on 1PPy/PVDF-HFP.

3.2. Silane effect on 1%PPy/PVDF-HFP dielectric properties

To enhance interfacial polarization, silane contents are considered. The 1, 5, 10, and 20% Silane on 1%PPy/PVDF-HFP are continuously prepared. As Figure 1 (d), dielectric constant at 10-10^3 Hz sharply increases from 111 (0%Silane/1PPy) to around 222 (1%Silane/1PPy), which seem 2 times differently. Dielectric constant, then, suddenly drops to 75 (5%Silane/1PPy), which is lower than the 0%Silane/1PPy. Afterward, there is a gradual decrease to 8.8 (20%Silane/1PPy) that is the lowest value. Surprisingly, this 1%Silane/1PPy can decrease their conductivity from 2.06x10^{-4} S/m of 0%Silane/1PPy to be 3.30x10^{-5} S/m. Afterward, the conductivity slightly increases at 5%Silane/1PPy before sharply decreased to 7.32x10^{-9} S/m at 20%Silane/1PPy. The trend of this conductivity is similar behavior with dielectric loss, matching with a prior study [2].

The larger insulating silane that covers on PPy has the ability to reduce PVDF-HFP conductivity as well as dielectric loss. The silane head (Si-Si with OH group) is bonded with the OH group on PPy filler surface, and the silane tail (Amino group) is completely connected with the PVDF-HFP polymer chain. It seems that the 1%silane on PPy filler is the suitable silane content that performs a good connector between polymer chain and PPy filler, resulting in the easiest switchable polarization and the largest dielectric properties. Surprisingly, this silane has ability to reduce conductivity as well as dielectric loss. Compared with thicker silane content, the stronger chemical bonding in thinner silane content is likely easier in switchable polarization, resulting in great dielectric properties [2].

3.3. Morphological PPy distribution and dispersion on treated PPy

Scanning Electron Microscopy (SEM), Hitachi TM 3030plus, is observed in the morphology on the sample surface. The PPy filler distribution is distinguishable between treated and untreated silane on 1%PPy in PVDF-HFP as showed in **Figure 2**. There are several big porosities on this sample by the DMF evaporation after drying. The PPy is noticeable as the small black dot. These dots have more noticeable and completely separated two phases on untreated PPy by silane (Figure 2 (a)). As a result, it seems a lot of PPy particles dispersing on PVDF-HFP. In contrast, when the treated PPy is connected to PVDF-HFP by silane, the separation between two phases is blurred. As a result, it seems less of PPy particles dispersing on PVDF-HFP.

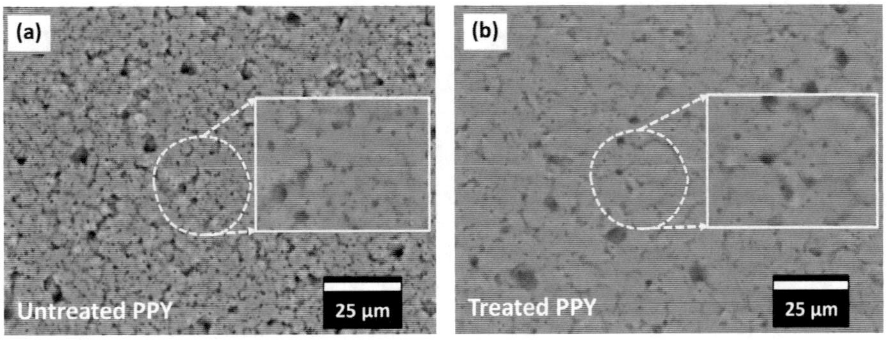

Figure 2 SEM images of untreated (a) and treated (b) PPy by a silane coupling agent.

4. Conclusion

The 0.25, 0.50, 0.75, and 1.00%PPy in PVDF-HFP polymer composites were fabricated by tape casting method. The dielectric constant, conductivity, and dielectric loss were increased with PPy content. Among them, the best dielectric constant was shown by 1.00%PPy in PVDF-HFP. Nevertheless, that condition also possessed the highest conductivity and dielectric loss. To reduce those two conductivity and dielectric loss, the 3-Aminopropyltriethoxysilane coupling agent was added into these composites. The silane content was varied of 1, 5, 10, and 20%. As a result, 1 %silane improved dielectric constant, and reduced both conductivity and dielectric losss. On the other hand, adding silane more than 1% could cover the filler to be thicker. As a consequence, the dipole moment is hard to polarize following the external electric field, resulting in decrease dielectric constant, conductivity along with dielectric loss. This silane content in this work has a potential role to increase dielectric properties for the advanced capacitor in the future.

5. References

[1.] Salea, A., et al., *The microstructure of negative electrocaloric Polyvinylidene fluoride-hexafluoropropylene copolymer on graphene loading for eco-friendly cooling technology.* Journal of Cleaner Production, 2020. **251**: p. 119730.
[2.] Zhang, Q., et al., *Enhanced dielectric tunability of Ba0.6Sr0.4TiO3/Poly(vinylidene fluoride) composites via interface modification by silane coupling agent.* Composites Science and Technology, 2016. **129**: p. 93-100.
[3.] Tiwari, V.K., et al., *Thin and surface adhesive ferroelectric poly(vinylidene fluoride) films with β phase-inducing amino modified porous silica nanofillers.* Journal of Polymer Science Part B: Polymer Physics, 2016. **54**(23): p. 2401-2411.
[4.] Xie, L., et al., *Core-shell Structured Hyperbranched Aromatic Polyamide/BaTiO3 Hybrid Filler for Poly(vinylidene fluoride-trifluoroethylene-chlorofluoroethylene) Nanocomposites with the Dielectric Constant Comparable to That of Percolative Composites.* ACS Applied Materials & Interfaces, 2013. **5**(5): p. 1747-1756.
[5.] Tsangaris, G.M., G.C. Psarras, and N. Kouloumbi, *Electric modulus and interfacial polarization in composite polymeric systems.* Journal of Materials Science, 1998. **33**(8): p. 2027-2037.

Acknowledgment
We would like to thanks the Thailand Center of Excellence in Physics (ThEP-61-PIP-PSU3), Science Achievement Scholarship of Thailand (SAST), including Physics Department, Faculty of Science, and Prince of Songkla University.

IOP Conf. Series: Materials Science and Engineering **773** (2020) 012021 doi:10.1088/1757-899X/773/1/012021

Preparation and ferroelectric properties of poly (vinylidene fluoride-hexafluoropropylene) (PVDF-HFP) filled with graphene-nanoplatelets film composites

S Chaipo[1], C Putson[1,*]

[1] Materials Physics Laboratory, Department of Physics, Faculty of Science, Prince of Songkla University, Hat Yai, Songkhla 90112, Thailand.

*Corresponding author: chatchai.p@psu.ac.th

Abstract. The poly(vinylidene fluoride-co-hexafluoropropylene), PVDF-HFP composite films with addition graphene-nanoplatelets (GPN) was prepared by using the tape casting solution method. The obtained composite films were stretched with help from thermal stretching machine at 80 $_{o}$C with 5 mm/min rate. Dielectric constant and hysteresis loop (PE-loop) between the stretching and non-stretching films at different fillers percentage were compared in this study. Dielectric constant was investigated by the LCR meter. The PE-loop was measured by the ferroelectric polarization loop test system. The experimental results showed that the dielectric constant of all samples increases with increasing fillers content, regardless of frequency. The dielectric constant of stretching composite films was higher than non-stretching composite films. In addition, the PE-loop shapes of the stretching films have slimmer than the non-stretching films regardless of filler content. However, the PE-loop produced the shape to be bigger loop with increasing filler content. The energy efficiency of obtained PVDF-HFP composite films will be discussed on their dielectric constant, dielectric loss, AC conductivity, and polarization performances for electric–capacitor materials applications.

1. Introduction

Poly(vinylidene fluoride-hexafluoropropylene) (PVDF-HFP) is widely investigated due to its exceptional ferroelectric properties. These properties are at the origin of various applications, especially in the field of sensor, transducers, actuator devices, and electric–capacitor materials [1],[2]. The β - phase of P(VDF-HFP) exhibited the strongest ferroelectric properties. The increasing of β - phase transition of polymer has been investigated on several techniques. Stretching films at a temperature below the crystal phase transition temperature resulted in a significant increase in the β – phase [3]. However, the dipolar moments of polymer are randomly oriented. The poling field will induce a preferred orientation of the dipoles along the field direction that can generate the β –phase regions. On the other hand, the β -phase can be enhanced by conductive particle fillers [4]. Graphene-nanoplatelets (GPN) have been recognized due to its excellent conductivity, thermal, mechanical, electrical properties, and its possibility of mass production at low cost. In fact, GPN are layered graphene nano - crystals in the structure of platelets stacked by van der Waal's forces. Thus, the ferroelectric and dielectric properties of PVDF-HFP can be enhanced with increasing GPN content. In this work, the PVDF-HFP blend with GPN in DMF solvent was prepared and synthesized film by the solution casting method. The dielectric constant, AC conductivity, dielectric loss factor, PE-loop and energy efficiency were assessed by using LCR meter and the ferroelectric polarization loop test system. The obtained PVDF-HFP composites with and without stretching various content fillers was studied and discussed on the dielectric and ferroelectric properties.

2. Experimental

2.1 Material preparation

Polyvinylidene fluoride-hexafluoropropylene (PVDF-HFP, Solef 11010/1001, purchased from Solvay Solexis, Belgium) filled with graphene-nanoplatelets (GPN, 806633 from Sigma-Aldrich). The polymeric films were prepared by the solution casting method. Firstly, the GPN (conditions of 1, 2, 3 and 4 %wt.) was dissolved and homogenized in N, N-dimethylformamide (DMF, 99% purity, purchased from RCI Labscan Limited, Thailand), and dispersed by the ultrasonic probe (Hielscher UP400S) for 20 min. Subsequently, the obtained GPN was mixed with PVDF-HFP solution and continuously stirred for 3h at room temperature. The solutions were cashed on the glass plate and annealed at 80°C for 12 h for evaporation of the solvent. And then, the films were taken off from the glass plate. Finally, the composite films were stretched with the thermal stretching machine at 80 $_{o}$C with rate of 5 mm/min.

2.2 Dielectric characterization

The dielectric constant, AC conductivity and dielectric loss factor of films were investigated by the LCR meter (IM 3533 HIOKI) with frequency of 1 - 10$_5$ Hz at room condition. The dielectric constant and AC conductivity were calculated by equation (1) and (2), respectively:

$$\varepsilon_r = \frac{C_p t}{\varepsilon_0 A} \quad (1)$$

$$\sigma = \frac{Gt}{A} \quad (2)$$

where ε_r and σ are dielectric constant and AC conductivity and C_p, t, A, ε_o, G are electrical capacitor, the thickness of film (Samples thickness is 100 ± 3 μm in all case), the area of electrode, the permittivity of air (8.853 x 10-$_{12}$ F m-$_1$) and the conductance, respectively.

2.3 Ferroelectric characterization: Au electrodes were sputtered on the both surface of films. The ferroelectric properties was analysed by Polarization-Electric filed (P-E) loop test system with the electric fields range of -40 to 40 MV/m at room temperature. The energy density (U_e), and energy efficiency (η) can be calculated from the curve of PE-loop as Equation (3) and (4) [5], respectively. For loss energy density (U_l) was calculated in the closed area in PE-loop:

$$U_e = \int_{Pr}^{Pmax} EdP \quad (3)$$

$$\eta(\%) = \left(\frac{Ue}{Ue+Ul}\right) \times 100 \quad (4)$$

where E, P_r and P_{max} were the electric field, the remnant polarization, the maximum polarization, respectively.

3. Results and discussion

3.1 Dielectric properties: In Figure 1(a) shows the dielectric constant of all samples increases with increasing content of fillers. The dielectric constant of stretching PVDF-HFP films is higher that of non-stretching films, regardless of used content of fillers. For examples, the dielectric constant of stretching and non-stretching PVDF-HFP with GPN 4%wt was 2.99 and 1.86, respectively. The micro-polarization contributions in case of the heterostructure can be increased which is strongly relevant to the interfacial polarization between polymer matrix and fillers. In fact, the GPN exhibited the good conductivity, thus there are enhancement of charge accumulation in polymer matrix when the GPN was used as fillers.

Furthermore, it may be attributed that its crystallinity of polymer chain was increased when the composites films were stretched, lead to enhancement of the β – phase region. While, the AC conductivity of stretching PVDF- HFP films is not more change when compared with non – stretching films. Figure 1(b-c) exhibited the AC conductivity of PVDF-HFP composites increases with increasing GPN content. The dielectric loss factor had the same trend with the AC conductivity, can assume by , the dielectric loss factor (tanδ) depends on AC conductivity (σ) [6].

$$\tan\delta = \frac{\varepsilon_r''}{\varepsilon_r'} + \frac{\sigma}{\omega\varepsilon_0\varepsilon_r'}$$

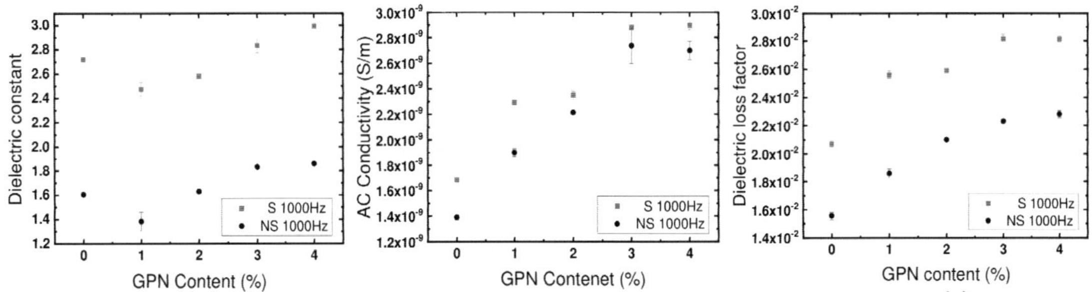

(a) (b) (c)

Figure1. (a) The dielectric constant, (b) AC conductivity and (c) The dielectric loss factor of PVDF-HFP filled with GPN compared between stretching film (S: Red square symbol) and non-stretching film (NS: Black circle symbol) at 1000Hz.

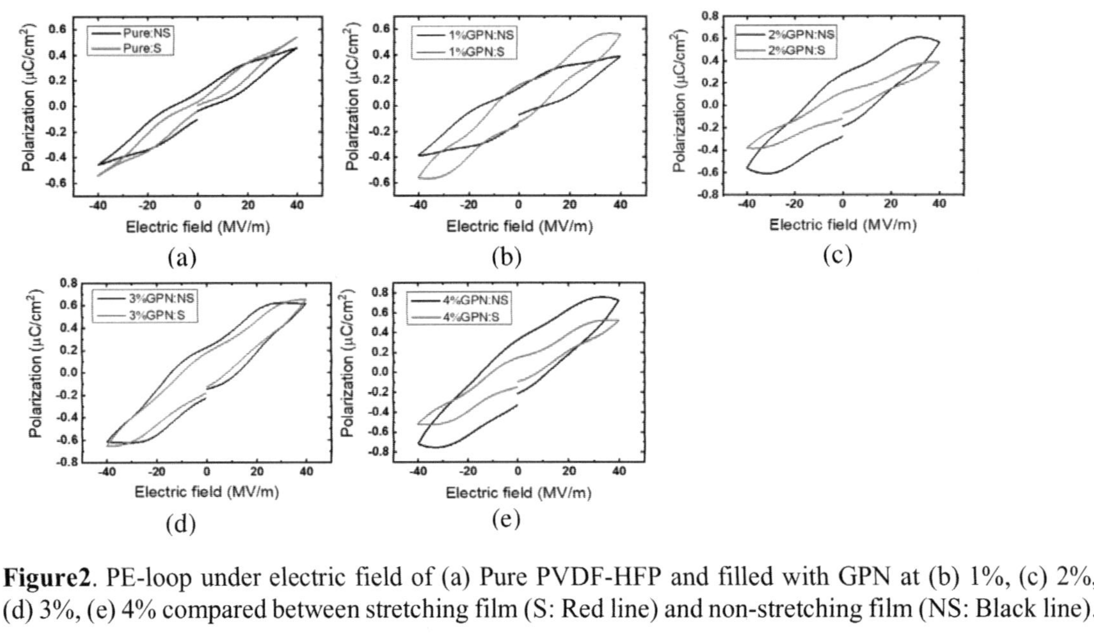

Figure2. PE-loop under electric field of (a) Pure PVDF-HFP and filled with GPN at (b) 1%, (c) 2%, (d) 3%, (e) 4% compared between stretching film (S: Red line) and non-stretching film (NS: Black line).

3.2 Ferroelectric properties: The shape of the hysteresis loop (PE-loop) as shown in Figure 2(a-e). The loss energy density (U_l) of the stretching films have less than the non-stretching films, regardless of filler content. From PE-loop can evaluate the energy efficiency (η) from equation (4). the η of stretching film is higher than non-stretching film. However, the neat film stretched had the highest efficiency at 77% (Figure 3). It may be attributed that the addition of GPN related to crystalline size of PVDF-HFP composite films [7]. Agglomerate of GPN particle resulted in increasing of U_l and reduction U_e, so the η of the composite film had the decreasing trend.

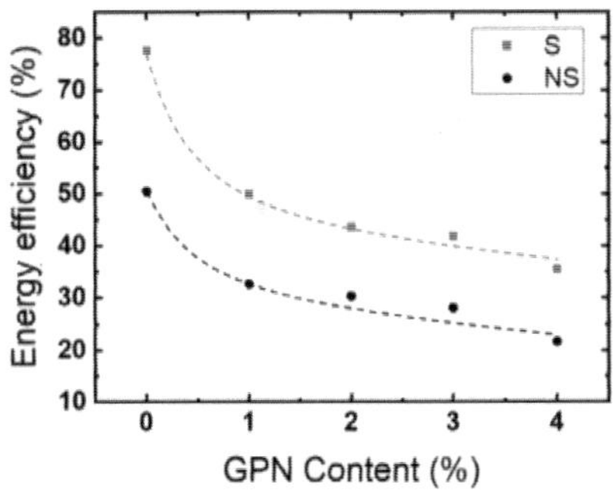

Figure3. The energy efficiency of pure PVDF-HFP and PVDF-HFP filled with GPN 1%, 2%, 3% and 4% compared between stretching film (S) and non-stretching film (NS).

4. Conclusions

This work demonstrates the enhancement of the dielectric constant, AC conductivity and dielectric loss factor of the stretching PVDF-HFP film by GNP particle. In addition, the shape and volume of the hysteresis loop or the loss energy density of the stretching films have less than the non-stretching films, regardless of filler content. It may be attributed that the changing of hysteresis loop related to the crystallinity of PVDF-HFP composite films when GNP fillers was used.

5. References

[1] X. He, K. Yao, and B. K. Gan, "Phase transition and properties of a ferroelectric poly(vinylidene fluoride-hexafluoropropylene) copolymer," *J. Appl. Phys.*, vol. 97, no. 8, 2005.

[2] M. Wegener, W. Künstler, K. Richter, and R. Gerhard-Multhaupt, "Ferroelectric polarization in stretched piezo- and pyroelectric poly(vinylidene fluoride-hexafluoropropylene) copolymer films," *J. Appl. Phys.*, vol. 92, no. 12, pp. 7442–7447, 2002.

[3] J. Gomes, J. S. Nunes, V. Sencadas, and S. Lanceros-Mendez, "Influence of the β-phase content and degree of crystallinity on the piezo-and ferroelectric properties of poly(vinylidene fluoride)," *Smart Mater. Struct.*, vol. 19, no. 6, 2010.

[4] L. Yang *et al.*, "Graphene enabled percolative nanocomposites with large electrocaloric efficient under low electric fields over a broad temperature range," *Nano Energy*, vol. 22, pp. 461–467, 2016.

[5] T. F. Zhang *et al.*, "Enhanced electrocaloric analysis and energy-storage performance of lanthanum modified lead titanate ceramics for potential solid-state refrigeration applications," *Sci. Rep.*, vol. 8, no. 1, pp. 1–12, 2018.

[6] Ardimas, C. Putson, and N. Muensit, "High electromechanical performance of modified electrostrictive polyurethane three-phase composites," *Compos. Sci. Technol.*, vol. 158, pp. 164–174, 2018.

[7] Y. Zhu, P. Jiang, Z. Zhang, and X. Huang, "Dielectric phenomena and electrical energy storage of poly(vinylidene fluoride) based high-k polymers," *Chinese Chem. Lett.*, vol. 28, no. 11, pp. 2027–2035, 2017.

Acknowledgement

This work was supported by Development and Promotion of Science and Technology Talents Project (DPST), Institute for the promotion of teaching science and technology Thailand (IPST) and department of Physics Faculty of Science, Prince of Songkla University (PSU), Thailand.

IOP Conf. Series: Materials Science and Engineering **773** (2020) 012022 doi:10.1088/1757-899X/773/1/012022

Synthesis of Carbon Dots from the Biomass Products for Supercapacitor Applications

A Tibodee[1], W Hirunpinyopas[2], K Jeamjumnunja[2], and W Sirisaksoontorn[1,*]

[1] Department of Chemistry and Centre of Excellence for Innovation in Chemistry, Faculty of Science, Kasetsart University, Chatuchak, Bangkok 10900, Thailand
[2] Department of Chemistry, Faculty of Science, Kasetsart University, Chatuchak, Bangkok 10900, Thailand

*Corresponding author: fsciwks@ku.ac.th

Abstract. Biomass-derived CDs were prepared via a green one-step synthetic method. The orange juice biomass was used as a CD precursor. The resulting CDs were mixed with rGO under the hydrothermal condition to prepare rGO/CD composites with the variation of rGO:CD mass ratio. The composites were characterized through both physical and electrochemical methods. Moreover, the capacitive property of rGO/CD composites in 2 M KOH aqueous electrolyte was studied by cyclic voltammetry, galvanostatic charge-discharge measurements, and electrochemical impedance spectroscopy. The rGO/CD composite with a 4:1 mass ratio exhibits the high specific capacitance with 184.8 F g^{-1} at a current density of 0.5 A g^{-1}. The rGO/CD 4:1 composite retains 97.3% of its initial capacitance after 1000 cycles at the current density of 10 A g^{-1} in 2 M KOH electrolyte solution.

1. Introduction

Supercapacitors (SCs), also known as ultracapacitors or electrochemical capacitors, have many advantages for energy storage due to their high power densities, long cycle life, low maintenance cost, and environmentally eco-friendly nature. In general, SCs can be classified into three categories such as the electrical double-layer capacitors, pseudocapacitors, and hybrid capacitors. Each class is dependent upon its unique mechanism for storing charges [1]. In terms of carbon materials, carbon dots (CDs), a zero-dimensional carbon material, have attracted tremendous interest owing to their outstanding electronic properties, quantum confinement, and edge effects. CDs are used for several practical applications such as supercapacitors, batteries, optoelectronics, and biomedical field. Recently, several carbon sources, including small chemical molecules, have been employed for CD synthesis. One interesting challenge is to use CDs obtained from the sustainable resource. Biomass products are considered to be one of the most competent candidates for CD synthesis. Advantages of biomass energy are in terms of reducing chemical exposure, reducing waste, cheap and eco-friendly synthesis, and abundant sources [2].

Herein, rGO/CD composites were prepared by using a hydrothermal method and their electrochemical properties were systematically investigated. CDs were synthesized by using orange juice as a biomass-derived source. The effect of rGO:CD ratio on the electrochemical performance of the composites was also presented.

2. Experimental

2.1. Preparation of CDs by using orange juice
40 mL of the pulp-free orange juice and 300 µL of ethylenediamine were mixed into a 125 mL Teflon-lined stainless steel autoclave and heated at 160°C for 4 h. After cooling, the dark brown solution was centrifuged at 10000 rpm for 30 min to remove large particles. The dark brown supernatant was collected and filtered through a 0.22-µm syringe filter. After that, the solution was washed with excess dichloromethane to extract unreacted organic species. Subsequently, the yellowish brown aqueous solution was dialyzed in DI water using a tubular dialysis bag with the molecular weight cut-off of 3.5 kDa for 24 h. The purified CD aqueous dispersion was pre-concentrated and freeze-dried.

2.2. Preparation of rGO/CD composites
GO was prepared by using modified Hummers' method [3]. Then, 100 mg of as-prepared GO was dissolved in aqueous solution and sonicated for 1 h. Variable amounts of CD were added to GO solution with continuous stirring to establish a homogeneous solution. After that, the reaction solution was hydrothermally heated in the autoclave at 180°C for 9 h. After cooling, the rGO/CD composites were centrifuged at 10000 rpm for 10 min and then washed with DI water. Finally, the composites were vacuum-freeze dried. The rGO/CD composites with different rGO:CD mass ratio of 6:1, 4:1, 2:1, and 1:1 were systematically denoted as rGO/CD 6:1, rGO/CD 4:1, rGO/CD 2:1, and rGO/CD 1:1, respectively.

2.3. Electrochemical measurements
The electrochemical performance of all samples was evaluated by using a conventional three-electrode system. Three-electrode setup was composed of a working electrode (electroactive material), reference electrode (Ag/AgCl in 3 M KCl), and counter electrode (Pt sheet) in a 2 M KOH aqueous electrolyte. The active material was prepared by mixing rGO/CD, PVDF binder, and conducting carbon black with 80:10:10 mass ratio. The specific capacitance can be calculated from the galvanostatic discharge of a supercapacitor by using the following equation; $C_s = \frac{I\Delta T}{m\Delta V}$, where C_s is the specific capacitance (F g_{-1}), I is the discharge current (A), Δt is the discharge time (s), ΔV is the discharge potential window (V), and m is the mass of active material (g).

2.4. Material characterization
The morphology of all samples was characterized by scanning electron microscopy (SEM, QUANTA 450) and transmission electron microscopy (TEM, JEM-3100F). The X-ray photoelectron spectrum was recorded using the PHI 5000 Versa Probe II XPS system (ULVAC-PHI, Japan) with Al Kα (1486.6 eV) as an X-ray source. Raman spectra were collected on a Raman microscope (InVia, Renishaw) with a laser excitation at 514 nm.

3. Results and discussion
As shown in Figure 1a, the spherical CDs were mainly observed in TEM with monodisperse particles and lattice fringe of 0.21 nm, corresponding to the (1 0 0) plane of graphitic carbon [4]. The size distribution of CDs ranges from 2.5 - 5.5 nm with the average size of 4.0 nm. Figure 1b presents a high-resolution C 1s XPS spectrum of CDs. The spectrum was deconvoluted into five distinct carbon components at the binding energy of 284.0, 285.5, 287.2, 288.4, and 291.7 eV, which is consistent with the presence of C–C/C=C, C–O/C–N, C=O, O–C=O, and π- π* transition, respectively [5]. It suggests that the hydrophilic group is abundant in the CD structure. Moreover, the morphology of rGO in Figure 2a exhibits crumpled sheets with random aggregation. Upon incorporation of CDs, the composite exhibits more open structure and the aggregation of rGO sheets was prevented as shown in Figure 2b. The TEM image of selected rGO/CD 1 : 1 in Figure 2c exhibits transparent thin rGO nanosheets with small CD aggregates.

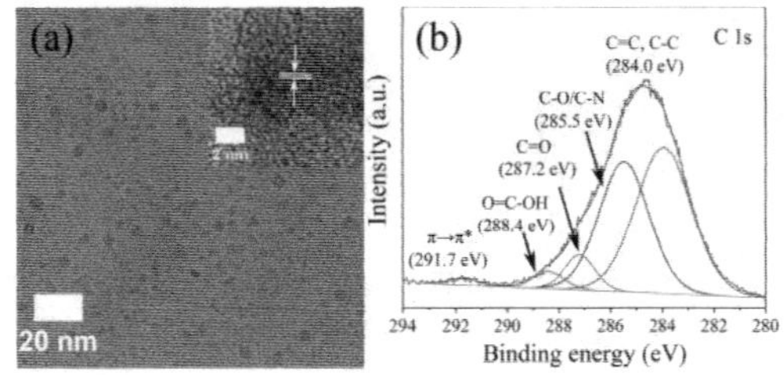

Figure 1. (a) TEM and high resolution (inset) images and (b) C1s XPS spectrum of CDs.

Figure 2d reveals the Raman spectra of GO, rGO, and rGO/CD 4:1. For GO, the I_D/I_G ratio was calculated to be 1.76, which suggests a relatively low graphitic sp^2 carbon in the structure. After reduction, the I_D/I_G ratio of rGO is decreased to 1.61. For rGO/CD 4:1, the I_D/I_G ratio is slightly lower than that of rGO, indicating that the composite contains the larger extent of graphitic-sp^2 carbon.

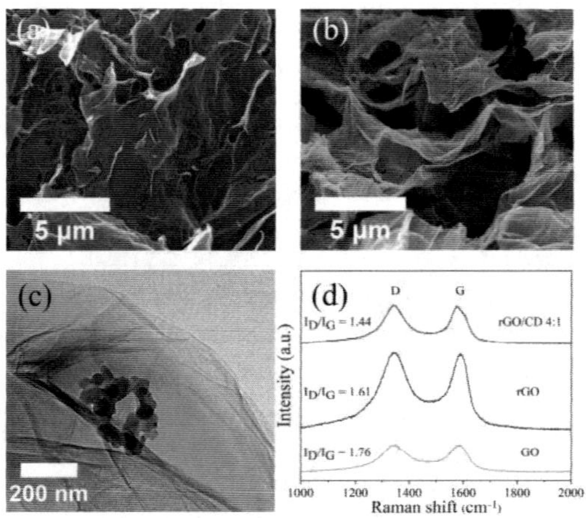

Figure 2. SEM images of (a) rGO and (b) rGO/CD 4:1 (c) TEM image of rGO/CD 1:1 and (d) Raman spectra of GO, rGO, and rGO/CD 4:1.

Figure 3a shows the CV plots of all samples at a scan rate of 200 mV s^{-1}. The CV curve of rGO is distorted from the rectangular shape. The reversible anodic and cathodic humps between -0.4 and -0.8 V originate from the combined contribution of the electrical double layer and faradaic redox capacitances. In comparison with rGO, CV curves of all as-prepared rGO/CD also show the distorted rectangular shape. The area under the CV curves of rGO/CD 4:1 is slightly larger than that of the other samples, suggesting the enhanced specific capacitance of the composite. Furthermore, the observed current of rGO/CD 4:1 in Figure 3b gets increased with increasing the scan rate, which is ascribed to the increasing concentration of electroactive ions at the electrode-electrolyte interface at the higher scan rate. The galvanostatic charge/discharge curves of electrodes with different rGO/CD mass ratio are compared at a current density of 0.5 A g^{-1} as shown in Figure 3c. The specific capacitance of rGO is 171.3 F g^{-1}. Meanwhile, the rGO/CD composites provide the enhanced specific capacitance with 183.8 F g^{-1} for rGO/CD 6:1 and 184.8 F g^{-1} for rGO/CD 4:1. However, the specific capacitance declines to 158.7 F g^{-1} for rGO/CD 2:1 and 140.9 F g^{-1} for rGO/CD 1:1 with an exceeding amount of CDs. This improved electrochemical behaviour is due to the positive effect of decorated CDs. The presence of CDs can shorten the transport pathways for electrons and ions and provides accessible edges [6]. However, the exceeding amount of CDs has an adverse impact on the electrochemical performance because of the blockage of ion channels [7]. Figure 3d illustrates the Nyquist plots of rGO and rGO/CD 4:1. R_s of the composite is lower than that of rGO, suggesting the lower intrinsic resistance. Nevertheless, the semicircles of both samples at the high frequency region were not clearly observed, indicating that both samples have a small electron transfer resistance and the mass transfer dominates on the Nyquist plot. Figure 3e illustrates the cyclability performance of rGO and rGO/CD 4:1 at the current density of 10 A g^{-1}. The result suggests that rGO/CD 4:1 can retains 97.3% of its initial capacitance, whereas the rGO only maintain 93.4% of its initial capacitance after 1000 cycles. The improved electrochemical stability of rGO/CD 4:1 composite is ascribed to the stable porous structure, which is beneficial for the ion transport during the charge/discharge process.

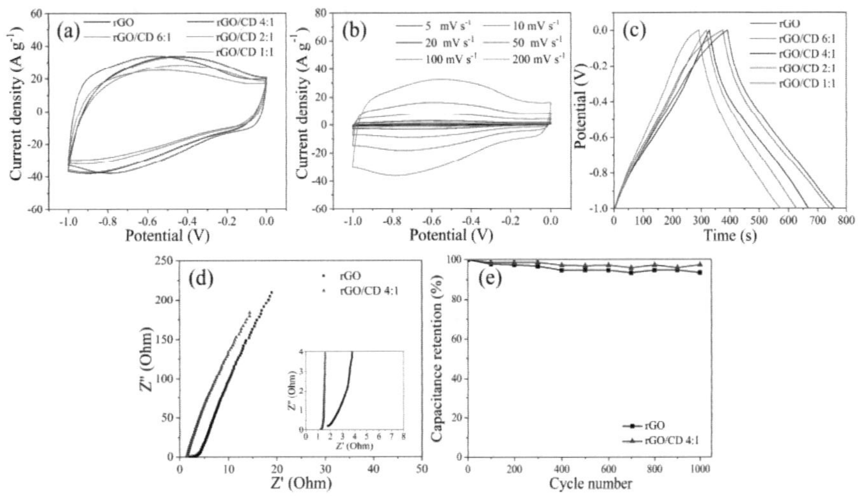

Figure 3. CV curves of (a) rGO and rGO/CD composites at the scan rate of 200 mV s^{-1}, (b) rGO/CD 4:1 at different scan rates, (c) Galvanostatic charge/discharge curves of rGO and rGO/CD composites, (d) Nyquist plots of rGO and rGO/CD 4:1 and their magnified portion at high frequency (inset), and (e) Capacitance retention of rGO and rGO/CD 4:1 composite at the current density of 10 A g^{-1}.

4. Conclusions

We successfully synthesized CDs from orange juice biomass. The average particle size of CDs is 4.0 nm and CDs were used to prepare rGO/GO composites via facile hydrothermal treatment. The addition of CDs could hinder the rGO re-stacking, increase additional pseudo-capacitance, and enhance the electron transport. The rGO/CD 4:1 exhibits the superior specific capacitance of 184.8 F g^{-1} at the current density of 0.5 A g^{-1} and retains high capacitance retention of 97.3% after 1000 cycles.

5. References

[1] Mao L, Guan C, Huang X, Ke Q, Zhang Y and Wang J 2016 *Electrochim. Acta.* **196** 653.
[2] Maneerung T, Liew J, Dai Y, Kawi S, Chong C and Wang C-H 2016 *Bioresour. Technol.* **200** 350.
[3] Tibodee A, Pannak P, Akkarachaneeyakorn K, Thaweechai T and Sirisaksoontorn W 2019 *Mater. Chem. Phys.* **235** 121755.
[4] Ding H, Ji Y, Wei J-S, Gao Q-Y, Zhou Z-Y and Xiong H-M 2017 *J. Mater. Chem. B* **5** 5272.
[5] Xu M, He G, Li Z, He F, Gao F, Su Y, Zhang L, Yang Z and Zhang Y 2014 *Nanoscale* **6** 10307.
[6] Xu Y, Feng Y, Li X, Hu G, Luo Y Sun L, Tang T, Wen, J wang J and Li M 2017 *Int. J. Electrochem. Sci.* **12** 8820.
[7] Dang Y-Q, Ren S-Z, Liu G, Cai J, Zhang Y and Qiu J 2016 *Nanomaterials* **6** 212.

Acknowledgement

This research was supported, under the Strengthening and Developing New Researcher Plan, in conformance with the Research and Innovation of Graduate Study Strategy of the National Research Council of Thailand (NRCT) as of the fiscal year 2019.

An accelerated biodegradation of Poly(lactic acid) by inoculation of *Pseudomonas geniculate* WS3 combined with nutrient addition

Y Boonluksiri[1,*], B Prapagdee[2], and N Sombatsompop[1]

[1] Polymer PROcessing and Flow (P-PROF) Research Group, Division of Materials Technology, School of Energy, Environment and Materials, King Mongkut's University of Technology Thonburi (KMUTT), Thungkru, Bangkok 10140, Thailand.
[2] Laboratory of Environmental Biotechnology, Faculty of Environment and Resource Studies, Mahidol University, Salaya, Nakhon Pathom 73170, Thailand.

*Corresponding author: yeiangchart@gmail.com

Abstract. This study aims to develop an efficient method for accelerating the biodegradation rate of Polylactic acid (PLA). A combined use of PLA-degrading bacterium, *Pseudomonas geniculate* WS3 and nitrogen source or enzyme inducer to accelerate biodegradation of PLA was proposed. PLA films were prepared and submerged in basal salt medium (BSM) amended with ammonium sulfate, soytone, sericin or sodium lactate and inoculated with *P. geniculate* WS3 for 30 days. The results showed that the highest percentage of PLA film-weight loss was found in the treatment of soytone addition, followed by sodium lactate addition. PLA films in culture broth with *P. geniculate* WS3 and soytone were cracked and broken down into small fragments within 20 days. In addition, increasing the lactic acid content as a monomer of PLA in culture broth was directly correlated with increasing the percentage of PLA film-weight loss. It could be concluded that a combined use of *P. geniculate* WS3 and soytone exhibited a high potential to significantly accelerate the PLA biodegradation under the submerged condition.

1. Introduction

Polylactic acid (PLA) has been intensively used, especially in form of single-use biodegradable products and PLA biodegradation is relatively slow because PLA-degrading microorganisms in the environment are limited [1-3]. The important approaches to accelerate PLA biodegradation are biostimulation and bioaugmentation techniques. Biostimulation technique refers to addition of nutrients, water and oxygen to stimulate naturally occurring microbial populations and enhance microbial degradation activities [4-5]. Some nutrients and degradable enzyme inducers to enhance bacterial growth and enzyme production have been reported, including ammonium sulfate (inorganic nitrogen) [6], soybean meal (organic nitrogen) [6], sericin (enzyme inducer) [7] and sodium lactate (enzyme inducer) [8]. In addition, the efficiency of bioaugmentation with bacteria on accelerating PLA biodegradation in the soil has been stated by several studies [9-11].

A combined use of both biostimulation (the addition of nitrogen and inducers) and bioaugmentation (a specific PLA-degrading bacterium and/or microbial consortium) to accelerate biodegradation of PLA is lack of studies. Therefore, this study aims to develop an efficient method for accelerating biodegradation of PLA. A combined use of PLA-degrading bacterium, *Pseudomonas geniculate* WS3 [12] and nitrogen source or enzyme inducer to accelerate biodegradation of PLA was studied.

2. Experimental

2.1. Materials and chemicals
Polylactic acid (PLA; Ingeo™ biopolymer, 2003D grade) was purchased from NatureWorks, USA. Ammonium Sulfate was obtained from Ajax Finechem, Thailand. Soytone (Bacto™ Soytone) were purchased from BD Biosciences, USA. Sodium lactate (Sodium L-lactate) was purchased from Sigma-Aldrich, USA.

2.2 Biodegradation of PLA film under submerged condition
P. geniculate WS3 inoculum was prepared by culturing in nutrient broth (NB) and incubated overnight at 30 °C with shaking at 150 rpm. An overnight inoculum was centrifuged at 4 °C, 8000 rpm for 30 min and washed with a 50mM sterile phosphate buffer. Cell pellet of WS3 was suspended in a basal salt medium (BSM) broth amended with 0.1% ammonium sulfate [6], 1% soytone (digested soybean meal) [6], 0.1% sericin [7] or 0.2% sodium lactate [8] to give a cell density of OD_{600} ~ 0.2. PLA films were formed by solvent casting technique, with a dimension of 20 x 30 x 0.1 mm. The surface sterilized PLA films were placed in the culture of WS3 in BSM broth and incubated in a rotary shaker at 30 °C, 150 rpm for 30 days. PLA films and the culture broth were collected at 10, 20 and 30 days, respectively for measuring film weight, pH, lactic acid content and viable bacterial cells.

3. Results and discussion

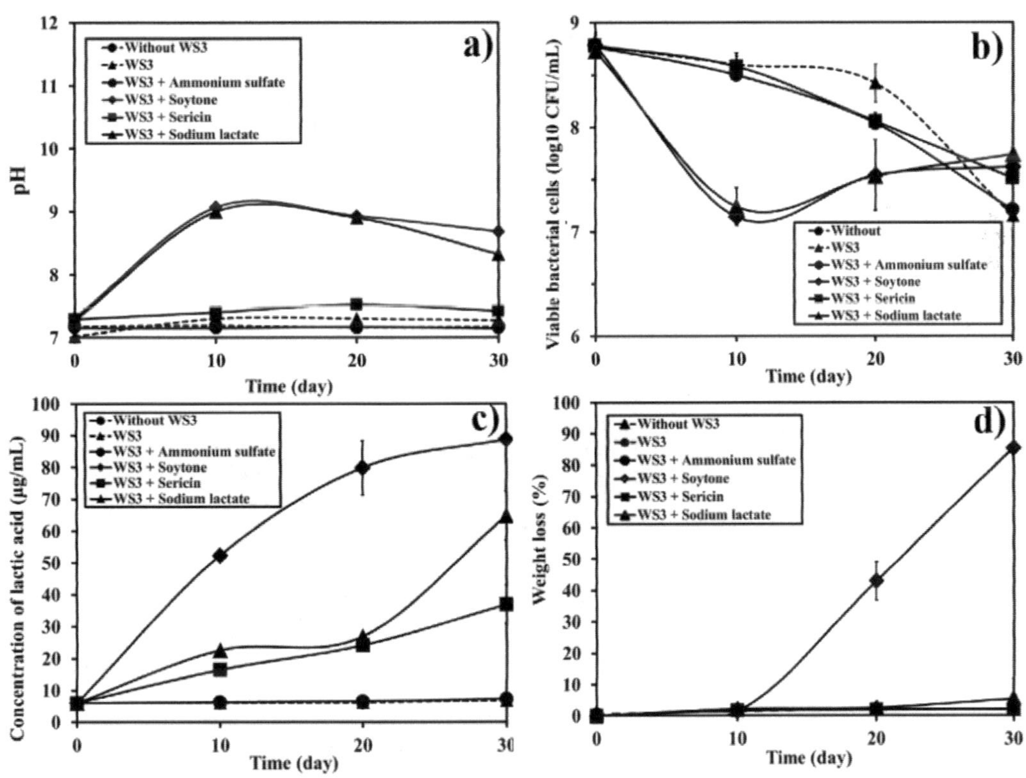

Figure 1. Biodegradation of PLA films after submerged in BSM with WS3 amended with ammonium sulfate, soytone, sericin or sodium lactate at 10, 20 and 30 days: a) pH value, b) viable bacterial cell, c) concentration of lactic acid, d) weight loss.

The results of the changes of pH, number of viable bacterial cells, lactic acid content and film-weight loss in the culture broth of *P. geniculate* WS3 amended with nitrogen source or enzyme inducer are shown in Fig. 1. It was found that pH value in the culture broth of WS3 amended with soytone (Fig. 1a) dramatically increased in the first 10 days because chemical hydrolysis of PLA generated hydroxide ion and it was slightly decreased in 20 and 30 days because of the generation of lactic acid which was corresponded with the study of Bubpachat et al. [12]. The change of pH affected the number of viable bacterial cells as shown in Fig. 1b, showing that the number of viable bacterial cells of WS3 in the culture broth amended with soytone dramatically decreased in the first 10 days because the medium turned to be alkaline and it slightly increased in 20 and 30 days. It can be explained that WS3 can degrade PLA structure and PLA turned to be lactic acid which was assimilated by bacteria [12-13]. The contents of lactic acid in the culture broth of WS3 amended with soytone (Fig. 1c) sharply increased and higher than that of other amended substances. It would be suggested that the addition of soytone promote the PLA biodegradation of WS3. In addition, increasing lactic acid content in culture broth of WS3 amended with soytone was directly correlated with increasing percentage of PLA film-weight loss (Fig. 1d). The percentages of PLA film-weight loss in the treatment of WS3 and soytone addition sharply increased higher than other treatments, particularly at 20 and 30 days of incubation periods. This can be explained that the degraded PLA by biochemical activity can be dissolved in medium [14], resulting in decreasing PLA mass.

Table 1. Digital photographs of PLA films in BSM broth inoculated with WS3 and amended with different substances at 10, 20 and 30 days of incubation period

Treatment	Time (days)		
	10	20	30
Without WS3	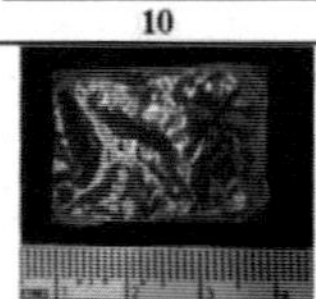	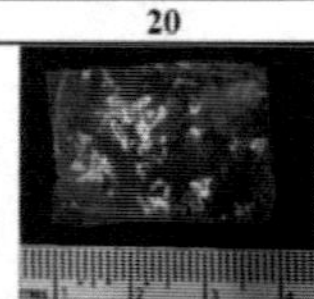	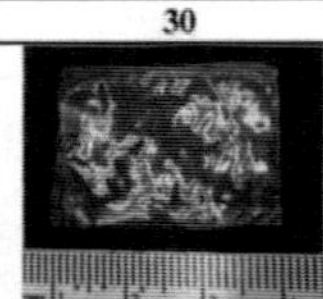
WS3	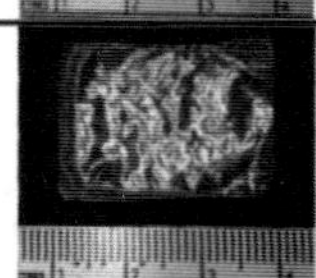	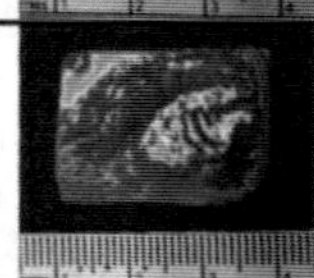	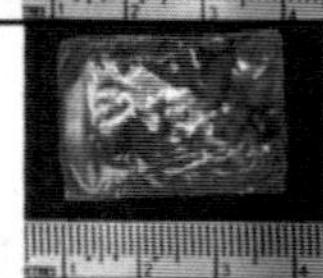
WS3 + Ammonium sulfate	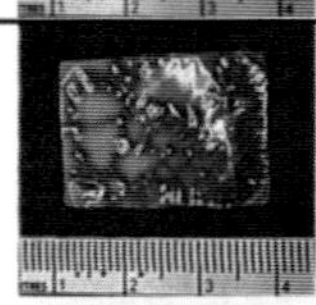	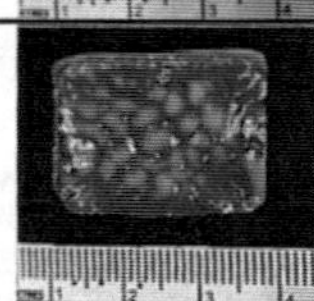	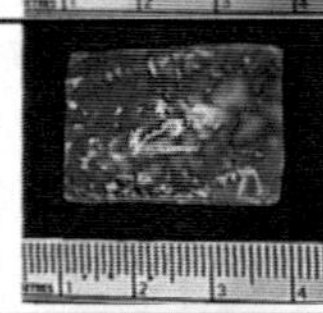
WS3 + Soytone	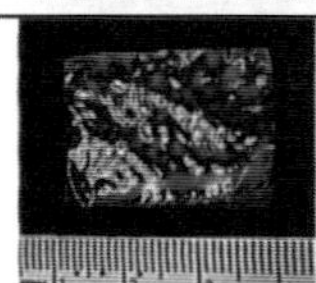		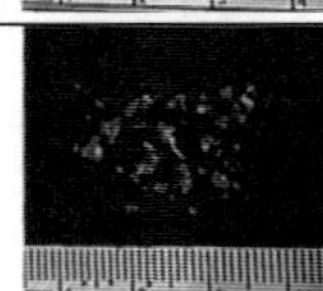
WS3 + Sericin	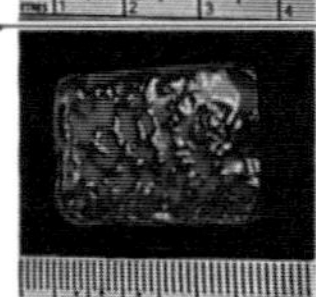	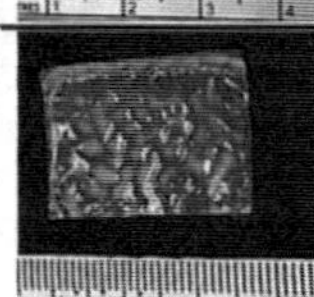	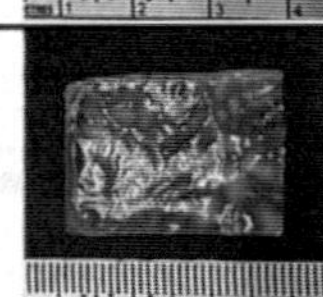
WS3 + Sodium lactate	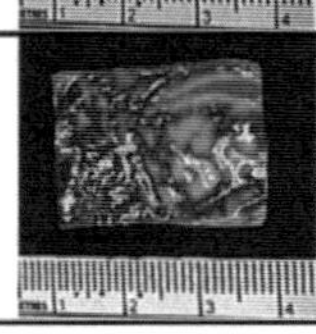	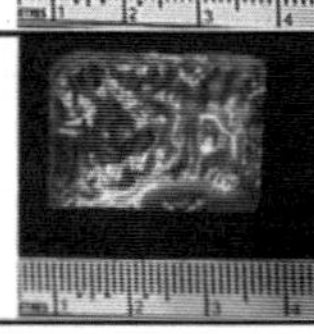	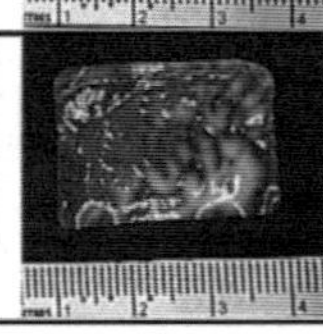

As shown in Table 1, it was found that PLA films in the culture broth of WS3 amended with soytone were cracked and broken down into small fragments. This can be explained that PLA films lost their transparency in the first 10 days because of chemical hydrolysis, after that PLA films became opaque, fragile and disintegrate due to enzymatic hydrolysis [14].

4. Conclusions

The highest percentage of PLA film-weight loss was found in the treatment of soytone addition, followed by sodium lactate addition. PLA films in culture broth with *P. geniculate* WS3 and soytone were cracked and broken down into small fragments within 20 days. Increasing the lactic acid content as a monomer of PLA in culture broth was directly associated with increasing the PLA film-weight loss. A combined use of *P. geniculate* WS3 inoculation and soytone could accelerate the PLA biodegradation rate under the submerged condition.

5. References

[1] Pranamuda H, Tokiwa Y, Tanaka H 1997 Appl. Environ. Microbiol. **63** 1637.
[2] Tokiwa Y, Calabia BP 2007 J. Polym. Environ. **15** 259.
[3] Ahn HK, Huda MS, Smith MC, Mulbry W, Schmidt WF, Reeves JB 2011 Bioresour. Technol. **102** 4930.
[4] Scow KM, Hicks KA 2005 Curr. Opin. Biotechnol. **16** 246.
[5] Couto MNPFS, Monteiro E, Vasconcelos MTSD 2010 Environ. Sci. Pollut. Res. **17** 1339.
[6] Lotrakul P, Dharmsthiti S 1997 World J. Microbiol. Biotechnol. **13** 163.
[7] Terada S, Nishimura T, Sasaki M, Yamada H, Miki M 2002 Cytotechnology **40** 3.
[8] Satti SM, Shah AA, Marsh TL, Auras R 2018 J. Polym. Envirin. **26** 3848.
[9] Apinya T, Sombatsompop N, Prapagdee B 2015 Int. Biodeter. Biodegr. **99** 23.
[10] Pattanasuttichonlakul W, Sombatsompop N, Prapagdee B 2018 Int. Biodeter. Biodegr. **132** 74.
[11] Satti SM, Shah AA, Auras R, Marsh TL 2017 Polym. Degrad. Stabil. **144** 392.
[12] Bubpachat T, Sombatsompop N, Prapagdee B 2018 Polym. Degrad. Stabil. **152** 75.
[13] Saadi Z, Rasmont A, Cesar G, Bewa H, Benguigui L 2012 J. Polym. Environ. **20** 273.
[14] Morawska M, Krasowska K 2017 Pol. J. Chem. Technol. **19** 44.

Acknowledgement
The authors would like to thank Thailand Science Research and Innovation (TSRI) under a Directed Basic Research Grant (DBG6280004).

IOP Conf. Series: Materials Science and Engineering **773** (2020) 012024 doi:10.1088/1757-899X/773/1/012024

Comparative X-ray shielding properties of bismuth oxide/natural rubber composites using a Monte Carlo code of PHITS

D Toyen[1] and K Saenboonruang[2,3*]

[1] Scientific Equipment and Research Division, Kasetsart University Research and Development Institute, Bangkok 10900, Thailand
[2] Department of Applied Radiation and Isotopes, Faculty of Science, Kasetsart University, Bangkok 10900, Thailand
[3] Specialized center of Rubber and Polymer Materials in agriculture and industry (RPM), Faculty of Science, Kasetsart University, Bangkok, 10900, Thailand

*Corresponding author: kiadtisak.s@ku.th

Abstract. Natural rubber (NR) is a material for potential development as a flexible X-ray shielding material, subject to improvement in its attenuation ability as one of the necessary measures to ensure safety for radiation-related workers and the general public. Consequently, this work used Monte Carlo coding and the Particle and Heavy Ion Transport System (PHITS) to simulate the X-ray shielding properties of NR composites that were added with Bi_2O_3 particles at different Bi_2O_3 contents (0–50 wt% in 10-wt% increments) and different material thicknesses (2, 4, and 6 cm). The X-ray source used for the simulation was a point source with energy ranges of 0.05, 0.1, 0.5, 5.0, and 15.0 MeV, respectively, and the detector was assumed to have 100% detection efficiency. The simulated results showed that at lower energies of X-rays (0.05 and 0.1 MeV), the values of X-ray transmission substantially decreased with increasing filler contents and material thicknesses. However, at higher energies of X-rays (0.5, 5.0, and 15.0 MeV), changes in the X-ray transmission ratios for different filler contents and material thicknesses were not as pronounced as for the lower-energy X-rays due to the high penetration ability of the high-energy X-rays. In particular, the highest attenuation ability of the materials obtained in this work was in the NR composites with 50 wt% Bi_2O_3, which had the values of half value layer (HVL; the thickness required to attenuate the incoming X-ray intensity by 50%) of 0.002, 0.004, 0.324, 0.990 and 1.540 m for X-ray energies of 0.05, 0.1, 0.5, 5.0, and 15.0 MeV, respectively. Other parameters including the linear attenuation coefficients (μ) and HVL of all filler contents were also reported and thoroughly discussed in this work.

1. Introduction

X-ray technology has many uses such as in agriculture, in food preservation, in industry, and especially in medical fields for diagnosis and therapy [1]. Although X-ray radiation is used in many useful applications, researchers, technicians, and the general public are at risk of receiving excessive doses of X-rays that are harmful to human health, cause symptoms such as skin burn and vomiting within hours, and can also result in long-term health effects such as memory loss, cancer, and cardiovascular disease [2].

Consequently, it has become necessary to evaluate the risks and to reduce the hazards from excessive X-ray radiation exposure. In principle, X-ray attenuation requires their energy transfer into matter encountered through two main interactions: the photoelectric effect and Compton scattering by interacting with heavy elements such as lead or lead compounds [3]. Examples of such materials are lead/polyethylene composites or lead/glass composites [4]. However, while lead is useful and effective in X-ray radiation reduction, it is considered a hazardous chemical, particularly for the manufacturer, especially when it is accidentally inhaled or ingested [5]. In this work, we aimed to develop flexible, lead-free, X-ray-shielding materials that are safe and environmental-friendly based on natural rubber (NR) composites that could provide flexibility, resistance to abrasion, and highly efficient X-ray attenuation, with the addition of bismuth (III) oxide (Bi_2O_3). The selection of Bi_2O_3 used in this work was based on the fact that bismuth has a high atomic number (Z) of 83 (greater than lead, which has Z = 82), high density, a high melting point, and also is highly effective in X-ray attenuation.

This work aimed to simulate the shielding properties of NR composites having different Bi_2O_3 contents, different material thicknesses, and different X-ray energies (0.05, 0.1, 0.5, 5.0, and 15.0 MeV) by using Monte Carlo coding of a particle and heavy ion transport code system (PHITS) to simulate and to measure important properties, including the X-ray transmission ratio ($\frac{I}{I_0}$), the linear attenuation coefficient (μ), and the half-value-layer (HVL) to assess the capabilities of X-ray attenuation of the NR composites and to design X-ray shielding materials for use in actual experiments.

2. Experimental

PHITS is a general-purpose Monte Carlo particle transport simulation code using several nuclear reaction models and nuclear data, of which the program has been used in various research fields, including in accelerator technology, in medical physics, in geosciences, in facility design, and in radiation protection. In this work, PHITS was used to simulate X-ray shielding properties of Bi_2O_3/NR composites that were added with different Bi_2O_3 contents (0–50%wt) in air. The sizes of all samples used in the simulation were 20 cm × 20 cm with thicknesses varied from 2, 4, to 6 cm, respectively. X-ray point source had a 0.005-cm diameter and its emitted energies varied from 0.05 MeV to 0.1, 0.5, 5.0, and 15.0 MeV, respectively. The X-ray source was located 5 cm away from the surface of NR composites. In order to correctly estimate the X-ray shielding properties of the samples, the detector was set to have 100% detection efficiency and located 20 cm away from the surface of NR composites in the opposite side of the X-ray source. The setup of the simulation is shown in Figure 1.

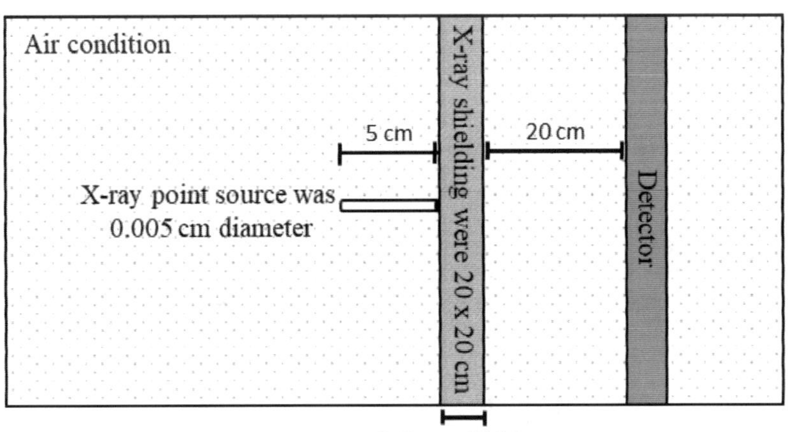

Figure 1. Setup of the simulation of Bi_2O_3 composites by using PHITS.

From the simulation, the X-ray shielding properties ($\frac{I}{I_0}$, μ, and HVL) could be calculated using Eq. 1 and Eq. 2:

$$I = I_0 e^{-\mu x} \tag{1}$$

$$HVL = \frac{In(2)}{\mu} \tag{2}$$

where I_o is the initial intensity of the X-ray beam, I is the intensity of the X-ray after passing through the NR composites, μ is the linear attenuation coefficient of shielding calculated from the material thicknesses of 0.5, 1.0 and 1.5 cm in order to reduce the effect of build-up factors, and x is the thickness of the NR composites.

3. Results and discussion

The results from the simulation are shown in Fig. 2 and Table 1, which showed that, at 0.05 and 0.1 MeV, the values of X-ray transmission substantially decreased with increasing filler contents and material thicknesses due to the greater ability of X-ray interaction and the impact of the photoelectric effect or Compton scattering with bismuth atoms in the material to attenuate and to reduce X-ray energies and intensities. However, at higher X-ray energies (0.5, 5.0, and 15.0 MeV), the changes in the X-ray transmission ratios for different filler contents and material thicknesses were not as pronounced as for the lower X-ray energies. This behaviour was observed because of the much increases in penetrating ability of the high-energy X-rays [6].

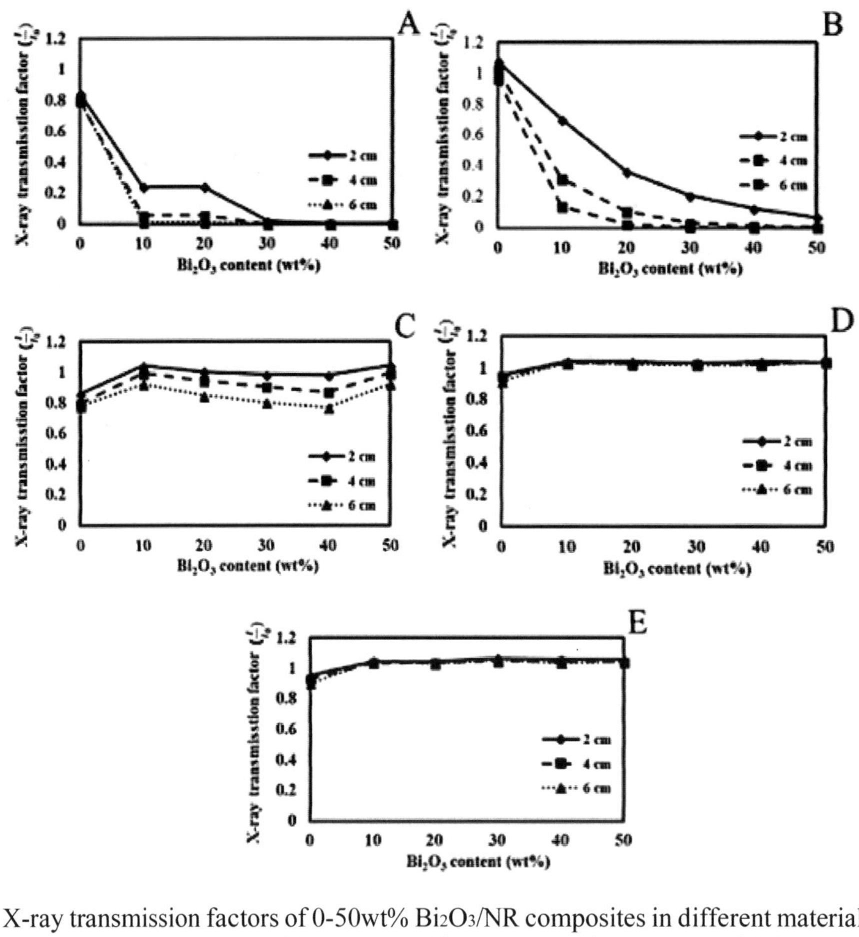

Figure 2. X-ray transmission factors of 0-50wt% Bi_2O_3/NR composites in different material thicknesses (2, 4 and 6 cm) and (a) 0.05-MeV, (b) 0.1-MeV, (c) 0.5-MeV, (d) 5-MeV, and (E) 15-MeV incoming X-rays.

The ability of the X-rays to penetrate the Bi_2O_3 composites being dependent on their energy, at the lower energies (0.05, 0.1 MeV) of X-rays had a greater probability of being absorbed by Bi_2O_3 when the values of μ clearly increased with increasing Bi_2O_3 content (0-50 wt%). However, at higher X-ray energies (0.5, 5.0, and 15.0 MeV) had a greater probability of penetrating the composites and the differences in values of μ and *HVL* for different energies were not as pronounced as in the case of lower energies. This could be because the probability of X-ray interaction with an orbital electron of Bi atoms is the largest when the X-ray energy equals the binding energy of bismuth atom and the probability tends to decrease when the X-ray energy becomes larger than the binding energy [7].

Table 1. μ and *HVL* results of Bi_2O_3/NR composites.

X-ray energies (MeV)	Linear attenuation coefficient (m^{-1})					
	Bi_2O_3 content (wt%)					
	0	10	20	30	40	50
0.05	5.48	68.74	82.18	194.66	252.87	315.22
0.1	1.82	26.37	67.48	70.55	136.19	168.35
0.5	2.14	2.13	2.01	2.65	2.48	2.14
5.0	0.62	0.58	0.37	0.67	0.38	0.70
15.0	0.42	0.43	0.59	0.34	0.37	0.45
X-ray energies (MeV)	**Half value layer (m)**					
	Bi_2O_3 content (wt%)					
	0	10	20	30	40	50
0.05	0.126	0.010	0.008	0.004	0.003	0.002
0.1	0.381	0.026	0.010	0.010	0.005	0.004
0.5	0.324	0.325	0.345	0.262	0.279	0.324
5.0	1.118	1.195	1.873	1.035	1.824	0.990
15.0	1.650	1.612	1.175	2.039	1.873	1.540

4. Conclusions

In this study, Bi_2O_3/NR composites provided effective shielding against low energy X-rays (energy less than 0.1 MeV). Furthermore, it was found that the X-ray attenuation property improved with increasing Bi_2O_3 content and material thickness.

5. References

[1] Shepp L. A. and Kruskal J. B. 1978. *The American Mathematical Monthly* **85(6)** 420-439.
[2] Box H. C 1977 in Radiation Effects, *Academic Press*, 201-221.
[3] Fernández J 2000 in Microscopical X-Ray Fluorescence Analysis, *New York Press*, 17-62.
[4] Mahmoud ME, El-Khatib AM, Badawi MS, Rashad AR, El-Sharkawy RM, Thabet AA. 2018 *Radiation Physics and Chemistry* **145** 160-73.
[5] Wani AL, Ara A, Usmani JA. Lead toxicity 2015 *Interdisciplinary toxicology* **8(2)** 55-64.
[6] Singh K.. J. 2014. *Radiation Physics and Chemistry* **96** 153-157.
[7] Als-Nielsen J. 2008. *Comptes Rendus Physique.* **9(5)** 479-486.

Acknowledgement
The authors acknowledge technical support from the Thailand Institute of Nuclear Technology (TINT), the Japan Atomic Energy Agency (JAEA), and from the Kasetsart University Research and Development Institute (KURDI), and the Specialized Center of Rubber and Polymer Materials for Agriculture and Industry (RPM), Faculty of Science, Kasetsart University, Bangkok, Thailand for publication support.

Monte Carlo simulations of nano-rod filler in stretched polymer nanocomposites

N Kerdkaen[1,2,3], T Sutthibutpong[2,3,4,*], S Phongphanphanee[2,5,6], S Boonchuay[1,6], and J Wong-ekkabut[1,2,3,6,*]

[1] Department of Physics, Faculty of Science, Kasetsart University, Bangkok 10900, Thailand.
[2] Computational Biomodelling Laboratory for Agricultural Science and Technology (CBLAST), Faculty of Science, Kasetsart University, Bangkok 10900, Thailand.
[3] Thailand Center of Excellence in Physics (ThEP Center), Ministry of Higher Education, Science, Research and Innovation, Bangkok 10400, Thailand.
[4] Department of Physics, Faculty of Science, King Mongkut's University of Technology Thonburi (KMUTT), Bangkok 10140, Thailand.
[5] Department of Material Science, Faculty of Science, Kasetsart University, Bangkok 10900, Thailand.
[6] Specialized Center of Rubber and Polymer Materials for Agriculture and Industry (RPM), Faculty of Science, Kasetsart University, Bangkok 10900, Thailand.

*Corresponding E-mail: jirasak.w@ku.ac.th, thana.sut@mail.kmutt.ac.th

Abstract Conductive polymer nanocomposites material (PNCs) is one type of alternative polymeric materials to replace high-cost intrinsically conductive polymers (ICPs). PNC is a composite of electrical insulative polymer matrix and electrically conductive filler in which it is made by less complicated synthesis protocols but has similar quantitative conductive properties to existing conducting polymers. Therefore, PNC is a candidate for many applications, such as, light-emitting diodes, flexible electrodes, batteries and strain sensors [1]. In this study, an in-house Monte-Carlo simulation was used to investigate the percolation paths of the 3D model of nanorod filler network in the polymer lattice and estimate the nanorod concentration at percolation threshold [2]. The model also includes the nanorod filler orientation angles. We then focused on the effects of stretching lattice on the percolation threshold. The dimension of lattice length was varied with constant volume for each simulation system (incompressible material). Results of simulations showed that the percolation thresholds decreased when increasing the lattice stretching and the nanorod orientation angles have been confined by finite lattice dimension which shows there is the effect of orientations angle on the percolation threshold. Our finding will be a useful guideline for designing polymer nanocomposite as a switching sensor.

1. Introduction

Intrinsically conductive polymers (ICPs) are a class of polymers whose conductivity is an intrinsic property based on their chemical nature [3]. ICPs can be used in many applications but they also are expensive and have complex chemical synthesis routes. Polymer nanocomposites (PNCs) are alternative polymeric materials to replace high-cost ICPs. A PNC is a mixture of two materials, where the matrix is a polymer and the filler consists of nanoparticles [4]. PNCs possesses a lower-cost and least-complex synthesis route than ICPs. The quantitative conductive property of PNCs can be adjusted to the level similar to the existing ICPs. In insulator PNCs with conductive nanorod filler, the percolation path has been established within nanorod network. As the concentration of filler is increased, nanorods interact with its neighbors and form the conductive network for transporting electrons across the material. Therefore, PNCs can transit between insulator and conductor and the concentration that transition occurred is known as the percolation threshold [5]. In this study, we used Monte Carlo simulation [6] to investigate the percolation path of the 3D model of nanorod filler network in the polymer lattice and to estimate the nanorod concentration at percolation threshold. We then investigated the effect of lattice dimension on the percolation threshold. A lattice system was set up by generating a "nanorod" model by a 3D rigid spherocylinder-shaped model, represented the conductive nanofillers. The position and orientation of nanorods were filled with uniform-random function. The electrical percolation path in a lattice was determined. The percolation probability was also evaluated by the ratio of the number of percolated lattice and the number of simulations.

2. Methodology

2.1. Model and Intersection of Nanorod

Computer simulation techniques have been used in many fields of research as a "virtual experiment" to observe behavior of systems and reduce the cost from real experiment. However, large time consuming is required for computer simulation techniques. The models are needed to be simplified to reduce system complexity and computational time [2]. In this study, polymer nanocomposites consist of insulating polymer matrix and conductive nanorod filler. Insulating polymer matrix was represented by a fixed volume with closed cuboid lattice. Conductive nanorod fillers were represented by 3D spherocylinder-shaped particles [7]. Each rod consists of an impenetrable inner hard-core to represent the nanorod material and the penetrable outer soft-shell to represent the effective range of electron tunneling effect [7]. If the soft-shell of a rod is overlapped with the hard-core of another rod, the rods are considered to be electrical contact.

2.2. Percolation Path Search

Percolation theory has been successful and widely accepted for describing the conductive transition of polymer nanocomposite. The PNCs exhibit a transition from "insulator" to "conductor" when the filler concentration exceeds the critical value, so-called percolation threshold [8]. We used Monte Carlo method to solve the percolation threshold. Our in-house program starts with filling a closed cuboid lattice with randomly-positioned 3D spherocylinder-shaped particles and observe the orientations angle of each nanorod filler. After that, the percolation paths of nanorod filler network in the lattice are investigated. The simulations were performed 3000 lattices in each simulation parameter to calculate the probability of having percolation path in the lattice with Monte Carlo method. The percolation probability will be related to overall conductive property of PNCs.

2.3. Simulation Setup

Our simulation program requires several control parameters. In this study, hard-core part of a nanorod filler is 20 arbitrary unit (AU) in length and 2 AU in diameter. Soft-shell thickness (t) of a nanorod filler is 0.8 AU. Filler concentration was varied from 0 to 12 % by volume with 0.4 intervals to find the percolation threshold. The lattice was defined by a cuboid shape. The width was kept fixed at 100 AU. The height was varied by 100, 80, 60, 40 and 20 AU, corresponding to the lengths of 100, 125, 167, 250 and 500 AU, respectively (shown in Figure1). Therefore, the volume of lattice is constant (incompressible material).

IOP Conf. Series: Materials Science and Engineering **773** (2020) 012025 doi:10.1088/1757-899X/773/1/012025

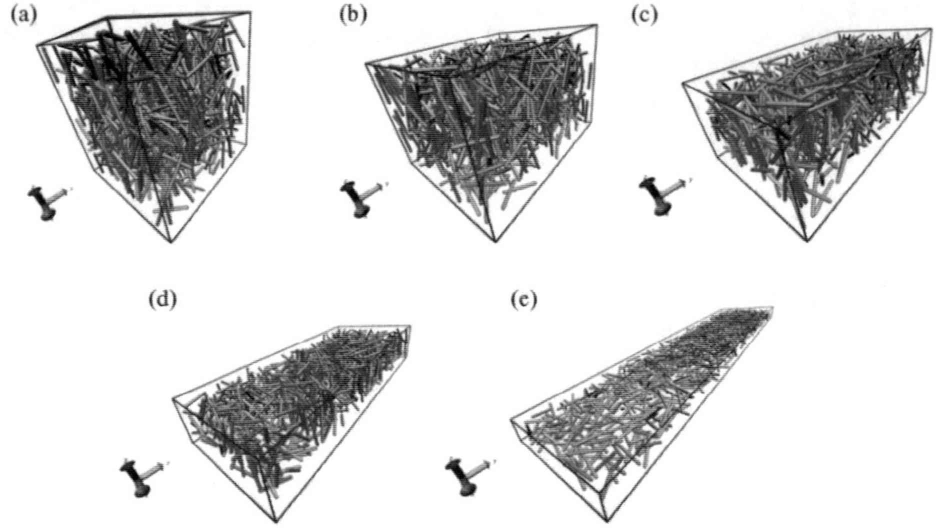

Figure 1. The lattice filling with nanorod filler at the lattice height of (a) 100 AU, (b) 80 AU, (c) 60 AU, (d) 40 AU and (e) 20 AU. Visualizations were done by Visual Molecular Dynamics(VMD)[9]

3. Result and discussion

3.1. Percolation Probability

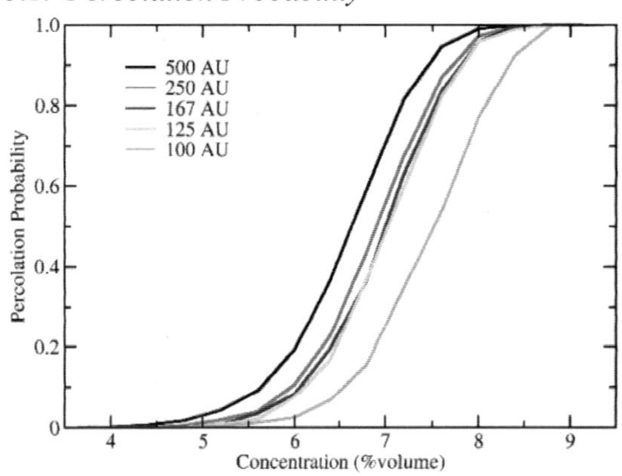

Figure 2. Percolation probability as a function of nanorod filler concentration

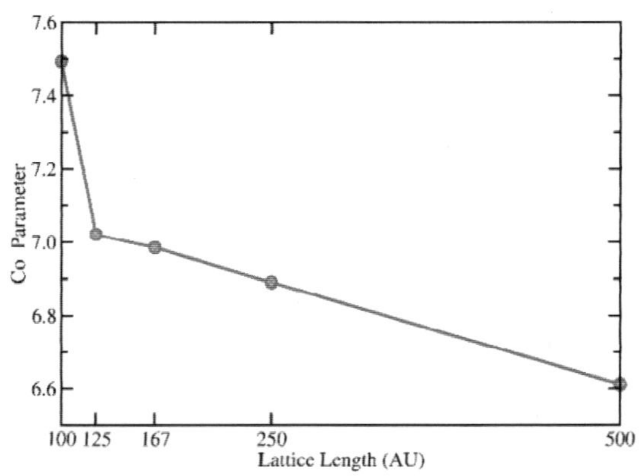

Figure 3. Percolation thresholds for different lattice lengths

The percolation probabilities with various lattice dimensions and constant lattice volume are shown in Figure 2. Percolation probability is dependent on nanorod filler concentration. The curves are S-shape and agreed well with logistic function. The curve of the lattice with long length was begun to rises up at lower concentration than the one with short lattice length. We fitted the percolation probability curves with a logistic curve and obtained the percolation threshold (C_o) parameter for the concentration value of the curve midpoint. Therefore, the percolation threshold, C_o was found approximately 7.4 %volume in the cube lattice of 100 AU and it is decreasing to 6.6 %volume when the lattice length increased to 500 AU. Percolation thresholds in this work are larger than in the previous works [5, 6] of approximately 3 - 7 %volume. Note that: the previous models were considered very thinned lattice [5] and large aspect ratio of nanorod [6].

3.2. Nanorod Orientation Angle

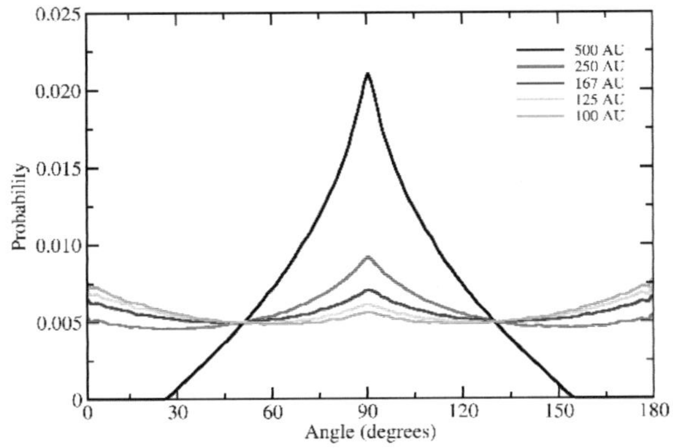

Figure 4. Distribution of nanorod filler orientations in the different lattice length at filler concentration equal to 7.2 %volume.

Orientation angles were defined by the angles between the nanorod vector and the Z-axis (parallel to lattice height). Finite size effect on the orientation of nanorod was observed in our simulation. The nanorods were confined within the long lattice length (the thinned lattice). Figure 4 shows that the probability of orientation at 90 degrees (perpendicular to the Z-axis) is higher than other angles and further increased as the increase of lattice length. At the lattice length 500 AU (with lattice height 20 AU, equal to the length of a rod), the probability of orientation at 90 degrees became significantly higher than other lattice lengths and two times higher than probability at 40 AU as seen Figure 2. The results showed that nanorod fillers are oriented perpendicular to the Z-axis when lattice was thinned in which indicated the more probability of nanorod connect to each other, the easier percolation paths in lattice form.

4. Conclusions

Monte Carlo simulations of the 3D rigid spherocylinder-shape filler model in closed cuboid lattice of polymer were used to estimate nanorod concentration at the percolation threshold. Our simulation results can show the relationship between percolation threshold and lattice length. The percolation threshold was predicted. Moreover, the percolation probability is increased as the increase of lattice length. The simulation results in this study help to understand the stretched conditions of the polymer nanocomposite with conductive nanorod filler.

5. References

[1] S. Paszkiewicz, A. Szymczyk. *Nanomaterials and Polymer Nanocomposites*. Elsevier (2019)
[2] Belashi Azita. *Percolation modeling in polymer nanocomposites*. Theses and Dissertations, Paper 525 (2011)
[3] Waldfried Plieth. *Electrochemistry for Materials Science*. Elsevier Science. 2008
[4] Amanda Dantas de Oliveira and Cesar Augusto Gonçalves Beatrice. *Polymer Nanocomposites with Different Types of Nanofiller*. Nanocomposites Recent Evolutions. IntechOpen. 2019
[5] Soto M, Esteva M, Martínez-Romero O, Baez J, Elías-Zúñiga A. Modeling Percolation in Polymer Nanocomposites by Stochastic Microstructuring. Materials (Basel). 2015;8(10):6697–6718. Published 2015 Sep 30. doi:10.3390/ma8105334
[6] Gu, Heng. *Three-dimensional Modeling of Percolation Behavior of Electrical Conductivity in Segregated Network Polymer Nanocomposites Using Monte Carlo Method.* Advances in Materials. 5. 1. 10.11648/j.am.20160501.11. 2016
[7] N Kerdkaen et al 2019 IOP Conf. Ser.: Mater. Sci. Eng. 526 012006
[8] Xiaojuan Ni et al 2018 Nanotechnology 29 07540
[9] Humphrey, W., Dalke, A. and Schulten, K., *VMD - Visual Molecular Dynamics*. J. Molec. Graphics 1996, 14.1, 33-38.

IOP Conf. Series: Materials Science and Engineering **773** (2020) 012026 doi:10.1088/1757-899X/773/1/012026

Effect of zeolite types on properties of polybutylene succinate/polylactic acid films

R Preedanorawut1, PThreepopnatkul* and A Sittatrakul

Department of Materials Science and Engineering, Faculty of Engineering and Industrial Technology, Silpakorn University, NakhonPathom, 73000, Thailand

*Corresponding author: poonsubt@yahoo.com

Abstract. In the current study, the effect of modified zeolite using (3-aminopropyl) triethoxysilane in polybutylene succinate (PBS) and polylactic acid (PLA) blend was investigated. Two types of modified zeolite i.e., zeolite 5A and 13X at 3wt% of polymer blend between PBS and PLA were mixed together in twin-screw extruder and thin-films were produced by cast-film extruder. The thickness of each film is between 50 – 70 micron. Mechanical properties, thermal properties, morphological properties and permeability of oxygen, carbon dioxide as well as water vapour were investigated. Adding of zeolite 5A into PBS/PLA blend was found to increase more tensile strength and Young's moduluswith the comparison to zeolite 13X whereas the zeolite 13X and 5A had increased the percentage of elongation at break more than PBS/PLA blend. The zeolite 5A and 13X tended to increase the thermal stability of the composite films. Gas permeation results showed that PBS/PLA with zeolite 5A allowed the permeation of carbon dioxide and oxygen more than 13X in composite films. Moreover, water vapour transmissionrate of PBS/PLA with zeolite 5A was higher than the one with zeolite 13X.

1. Introduction

Currently, Thailand has produced and exported agricultural products, such as fruits, with the export value up to 6,419,506,381 baht in 2018 [1] and tends to increase in each year. Due to the landscape and climate of Thailand are suitable for growing tropical fruits, the quality of Thai fruits is good. Therefore, it has a high demand by the international market.However, post-harvest handling is still not good in quality causing the fruit ripens before reaching the international market. There are many technologies in producing packaging that can delay fruit ripening such as active packaging or smart packaging [2].

Due to the population of the world is growing, the world has realized the limit of natural resources. Thus, humans develop plastic that will be used to replace natural resources, but the lifespan of common plastics put impact on the environment is too long, causing plastic waste. Biodegradable plastics are used to conserve the environment. Therefore, the researcher is interested in polybutylene succinate (PBS) and polylactic acid (PLA). The properties of PBS are high flexibility, similarpolyethylene (PE), but the disadvantage of PBS is the lack of transparency. So, PLA is mixed together with PBS because PLA has transparency, good mechanical properties, and good thermal stability [3]. The PBS/PLA blend film also has many desirable properties such as high flexibility.

As mention previously, PBS/PLA blend has a low ability to absorb ethylene and low permeability of carbon dioxide and oxygen gas. So the researcher is interested in improving its properties by adding zeolite, which has a porous structure and large cavity [4]. Each type of zeolite has a different ability to selective passing through of gases. The type of zeolite that can be selected passing through of carbon dioxide and oxygen must have the pore size that is larger than the molecular size of that gas type. For example, zeolite type that can be selected passing through by carbon dioxideare zeolite type 5A and 13X (pore size is about five and eight angstrom respectively) [4] because the zeolite has a pore size larger than the molecular size of carbon dioxide, which is 3.3 angstrom.

This research has aimed to developfood packaging film by adding zeolite and studies the effect of zeolite types (5A and 13X) on physical, mechanical, thermal and morphological properties. Moreover, the permeability of oxygen, carbon dioxide, water vapour of PBS/PLA film was investigated.

2. Experimental

The formulas of prepared film compoundsarePBS/PLA/Z5A and PBS/PLA/Z13X. (Z5A and Z13X means zeolite 5A and 13X respectively). The ratio of PBS to PLA is 60:40. Each zeolite was modified by using (3-aminopropyl)triethoxysilane in 2wt% of zeolite. A commercial grade PBS (FZ91PM, MFR 5 g/10min, density 1.26 g/cm_3) was produced from PTT MCC Biochem Co. Ltd. (Thailand). PLA (4043D, MFR 6 g/10min) was produced form NatureWork LLC Co. Ltd. (Thailand). Zeolite 5A, 13X and (3-aminopropyl)triethoxysilane were purchased from Sigma Aldrich Co. (Singapore). PBS, PLA and modified zeolite were blended together in twin-screw extruder at 165-180°C. Thin-films were produced by cast-film extruder at 155-175°C.

All 20 samples were prepared to test the mechanical properties. The percentage of elongation at break, tensile strength and Young's modulus of the films were measured according to ASTM D882 using Universal Testing Machine (5956; Instron Corp.). Thermal properties of the films was performed using Thermal Gravimetric Analyzer (TGA/DSC1, Mettler Toledo) under a nitrogen atmosphere. The water vapour transmission rate was determined using the Cup Method (ASTM E96). The permeability of carbon dioxide and oxygen were measured according to ASTM D3985 using Gas Permeability Tester (GDP-C, Bruggerfeinmechanik). The cross section of films was observed through a scanning electron microscopy (SEM). The films were immersed in liquid nitrogen after which they were broken. The films were sputtered coated with Pd/Au.

3. Results and discussion

As shown in Figs. 1a and 1b, adding zeolite 5A into PBS/PLA blend leads to increase tensile strength and Young's modulus more than zeolite 13X.The increase in tensile strength is a result of zeolite 5A allowing the PLA phase to be more compatible with PBS, as can be seen in Figs. 2b and 2c.In terms of Young's modulus, Young's modulus was increased by adding zeolite because zeolite is an inorganic compound consisting of alumina and silica that binds to a strong structure. Zeolite will reduce the mobility of the polymer chains, causing the polymer film to become more stiffness [5]. The zeolite 5A has a cubic shape and has a lower polarity when compared to the zeolite 13X[4]. The high polarity causes the zeolite to interact with each other more than zeolite interacts with the polymer. Therefore, the zeolite 13X is more easily agglomeration than zeolite 5A, resulting in the unevenly spreading of zeolite 13X in the polymer phase compared to zeolite 5A (as in Figs. 2b and 2c). PBS/PLA film with adding zeolite 13X had a lower Young's modulus than the PBS/PLA film adding with zeolite 5A. Meanwhile adding zeolite 13X and 5A had increased the percentage of elongation at break more than PBS/PLA blend in Figs 1c. The increasing in elongation at break is expected to be due to the addition of zeolite into PBS/PLA blend, making PLA more compatible with PBS. As a result, the PLA chain does not interfere with the PBS chain. Moreover,some parts of (3-aminopropyl)triethoxysilanemay be reacted with water [6]. Silanol groups that are formed by the hydrolysis reaction can coupling to form siloxane, which is flexible [7].

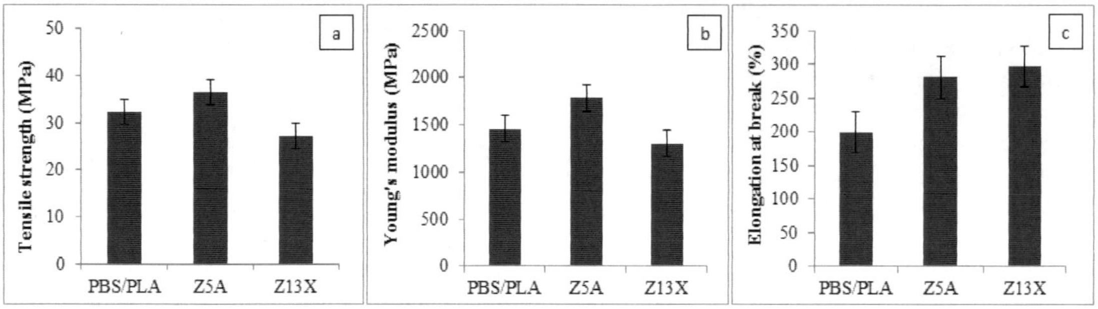

Figure 1. Mechanical properties of PBS/PLA and PBS/PLA/zeolite (5A and 13X). (a) Tensile strength (b) Young's modulus and (c) Elongation at break

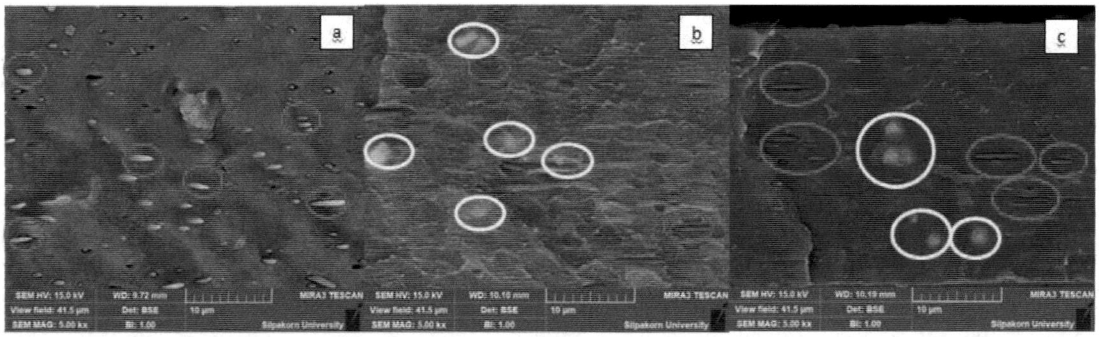

Figure 2. SEM micrograph of surface films (a) PBS/PLA (b) PBS/PLA/Z5A and (c) PBS/PLA/Z13X (the red circle is PLA phase in PBS and the yellow circle is zeolite in PBS/PLA blend)

From the analysis using the TGA technique, the thermal degradation behaviour of the composite and the blend films are shown in Figs.3. It is found that all films have two stages of degradation. The initial degradation temperature of the composite film is close to the degradation temperature of PLA. The second degradation temperature is close to the degradation temperature of PBS. The degradation of PLA and PBS are about 350 and 395 degree Celsius, respectively. However, considering the onset temperature, it is found that adding both of types of zeolites(5A and 13X) will increase the onset temperature. The results indicated that zeolite (5A and 13X) increase thermal stability.

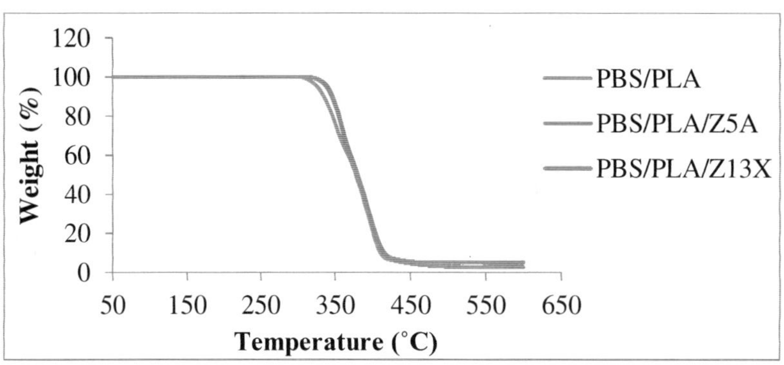

Figure 3. TGA thermogram of PBS/PLAFrom Figs.4a, 4b, it shows that adding zeolites reduce the permeation of carbon dioxide and oxygen because it increases path wayfor gas movement. Moreover, the zeolite structure is also porous, allowing the zeolite to absorb oxygen and carbon dioxide that are smaller than the zeolite's pores. Therefore, oxygen and carbon dioxide slowly move out of the films [8]. When considering the type of zeolite, PBS/PLA with zeolite 5A can allow the permeation of

oxygen and carbon dioxide faster than zeolite 13X. This is due to the zeolite 13X has a pore size larger than that of zeolite 5A, carbon dioxide and oxygen can be absorbed in the pores more.

The water vapour transmission rate of PBS/PLA/zeolite is higher than PBS/PLA because zeolites are highly polar compounds. It consists of silica and alumina and can allow water molecules to penetrate into it. Moreover, adding zeolite 5A makes the water vapour transmission rate more than zeolite 13X.Due to zeolite 13X has a larger pore size than that of zeolite 5A, its surface area can absorb more water molecules. Therefore, water molecules can take longer time to spread through the film with zeolite 13X more than the film with zeolite 5A [9].

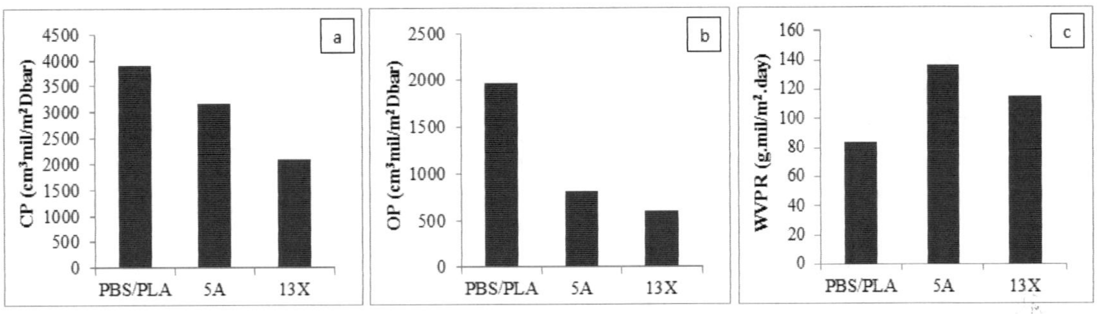

Figure 4. The permeation of PBS/PLA and PBS/PLA/zeolite (5A and 13X) (a) carbon dioxide (b) oxygen and (c) water vapour

4. Conclusions

In this research, the effect of zeolite on PBS/PLA properties was investigated. Adding zeolite 5A in PBS/PLA tended to increase the tensile strength, Young's modulus more than zeolite 13X because zeolite 5A is more evenly distributed and made PLA compatible with PBS better than zeolite 13X. From the TGA, PBS/PLA/zeolite (5A and 13X) make higher thermal stability. Moreover, PBS/PLA with zeolite 13X can absorb carbon dioxide and oxygen more than zeolite 5A whereas PBS/PLA with zeolite 5A allowed the water vapour transmission rate more than zeolite 13X due to the larger pore size of zeolite 13X.

5. References

[1] http://dataservices.mof.go.th/Dataservices/IETopExport; November 2019.
[2] JarosrojkulS 2000 *MTEC.* 60.
[3] Homklin R and Hongsriphan N 2013 *Energy procedia.***34** 871.
[4] Lamo N 2008 *Chulalongkorn University.* 111.
[5] Correia PR, Ramos IG, Viana A and Mascarenhas A2018 *J. Appl. Polym. Sci.* **135**45757.
[6] Demir H, Balköse D andÜlkü S 2006 *Polym. Degrad. Stabil.* **91** 1079.
[7] Sloane K and Rebecca C 2015 *Systems Biology in Toxicology and Environment Health.* 117.
[8]Farid A,LinnéaA, Steven O, Niklas H and BergströmL 2014 *J. Eur. Ceram. Soc.* **34** 1643.
[9] Kim KM, Oh HT, Lim SJ, Ho K, Park Y and Lee CH 2016 *J. Chem. Eng. Data.* **61** 1547.

Acknowledgement
It is a pleasure to acknowledge the financial support provided by Department of Materials Science and Engineering, Faculty of Engineering and Industrial Technology, Silpakorn University.

IOP Conf. Series: Materials Science and Engineering **773** (2020) 012027 doi:10.1088/1757-899X/773/1/012027

UV-protection property of Eri silk fabric dyed with natural dyes for eco-friendly textiles

N Rungruangkitkrai[1], R Mongkholrattanasit[2], P Phoophat[1], N Chartvivatpornchai[1], S Sirimungkararat[3], K Wongkasem[4], P Tuntariyanond[1], N Nithithongsakol[5] and R Chollakup[6,*]

[1]Department of Textile Science, Faculty of Agro-Industry, Kasetsart University, Bangkok, Thailand.
[2]Department of Textiles Chemistry Technology, Faculty of Industrial Textiles and Fashion Design, Rajamangala University of Technology Phra Nakhon, Bangkok, Thailand.
[3]Agricultural Biotechnology Research Center for Sustainable Economy, Khon Kaen University, Khon Kaen, Thailand.
[4]Faculty of Engineering, Mahasarakham University; Mahasarakham 44150, Thailand
[5]Natural Niche Co., Ltd., 10/39 Wutagas Road, Jomthong, Bangkok, Thailand.
[6]Kasetsart Agricultural and Agro-Industrial Product Improvement Institute, Kasetsart University, Bangkok, Thailand.

*Corresponding author : aaprmc@ku.ac.th

Abstract. Eri silk (*Samia ricini*) is a wild silkmoth which has been promoted and supported to culture for textile industry in Thailand. Especially for the recent decade, textile industry was interested to develop eri silk for casual eco-friendly textile style with functional properties. This research aims to study on UV protection property of eri silk fabric dyed with natural dyes from plant extracts and their washing fastness property. The processes included extracting the colorants from Thai plants i.e. natural indigo, Burma padauk bark, Neem bark, etc., mordanting the eri silk yarn with potassium aluminum sulphate, iron (II) sulfate and dyeing with their natural dyes to increase the washing fastness property. Eri silk fabrics were woven by the dyed eri yarn providing an excellent protection (UPF 50+) with a proper shade for making 3 prototype clothing products. The result shows that UPF property of eri silk fabric dyed with Burma padauk bark, Gloden shower seed, Neem bark, Andaman satinwood leave and mordanted with iron (II) sulfate presented highest UPF (>100) with good washing fastness. Three prototype products were designed with color matching with natural dyes and fabricated which were hat, scarf and dress. The first product, hat, made from the hand-weave fabric comprised of Ne 20/2 eri spun silk yarn as a warp yarn and Ne 20/2 eri spun silk yarn that dyed from Burma padauk bark as weft yarns. Second product was scarf which made from Ne 20/2 eri spun silk yarn as a warp yarn and Ne1.2 eri hand-reel silk yarn dyed from natural indigo and Burma padauk bark as weft yarns. The last product was dress which made from Ne 30/2 eri spun silk yarn that dyed from sappen as a warp yarn and Ne 30/2 eri spun silk yarn that dyed from indigo as weft yarns. These eco-friendly eri silk products exhibited excellent functional UV-protection property.

1. Introduction

Eri silk (*Samia ricini*) is a wild silkmoth which has a number of benefits and it is cultivated with the complete life cycle. The characteristic of eri silk are smooth, shiny and white. However, with its crimp characteristic, the spun yarns are often blended with other fibers such as cotton fiber [1]. The eri silk has the unique characteristic of lightness, soft-smooth feel, thermal property and additionally, it can easily absorb sweat and it is well ventilated. Dyeing and physical properties of the eri and cotton fiber blended yarns at different blend ratios were found differently depending on physical nature of eri silk and cotton fibers [2]. Moreover, there are large amount of cassava plantations in Thailand, accounting for 1,600 million square meters and the cassava leaves are wasted and do not create a value. Since cassava and castor leaves were feed for eri silkworm then production of eri silk can generate subsidiary income to farmers.

Natural dyes extracted from plants are non-toxic, non-hazardous, and environmentally friendly. Dyeing fiber with natural dyes not only adds further elegance to clothing but it also has special properties including bacteria [3] and UV protection [4]. However, the level of UV protective property depends on the plant and materials used in the process of production. Moreover, in these days, people are more concerned about functional properties of product like sun protection. This affected on purchasing fabric goods such as coats, scarves, hats and dresses of customers. As a result, an increasing trend on health and environmental awareness, together with consumer needs of UV protection goods, made our research group interested in the studies of the improvement on UV protection property of eri silk. Study of different natural dyes extracted from Thai plants in order to enhance its ability of UV protection for design the dyed eri silk products to be more appropriate use.

2. Experimental

2.1. Materials

Eri silk fabrics, used for the study of natural dye on UV protection, were woven by Ne 30/2 spun eri silk yarns as warp and weft yarns. Natural dyes were extracted at boiled condition from Gloden shower pods, Neem bark, Andaman satinwood leave, Burma padauk bark, Sappan and Acacia bark. Concerning for any eri silk products in this study, different yarns count of eri spun silk yarn and hand eri spun yarns were chosen to design diversity of woven fabrics.

2.2. Dyeing method

The eri spun silk yarns were dyed with extracted solution of these natural dyes at 100_oC at L:R of 1:20 for 1 hr. Then post mordant process was prepared by soaking these dyed eri fabrics at room temperature in $AlK(SO_4)_2$ and $FeSO_4$ solution (5 g/L). Finally, these dyed eri spun silk yarns were hand woven in plain weaving pattern by community enterprise group at Khonkan province based on matching color trends.

2.3. Characteristic

UV spectrophotometer was used to measure L, a*, b* values of the dyed eri silk with different natural dyes. Ultraviolet (UVR) transmission and the ultraviolet protection factor (UPF) of dyed eri silk spun yarns were determined according to AATCC138:2000 test method. Also, the color fastness to washing was characterized according to ISO 105-C01 A1S: 2012 standard test method.

2.4. Product development

Three prototype products of eri silk fabrics developed in this research were hat, scarf and dress depending on color and weaving structure.

3. Results and discussion

3.1. Effect of dyeing properties and UPF values of dyed eri silk fabrics

The color value (L, a*, b*) results were presented in Table 1. Eri spun silk yarns dyed with natural dyes which were Andaman satinwood leave, Golden shower pods, Neem bark, Sappan, Acacia bark and Burma padauk bark with an efficiency enhancement of post-mordanting by $AlK(SO_4)_2$ and $FeSO_4$. It was found that dyeing eri silk with Andaman satinwood leave was resulted in greenish-brown color. In addition, eri spun silk yarns dyed with Golden shower pods resulted in brown. With Neem bark, Sappan, Acacia bark and Burma padauk bark, the post-mordanted samples produced a reddish-brown color. After mordanting with both $AlK(SO_4)_2$ and $FeSO_4$, it gives the contrast results. The consequent color of mordanting with $AlK(SO_4)_2$ was shown with bright color which gave increasing value of a* or b*, while mordanting it with $FeSO_4$ resulted in decreasing value of L, a* or b* making color darken.

The Ultraviolet Protection Factor value (UPF) was used to determine the ability of UV protection of the fabric. It was found that without natural dyes and mordant, the UPF value of evaluated eri silk fabric was 35.08. Mordanting undyed eri silk with $AlK(SO_4)_2$ and $FeSO_4$ gave the UPF values of 49.9 and 65.3, respectively. It can be obviously observed that eri silk with natural dyes from Burma padauk bark, Andaman satinwood leave, Neem Bark and non-mordanting gave the results of UPF values of 201.94, 184.42 and 157.78, respectively ranking from the highest ability of protection. Since tannins from plant has an advantage on protecting against UV radiation, dyeing eri spun silk yarn with natural dyes enhances the ability of UV protection regarding an increase in tested value of UPF. The UPF value of samples shown in table 2 appeared to be more than 50 which indicated that the resistance of dyed eri silk was very strong giving the highest UV protection. Furthermore, the dyed eri silk with mordanting by $AlK(SO_4)_2$ and $FeSO_4$ gave higher UV protection property towards eri silk. However, mordanting with $FeSO_4$ tended to be more effective on UV protection property due to darken color shade.

3.2. Color fastness to washing test

According to Table 3, the results indicated that the color fastness to washing of eri silk fabrics with Sappan and Burma padauk bark extract was very good with rating of 4-5. In the same way, the fastness property to washing of those plants extract fabric with mordants ($AlK(SO_4)_2$ and $FeSO_4$) was also ranked as 4-5 which was considered as very good fastness property.

Table 1. Color values (L a* b*) of dyed eri silk fabrics by post-mordanting technique

Type of plants	Type of mordants	L	a*	b*	Color obtained	Type of plants	Type of mordants	L	a*	b*	Color obtained
Without	Without	87.45	-0.00	9.86		Sappan	Without	58.03	11.79	23.26	
	$AlK(SO_4)_2$	87.63	0.02	10.27			$AlK(SO_4)_2$	46.83	17.89	15.43	
	$FeSO_4$	67.88	2.34	11.67			$FeSO_4$	35.03	3.14	3.99	
Andaman satin wood leave	Without	78.40	1.60	19.36		Acacia bark	Without	51.39	10.92	13.30	
	$AlK(SO_4)_2$	70.95	1.24	24.25			$AlK(SO_4)_2$	48.56	9.46	14.21	
	$FeSO_4$	58.52	2.61	16.41			$FeSO_4$	34.94	2.68	4.69	
Gloden shower pods	Without	43.31	4.40	16.78		Burma padauk bark	Without	51.01	19.81	18.64	
	$AlK(SO_4)_2$	35.96	4.23	14.98			$AlK(SO_4)_2$	48.28	17.26	17.82	
	$FeSO_4$	32.80	2.09	9.92			$FeSO_4$	32.50	5.35	5.62	
Neem bark	Without	52.93	12.89	13.40							
	$AlK(SO_4)_2$	47.68	12.23	12.58							
	$FeSO_4$	35.68	4.16	5.91							

Table 2. UPF values and protection class of eri silk fabrics by post-mordanting technique

Type of plants	Type of mordants	UPF	UPF Protection class	Type of plants	Type of mordants	UPF	UPF Protection class
Without	Without	35.08	Very good	Sappan	Without	85.50	Excellent
	$AlK(SO_4)_2$	46.69	Excellent		$AlK(SO_4)_2$	583.23	Excellent
	$FeSO_4$	65.13	Excellent		$FeSO_4$	507.44	Excellent
Andaman satin wood leave	Without	184.42	Excellent	Acacia bark	Without	109.70	Excellent
	$AlK(SO_4)_2$	292.81	Excellent		$AlK(SO_4)_2$	312.87	Excellent
	$FeSO_4$	331.63	Excellent		$FeSO_4$	576.17	Excellent
Gloden shower pods	Without	147.28	Excellent	Burma padauk bark	Without	201.94	Excellent
	$AlK(SO_4)_2$	441.71	Excellent		$AlK(SO_4)_2$	717.99	Excellent
	$FeSO_4$	843.01	Excellent		$FeSO_4$	826.18	Excellent
Neem bark	Without	157.78	Excellent				
	$AlK(SO_4)_2$	204.57	Excellent				
	$FeSO_4$	262.82	Excellent				

Table 3. Color fastness to washing test at $40^{\circ}C$ ISO 105-C01 A1S: 2012

Type of plants	Color change	Color stain					
		Acetate	Cotton	Nylon	Polyester	Acrylic	Wool
Sappan	4/5	5	4	4/5	4/5	4	4/5
Sappan+ $AlK(SO_4)_2$	4/5	5	4	4/5	4/5	4	4/5
Sappan+ $FeSO_4$	4/5	5	4/5	4/5	4/5	4	4/5

Type of plants	Color change	Color stain					
		Acetate	Cotton	Nylon	Polyester	Acrylic	Wool
Burma padauk bark	5	5	4/5	4/5	4	3/4	4
Burma padauk bark + $AlK(SO_4)_2$	5	5	4/5	4/5	4/5	4	4/5
Burma padauk bark + $FeSO_4$	5	5	4/5	4/5	4/5	4	4/5

3.3. Development of three prototype products (hat, scarf and dress) with natural UV-protected dyes

The result was concluded that all UPF value tests of dyed eri silk from natural extract were above 50 indicating that natural dyes had the highest efficiency on UV protection. Accordingly, our research group selected the complement colors which were reddish-brown color from Sappan and Burma padauk bark and blue color in order to combine them and to create natural indigo dyed eri silk for prototype products with natural UV-protected dyes (Table 4). The natural indigo dye not only has high demand in the market but it also has an excellent ability on UV protection. As previous research, indigo dyeing in fabric provided significant efficiency on UV protection and moreover, its property raised after washing [5].

Table 4. Development of three prototype products with natural UV-protected dyes

Products	Descriptions	Pictures
Hat: the entire face, ears, and neck UV protection hat	Plain weave fabric made from the hand-weave fabric comprised of Ne 20/2 eri spun silk yarn as a warp yarn and Ne 20/2 eri spun silk yarn that dyed from Burma padauk bark as weft yarns (71x28 yarn per square inch). Ikat pattern was made from 20/22Dx4 mulberry silk filament yarn as weft yarn (71x41 yarn per square inch).	
Scarf	Plain weave fabric comprised of Ne 20/2 eri spun silk yarn as a warp yarn and Ne 2.1 eri hand-reel silk yarn dyed from natural indigo and Burma padauk bark as weft yarns.	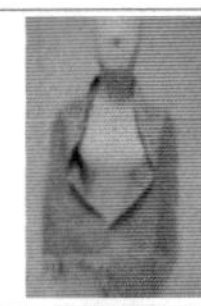
Dress with jacket: suitable pattern and texture for dress (casual dress for officers	Plain weave fabric comprised of Ne 30/2 eri spun silk yarn that dyed from sappen as a warp yarn and Ne 30/2 eri spun silk yarn that dyed from indigo as weft yarns (20x107 yarn per square inch)	

4. Conclusions

Eri silk yarns dyed with Andaman satinwood leave, Golden shower pods, Neem bark, Sappan, Acacia bark and Burma padauk bark had an excellent level on UV protection. Moreover, with post-mordanting by $AlK(SO_4)_2$ and $FeSO_4$, the UV protection property was observed in the UPF value. As a result, natural dyes together with post-mordanting were applied in order to UV protection for eri products which were hat, scarf and dress. The color of a reddish-brown color from Sappan and Burma padauk bark was matched with natural indigo in order to develop the products with a trendy stylish color and a greater quality on UV protection.

5. References

[1] Chollakup R, Suesat J and Ujjin S 2008 *Macromol. Symp* **264** 44.
[2] Chollakup R, Suesat J and Ujjin S 2010 *Color Technol* **126** 42.
[3] Karolia A and Khaitan U 2012 *Res. J. Text. Apparel* **16** 53.
[4] Mongkholrattanasit R, Klaichoi C, Rungruangkitkrai N and Sasivatchutikool N 2013 *Adv. Mater. Res* **821** 560.
[5] Sarkar KA 2004 *BioMed Central* **4** 1.

Acknowledgement

The authors are grateful to the Thailand Science Research and Innovation (TSRI) for financial support and Specialized Center of Rubber and Polymer Materials in Agriculture and Industry (RPM), Faculty of Science, Kasetsart University and Kasetsart University Research and Development Institute (KURDI) for the presentation support.

Apatite inducing ability on silk fabric and its ammonium gas adsorptivity

C Thongbai[1], S Buntham [2], C Ngamcharussrivichai [2,3] and D P Kashima[1,2,*]

[1] Research Unit of Advanced Ceramics, Department of Materials Science, Faculty of Science, Chulalongkorn University, Patumwan, Bangkok 10330, Thailand
[2] Center of Excellence on Petrochemical and Materials Technology, 7th floor, Chulalongkorn University Research Building, Patumwan, Bangkok 10330, Thailand
[3] Department of Chemical Technology, Faculty of Science Chulalongkorn University, Patumwan, Bangkok 10330, Thailand

*Corresponding author: dujreutai@gmail.com

Abstract. Silk fabric contains silk fibroin protein which consists of an amino group. This amino group can act as a nucleation site of hydroxyapatite by inducing the Ca^{2+} ions from surrounding solution at an appropriate pH. To increase the initial Ca^{2+} ions on the surface of silk fabric, the commercial grade silk fabric was drenched in $CaCl_2$ solution for 12 h, Na_2HPO_4 solution for 20 min, and $CaCl_2$ for 20 min, respectively. Finally, the fabric was rinsed by deionized water and the drenching processes were repeated for 7 times in order to increase the inducing ability of hydroxyapatite. XRD patterns showed the phases of hydroxyapatite coexisted with monetite. The sample drenched in $CaCl_2$ over 12 h gave more apatite peak intensity which was a result of increased Ca^{2+} ions over the surface and dot mapping technique with SEM showed the well distribution of both calcium (Ca) and phosphorus (P) on silk fabric. It is worth noting that hydroxyapatite induced on the surface of silk fabric can enhance the selectivity in ammonia gas adsorption. An ammonia gas adsorption resulted in better performance of silk with hydroxyapatite over raw silk fabric were displayed. These results are promising for development of Thai silk in term of superior ammonia gas adsorption characteristic.

1. Introduction

Silk fabric is a natural fabric from silkworm (*B. bombyx*). It has been known-well in beautifulness and unique textures. Silk fabric composed of fibroin protein which contain an amino group which is able to induce the positive ions via their amide group. Calcium ion, which is a positive charge, can be attracted to these amide group and acts as a nucleation site for hydroxyapatite (HA) formation under surrounding of phosphate anions under an appropriate pH [1].

HA is a famous material in both of biomedical and environmental fields due to its unique bioactivity, adsorptivity and stable even under the high temperature. Additionally, hydroxyapatite can be made as a good ammonium gas sensor because of its selectivity and high sensitivity to ammonia gas [2]. Moreover, it was reported that HA induced on the silk fiber via inducing mechanism showed a low crystallinity [3] which contain high surface area and suit for gas adsorption applications. From all mentioned above, this research aim at inducing HA on the silk fibers of Thai silk fabrics in order to increase the ammonia gas adsorption activity to Thai silk and fine tuning to optimize for an appropriate condition.

2. Materials and method

2.1. Materials preparation

Analytical grade of $CaCl_2 \cdot 2H_2O$ was purchased from Ajex finechem, Na_2HPO_4 from LOBO CHEMIE. Deionized water (DI) water was used as a solvent in all solutions. The commercial grade Thai silk fabric was kindly supplied from Jim Thompson Company Ltd. The commercial grade silk fabric 1.5 × 1.5 cm was firstly drenched in 1.5 M $CaCl_2$ solution, 20 mL for 12 h in order to gather a large number of Ca^{2+} on the amide group of Thai silk fabric. Secondly, it was drenched in 1M Na_2HPO_4 solution, 20 mL (with adjusting of pH to pH 7) for 20 min in order to let hydrogen phosphate group react to Ca^{2+} bound amide group on silk fiber. Afterward, the repeating of drenching process was done for 7 times as described in following: drenching in 1M $CaCl_2$ solution, 20 mL for 20 min and drenching again in 1M Na_2HPO_4, 20 mL for 20 min. The reaction was completed in both of room temperature (25°C) and 100°C to convert the undesired calcium orthophosphates to the desired HA. Finally, the processed fabric was air dried at room temperature for 8 h. The deposited powder on the fabric was collected for phase analysis by XRD characterization.

2.2. Characterization

Phase analyses of the samples prepared from 2.1 were characterized by X-ray diffractometer (XRD D8-Advance, Bruker AXS model D8, German) with Cu Kα (λ = 1.5406 Å) and microstructures were investigated by scanning electron microscope (SEM, JEOL JSM-6480LV).

2.3. Ammonium gas adsorption experiment

An adsorption test was carried out with ammonia detector (GasBadge pro, industrial scientific). An ammonia detector and processed fabric were put into the chamber with filter paper (**Figure 1**). The concentrated NH_4OH solution, 5 µL was dropped on the filter paper and then the chamber was sealed. The experiment was monitored within 3 h for measuring of the amount of reducing ammonia.

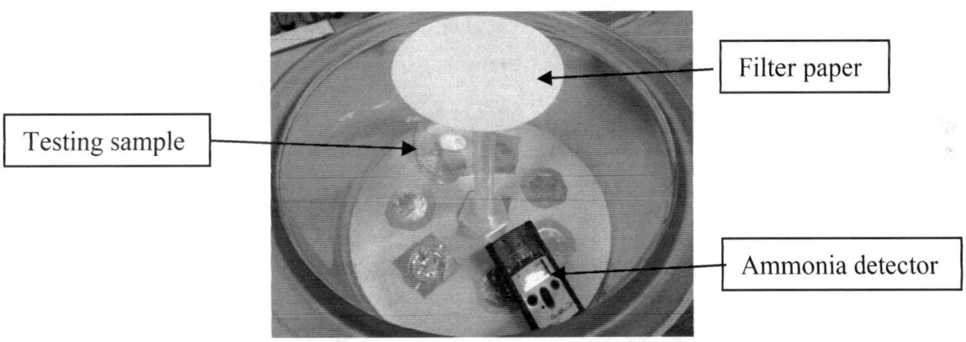

Figure 1. Ammonium gas adsorption experiment setup.

3. Results and discussion

3.1. Phase analysis

The XRD pattern showed the phase of hydroxyapatite coexisted with monetite on the surface of Thai silk fabric, monetite was the result of decreased pH from OH^- loss in brushite converting reaction as shown in **Figure 2**. The sample at 100°C gave more monetite intensity due to faster reaction resulted in rapidly decreasing of pH. This results also pointed out that drenching in $CaCl_2$ for 12 h in Ca^{2+} pregathering step gave higher intensity on both calcium phosphate because more Ca^{2+} ion was attracted on the silk fiber surface.

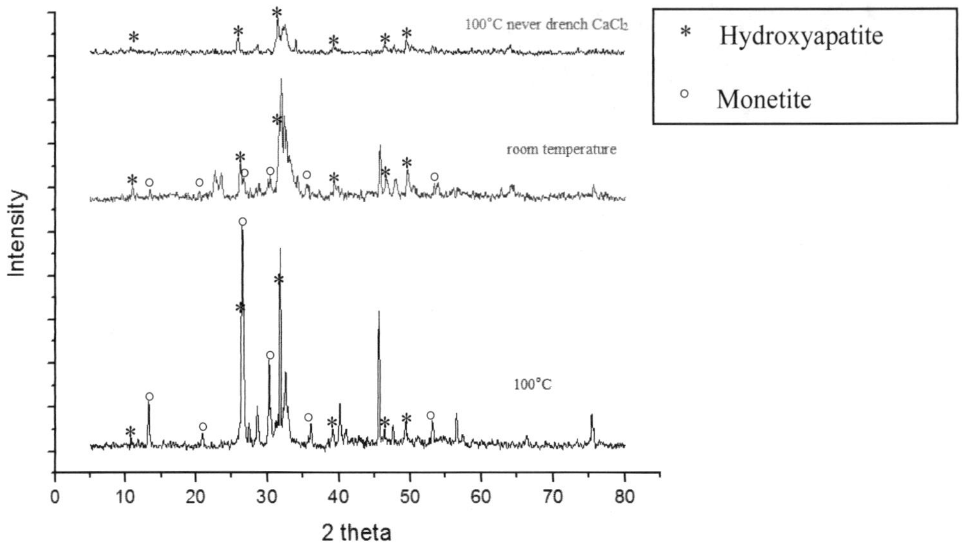

Figure 2. XRD pattern of processed silk at 100°C, room temperature and 100°C without pre-drenching step in $CaCl_2$.

3.2. Microstructure Investigation

SEM micrographs in **Figure 3** showed the hydroxyapatite tiny particles were induced on the surface of Thai silk fabric and dot mapping analysis showed the well distribution of Ca and P over the fabric suggested that the nucleation sites were distributed over the fabric as well.

Figure 3. SEM micrographs showed the hydroxyapatite induced on the silk fabric surface and dot mapping distribution of Ca (left) and P (right).

3.3. Ammonium gas adsorption

According to **Figure 4**, the HA-induced silk fabric showed the better performance for ammonia gas adsorption than the raw silk fabric which came from the results of increased surface area from HA on the surface. Although, the ammonia adsorption per gram of raw silk is 270.66 mmol/g which was more than HA-induced silk fabric (179.01 mmol/g) however, the HA-induced silk fabric could diminish the ammonia gas faster than the raw silk fabric. It is implied that ammonia adsorption site on HA-induced silk fabric was plenty enough compared to the raw silk fabric, and hydroxyapatite powder (peeled-off from the deposits of HA on HA-induced silk fabric).

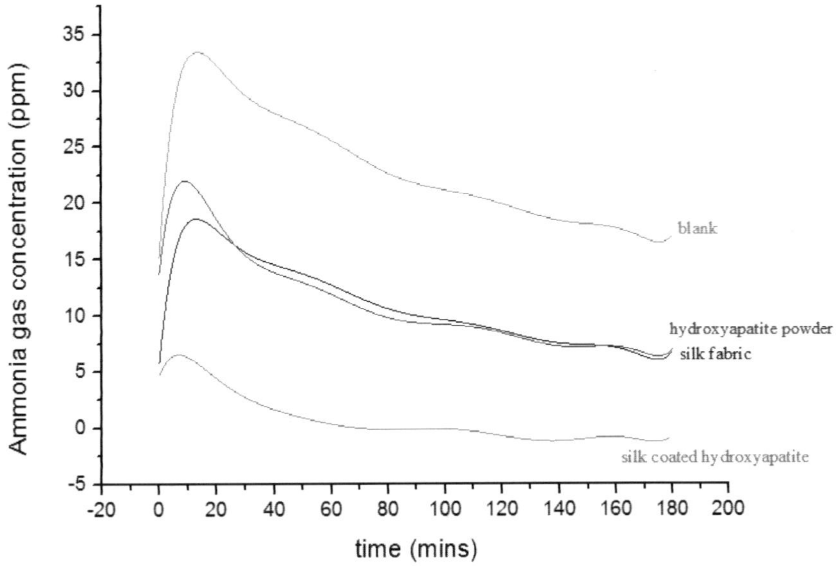

Figure 4. Ammonium gas adsorption results.

4. Conclusion

Hydroxyapatite could be induced onto Thai silk fabric and enhanced its ammonium gas adsorptivity due to the improved surface area. An inducing mechanism of hydroxyapatite on silk relied on amino group of fibroin protein which induced calcium ion to act as nucleation sites [4]. According to inducing mechanism, it is possible that hydroxyapatite on the surface is strongly attached to the Thai silk fabric and might be possible for cleaning by laundry process. By this technique it is promising to develop Thai silk with superior ammonium gas adsorption characteristic.

5. References

[1] X.D. Kong, *et al.*, Journal of crystal growth, (2004): 197-202.
[2] L. Huixia, *et al.*, The royal society of chemistry, (2015*): 3865-3874.*
[3] J. Zhao, *et al.*, Bone, (2009): 517-527.
[4] Y. Li, *et al.*, Applied surface science, (2008): 1681-1685.

IOP Conf. Series: Materials Science and Engineering **773** (2020) 012029 doi:10.1088/1757-899X/773/1/012029

The effect of non-rubber components on mechanical properties of TESPD silane coupling agent in silica-filled rubber compounds

S Kaewsikoun[1], S Kumarn[2], and J Sakdapipanich[1,2,*]

[1] Department of Chemistry and Center of Excellence for Innovation in Chemistry, Faculty of Science, Mahidol University, Bangkok 10400, Thailand
[2] Institute of Molecular Biosciences, Mahidol University, Nakhon Pathom 73170, Thailand

*Corresponding author: jitladda.sak@mahidol.ac.th

Abstract. Silica is reinforcing filler with high polarity, leading to the difficulty to get the homogenous mixing with a natural rubber (NR) compound, which normally becomes silica agglomeration. A silane coupling agent is applied in a silica-filled NR compound in order to enhance compatibility between silica and rubber. Moreover, non-rubber components, especially proteins, are believed to have competition with silane coupling agent during a silanization [1,2]. In this work, the interaction between the silane coupling agent and silica-filled NR materials was investigated to show the components influencing in the filler-filler and filler-rubber interaction as well as mechanical properties. The rubber types with different non-rubber components such as fresh NR, deproteinized NR (DPNR) and polyisoprene rubber (IR), absence of non-rubber constituents were mixed with *bis*-triethoxysilylpropyl disulfide (TESPD) and silica by using an internal mixer at high temperature. Payne effect of the unvulcanised rubber samples were characterized to study silica dispersion, respectively. A swelling test was used to determined crosslink density in the rubber samples after a vulcanisation system. In addition, the effect of non-rubber components on mechanical properties of rubbers was investigated to find the relationship between the crosslink density and filler-rubber network.

1. Introduction

Natural rubber (NR) latex is produced from *Hevea brasiliensis* tree. NR composes of *cis*-1,4 polyisoprene and non-rubber components, which play an important role in stabilization of colloidal rubber particles. The applications of NR materials are condom, gloves, tire and footwear due to high tensile strength, high abrasion resistance and high resilience. Raw rubber shows poor mechanical properties, so reinforcing filler and vulcanisation system are important to increase the strength and hardness. Silica is usually added in the rubber material because it can improve low heat build-up, low rebound and light colour of the rubber product [3,4]. However, silica is high polarity filler from silanol groups on its surface, that result in difficult compatibility with rubber (low polarity). Thus, a silane coupling agent is applied in the silica-filled rubber compound in order to improve the compatibility. The chemical structure of the silane coupling agent usually composes both of organic and inorganic parts. The organic parts such as hydrocarbon, amine and sulphur groups may give interaction or reaction with the unsaturated rubber chains. For the inorganic part, ethoxyl or methoxyl groups chemical bond with hydroxyl groups on silica surface, that is called silanization. However, the protein components in the NR chains can disturb the silane coupling agent work [1,2]. The aim of this research is to study the effect of the non-rubber components on physical and mechanical properties of *bis*-triethoxysilylpropyl disulfide (TESPD), commercially available silane coupling agent, in silica-filled rubber compounds, which is different rubber types such as NR, DPNR and IR.

2. Experimental

Table 1 presents the conditions for the mixing process. The mixing process of the rubber compound is separated into two steps in order to prevent premature vulcanisation. The raw rubber was masticated in an internal mixer (HAAKE™ (Rheocord 9000 Fisons) Banbury® rotor type) for 1 min under 150°C. Next, the half amount of ZnO (activator), stearic acid (co-activator), 6PPD (anti-oxidant) and silica were added and mixed for 5 min. TESPD (silane copling agent) was applied into the rubber compound for 1 min. After that, the leftovers of ZnO, stearic acid, 6PPD and silica is added into the rubber compound and mixed for 3 min. Finally, the ram of the internal mixer was brought up and down for 6 min. The rubber compound was taken out from the internal mixer and then, brought through a two-rolls mill (DR. Collin GMBH model W 100 T). CBS (fast-accelerator), DPG5 (slow-accelerator) and sulphur (curing agent) were applied into the rubber compound at 30°C for 2 min on each chemical substance with ten cutting. The rubber compound was stored at RT for 24 h. After mixing process, the complete vulcanisation was further carried out by an electrical hydraulic hot press at 160°C.

Table 1. Conditions for compounding in parts per hundred of rubber (phr).

No.	Chemical substances (phr)										
	NR	**DPNR**	**IR**	**Silica**	**ZnO**	**Stearic acid**	**6PPD**	**TESPD**	**Sulphur**	**CBS**	**DPG**
1	100	-	-	40	3	1	1	0	2	1	0.5
2	-	100		40	3	1	1	0	2	1	0.5
3	-	-	100	40	3	1	1	0	2	1	0.5
4	100	-	-	40	3	1	1	2.1	2	1	0.5
5	-	100		40	3	1	1	2.1	2	1	0.5
6	-	-	100	40	3	1	1	2.1	2	1	0.5

Rheological properties of the unvulcanised rubbers such as filler-filler interaction (Payne effect) were studied using Rubber processing analyser (RPA). Strain sweep test (0.5-1000%) was carried out at the test chamber temperature of 100°C and a constant frequency of 1 rad/s for the RPA instrument. The tensile properties of the vulcanised rubbers were analysed to study the strength and elongation by using a universal extensometer (Instron Model 5566) with a cross head speed of 500 mm/min and load cell of 1 kN. Next, crosslink density of the vulcanised rubbers was analysed using swelling test [5].

3. Results and discussion

The effects of non-rubber components on filler-filler interaction were analysed in comparison between unvulcanised NR, DPNR and IR compounds as given in **Figure 1.** The Payne effect, as filler-filler interaction of silica compounds with TESPD is lower than that of silica compounds without TESPD. Interestingly, the NR compound without TESPD presents the lowest filler-filler interaction because protein occurs interaction with silica through hydrogen bond, resulting in a reduction of silica-silica interaction [2]. However, after addition of TESPD, the filler-filler interaction of the NR compound was higher than that of the DPNR compound. This could be because of proteins in NR can disturb the reaction between silane coupling agent and silica from its shielding the silica surface through hydrogen bond [2]. Thus, it can be noted that there is a competition between the protein on rubber chain and the silane coupling agent in interaction or reaction with silica. For the IR compound, the filler-filler interaction exhibits the highest in both the absence and presence of TESPD. It can be noted that IR has only hydrocarbon chain, so it is considerably different polarity with silica, which resulting an agglomeration of silica. In addition, the absence of non-rubber components in the IR chain is caused for no interaction with silica to reduce silica-silica network. The silane coupling agent shows high efficiency for silanization in the DPNR compound which can be seen in low filler-filler interaction due to the absence of the competition with proteins and slightly different polarity. Moreover, it can be indicated that the silane coupling agent covalent bonded with silica can reduce silica agglomeration more than the hydrogen bond interaction between protein and silica.

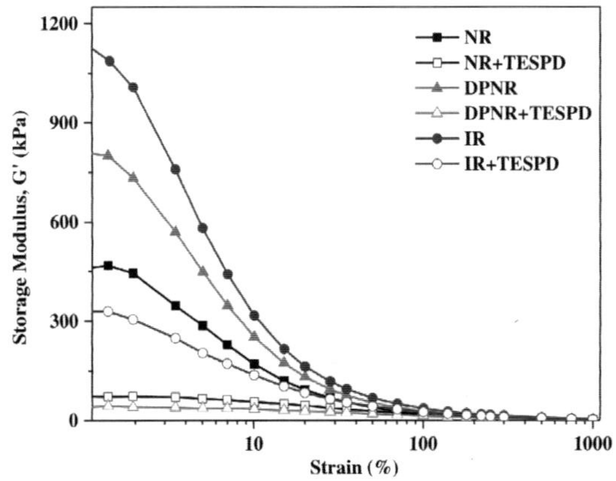

Figure 1. Payne effect of silica in NR, DPNR and IR unvulcanised compounds with and without TESPD.

Figure 2 presents the crosslink density and the tensile strength of silica in NR, DPNR and IR with and without TESPD after vulcanisation. All rubber vulcanised compound with TESPD shows higher degree of crosslink density more than the vulcanised compound without TESPD. The NR compound in both the absence and presence of TESPD presents the highest crosslink density owing to a natural accelerator from proteins in NR [1,6-8]. This can be noted that the DPNR compound has lower crosslink density than the NR compound. In the other hand, the crosslink density of IR compound is higher than that of the DPNR compound. This reason is a difference in microstructure and macrostructure between the IR and DPNR rubbers [9]. For tensile strength, the NR and DPNR compounds with TESPD have higher than for the compounds without TESPD, except for the IR compound. This could be explained by good silica dispersion and high crosslink density in the unvulcanised and vulcanised rubbers, respectively. Furthermore, nanomatrix structure formed by the non-rubber components and strain-induced crystallisation (SIC) caused by phospholipids are caused by performing the outstanding mechanical properties [10]. These reasons are confirmed in the DPNR vulcanised compound result, after removal of proteins from NR, the degree of crosslink and the tensile strength decrease. Even though the IR compound exhibits high crosslink density, the filler-filler interaction is poor. Thus, poor silica dispersion, the chemical composition, the absence of non-rubber of the IR can affect the mechanical properties.

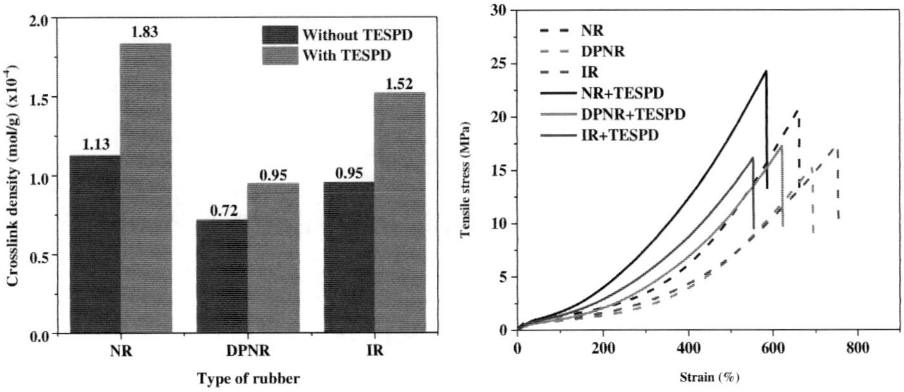

Figure 2. a) Crosslink density and b) tensile strength of TESPD and silica filled in (I) NR, (II) DPNR and (III) IR vulcanised compounds.

4. Conclusions

The non-rubber contents in rubber chains gave affect the silane coupling agent, which relates in physical and mechanical properties of TESPD in silica-filled rubber compounds. There is more silanization between TESPD and silica with absence of protein. On the other hand, the presence of non-rubber constituents can decrease silica agglomeration before addition of the silane coupling agent. The non-rubber components also have important role in enhancement of crosslink density, particularly the proteins. In addition, tensile strength of the rubber with the non-rubber contents presents outstanding from SIC, and nanomatrix structure.

5. References

[1] S. S. Sarkawi, W. K. Dierkes and J. W. M. Noordermeer, *Eur. Polym. J.* **49**, 10, 3199 (2013).
[2] S. S. Sarkawi, W. K. Dierkes and J. W. M. Noordermeer, *Eur. Polym. J.* **54**, 118 (2014).
[3] J. L. Leblanc, *Prog. Polym. Sci.* **27**, 4, 627 (2002).
[4] S. Prasertsri and N. Rattanasom, *Polym. Test.* **31**, 5, 593 (2012).
[5] N. Dechnarong, A. Nimpaiboon, S. Kumarn, P. Phinyocheep and J. Sakdapipanich, *Journal of Polymer Research* **25**, 1 (2017).
[6] K. Suchiva, T. Kowitteerawut and L. Srichantamit, *J. Appl. Polym. Sci.* **78**, 8, 1495 (2000).
[7] P. Posadas, M. A. Malmierca, A. Gonzalez-Jimenez, L. Ibarra, A. Rodriguez, J. L. Valentin, T. Nagaoka, H. Yajima, S. Toki, J. Che, L. Rong and B. S. Hsiao, *Express Polymer Letters* **10**, 1, 2 (2016).
[8] C. Kongkaew, W. Intiya, S. Loykulnant and P. Sae-oui, *KGK Kautschuk Gummi Kunststoffe* **70**, 37 (2017).
[9] S. Ostad-Movahed, K. Ansar Yasin, A. Ansarifar, M. Song and S. Hameed, *J. Appl. Polym. Sci.* **109**, 2, 869 (2008).
[10] Y. Zhou, K. Kosugi, Y. Yamamoto and S. Kawahara, *Polym. Adv. Technol.* **28**, 2, 159 (2017).

Acknowledgement

The authors would like to thank the Center of Excellence for Innovation in Chemistry (PERCH-CIC), Ministry of Higher Education, Science, Research and Innovation for the financial support. For support the rubber analyse machines, I would like to appreciate the Rubber Technology Research Centre (R-TEC). Sincere appreciation is extended to Thai Rubber Latex Co., Ltd. for generous supplies of NR latex.

Transmission of charge ion in Single-walled Carbon Nanotube

J Charoenpakdee[1], A Hutem[2], S Boonchui[1,*]

[1] Department of Physics, Faculty of Science, Kasetsart University, Bangkok 10900, Thailand.
[2] Physics Division, Faculty of Science and Technology, Phetchabun Rajabhat University, Phetchabun 67000, Thailand,

*Corresponding E-mail: fscistb@ku.ac.th

Abstract. In this paper, we investigate the effects of the chiral structure of single-wall carbon nanotube (SWCNT) on the probability that an electron is transmitted (or reflected) through SWCNT. We adopt the charge density calculated by using the density functional theory (DFT) method [New J. Phys. 18 (2016) 023029] to investigate the transmission of an electron through potential barriers from the charge density. We find that the alpha-electron density in SWCNT for the chiral structures (9,0), (9,2) and (9,3) causes resonant tunneling of electrons through potential barriers. For the chiral structures (9,1), (9,4) and (9,5), we obtain transmission and reflection as same as scattering at a finite potential barrier.

1. Introduction

Theoretical investigations were developed by several authors, focusing on the transport of fast charged particles through single-wall carbon nanotube (SWCNT). The hydrodynamic theory of a two-dimensional electron gas [1-2] has been used to derive energy loss of a probe electron when passing through or close by SWCNT. [3]. While the studies on results for stopping power in the linearized hydrodynamic model compared with the dielectric formalism in random-phase approximation had shown [4]. A close agreement between the two approaches is obtained for high speeds of a probe charged particle. The induced potential, the self-energy, and the stopping power for the examination charged particle moving parallel to the axis in CTNs had been derived. The numerical results show that the velocity dependents of these quantities are strongly affected by the damping factor, the nanotube radius, and the probe charged particle position. For fast molecular and cluster ion through solid and plasma targets had been studied [5]. For CNTs, problems in channeling H2+through SWCNTs were solved by the classical equation of motion, which involves both the repulsive continuum potential of SWCNT and the dynamical screened Coulomb repulsion between the two ions [6]. Strong vicinage effects are found in both the molecule self-energy and its stopping power when the speeds are high [7]. For other consideration, impact of the curvature on induced hybridization and modification of emission profiles for each chiral's index of SWCNT are showed that the maximum dipole vector circumference is obtained at the phase summation of π for the case of indirect transition [8] and electron transport in the chiral SWCNT and quantum dot coupling was showed in [9].

In the present paper, we adopt the charge density calculated by using the density functional theory (DFT) method [10] to investigate the transmission of an electron through potential barriers from the charge density.

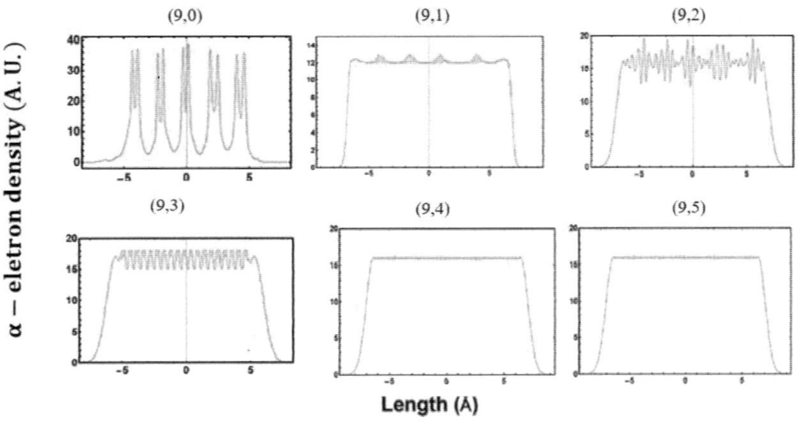

Figure 1. The α-electron density distribution along the tube axis of (9, m) for m = 0–5, calculated by using the density functional theory (DFT) method [10].

2. Methods

In the present, the computer simulation technique has been used in solutions to the one-dimensional Schrodinger equation with an arbitrary potential. A numerical method for quantum tunneling is essential for understanding many quantum transport phenomena. Instead, we have used the Crank-Nicolson method [11], which is stable and conserves probability. In this method, we evaluate the right-hand side of the Schroedinger equation at both times k_1 and k and take the average. Our discretized Schrodinger equation is then

$$\left(2 + 2\alpha + \frac{i\Delta t}{\hbar} V_j\right) \varphi_{j,k} - \alpha\varphi_{j+1,k} - \alpha\varphi_{j-1,k} = \left(2 - 2\alpha - \frac{i\Delta t}{\hbar} V_j\right) \varphi_{j,k-1} - \alpha\varphi_{j+1,k-1} - \alpha\varphi_{j-1,k-1} \tag{1}$$

when. $\alpha = i\hbar\Delta t / 2m\Delta x^2$. Eq. (1) can be rewritten in matrix form as

$$A\varphi_{j,k} = B\varphi_{j,k+1}, \tag{2}$$

and φ is the N-point array representing the wave function, k is the temporal index, and the matrices A and B are

$$A = \begin{pmatrix} 2 + 2\alpha + \frac{i\Delta t}{\hbar} V_1 & -\alpha & 0 & & 0 \\ -\alpha & 2 + 2\alpha + \frac{i\Delta t}{\hbar} V_2 & -\alpha & \cdots & \vdots \\ 0 & -\alpha & 2 + 2\alpha + \frac{i\Delta t}{\hbar} V_3 & & 0 \\ \vdots & & & \ddots & \vdots \\ 0 & 0 & 0 & \cdots & 2 + 2\alpha + \frac{i\Delta t}{\hbar} V_N \end{pmatrix}, \; B = \begin{pmatrix} 2 - 2\alpha - \frac{i\Delta t}{\hbar} V_1 & -\alpha & 0 & & 0 \\ \alpha & 2 - 2\alpha - \frac{i\Delta t}{\hbar} V_2 & -\alpha & \cdots & \vdots \\ 0 & -\alpha & 2 - 2\alpha - \frac{i\Delta t}{\hbar} V_3 & & 0 \\ \vdots & & & \ddots & \vdots \\ 0 & 0 & 0 & \cdots & 2 - 2\alpha - \frac{i\Delta t}{\hbar} V_N \end{pmatrix}. \tag{3}$$

Finally, to evolve the wave function forward by one-time step we perform the matrix multiplication

$$\varphi_{j,k} = A^{-1} B \varphi_{j,k+1} = C\varphi_{j,k+1} \tag{4}$$

where C is known as the Crank-Nicolson matrix.

The shape of the transmitted wave packet can, in principle, be determined by looking at the transmission amplitude for each of its energy eigenstate components following ref. [6]. We assume that the potential wall of the free electron moving though the SWCNT is proportional to the charge density of the alpha-electron density in SWCNT. We adopt the charge density calculated by using the density

IOP Conf. Series: Materials Science and Engineering **773** (2020) 012030 doi:10.1088/1757-899X/773/1/012030

functional theory (DFT) method [10] to investigate the transmission of an electron through potential barriers from the charge density.

3. Result and discussion

The results of the calculated probability of an electron through SWCNT, we obtain two characteristics of tunneling shown in figure 2-7 as resonant tunneling and rectangular potential. The electron transmission coefficient through the chiral structure depends on the alpha-electron density in SWCNT for the chiral structures (9,0), (9,2) and (9,3) causes resonant tunneling of electrons through potential barriers. For the chiral structures (9,1), (9,4) and (9,5), we obtain transmission and reflection as same as scattering at a finite potential barrier.

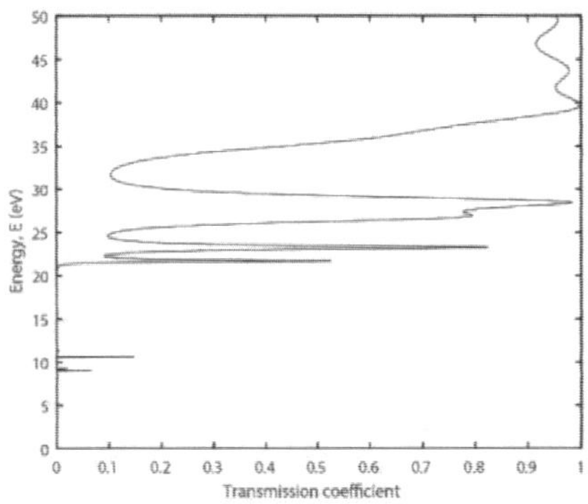

Figure 2. The electron transmission coefficient through the chiral structure (9,0) as a function of energy.

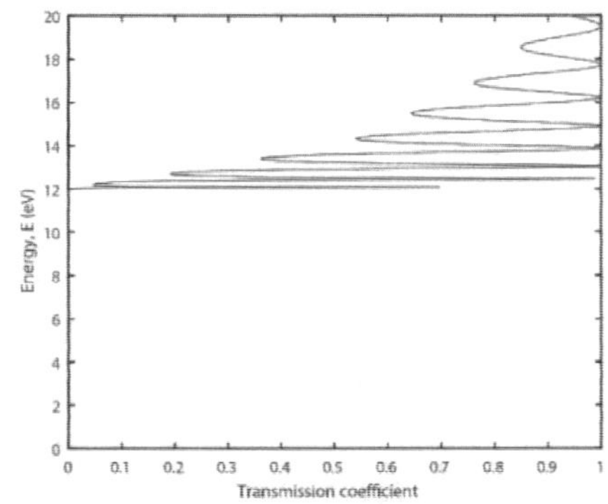

Figure 3. The electron transmission coefficient through the chiral structure (9,1) as a function of energy.

From Figure 2-4, the result of our simulation had predicted that the electron transmission coefficient through the chiral structure threshold is a complex phenomenon that compiles two quantum phenomena as quantum tunneling and quantum interference that show in low energy region. It refers to resonant tunneling in which the electron transmission coefficient through a structure is sharply peaked about certain energies in Figure 2. The total amplitude to transfer the electron through the chiral structure can be viewed as a sum of partial waves, executed several reflections in the central area.

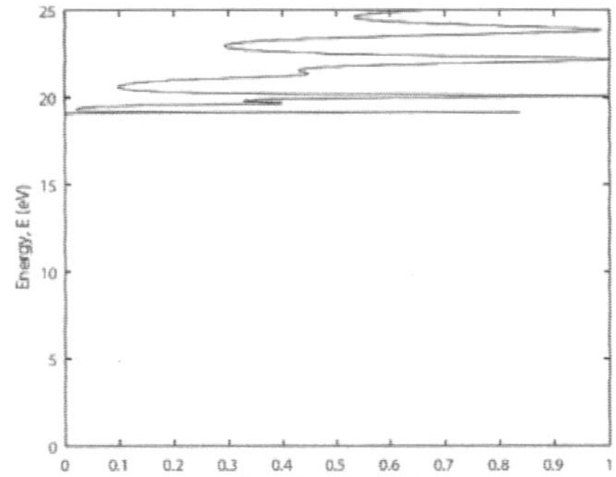

Figure 4. The electron transmission coefficient through the chiral structure (9,3) as a function of energy.

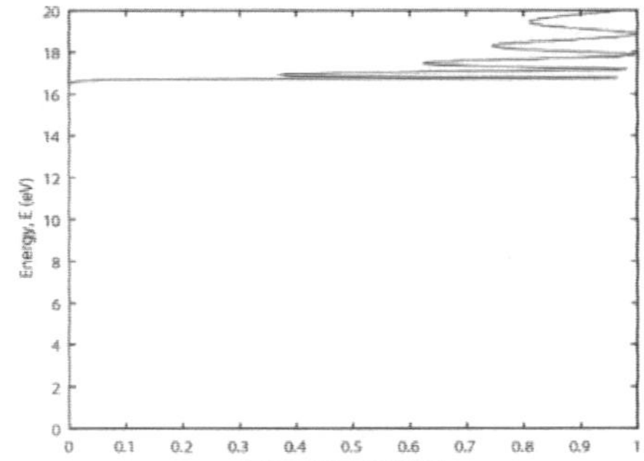

Figure 5. The electron transmission coefficient through the chiral structure (9,4) as a function of energy.

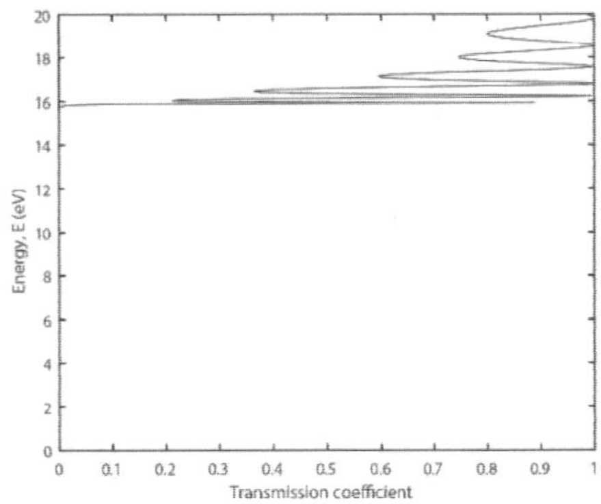

Figure 6. The electron transmission coefficient through the chiral structure (9,4) as a function of energy.

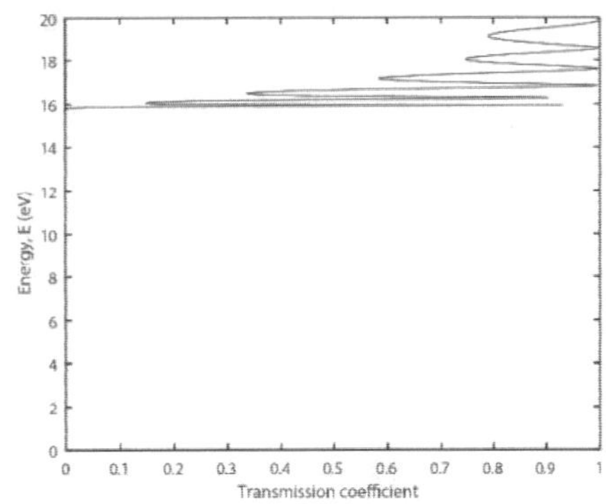

Figure 7. The electron transmission coefficient through the chiral structure (9,5) as a function of energy.

4. Conclusions

In this study, we investigate the dependence of the transmission coefficient of the electron through along the tube axis of the chiral structure (9, m) form 0 – 5. Our simulation had predicted the transmission coefficient threshold and show that the low energy electron is possible through the chiral SWCNT is decreasing while the chiral angle is increased.

5. References

[1] J. Cumings and A. Zettl, “Low-Friction Nanoscale Linear Bearing Realized from Multiwall Carbon Nanotubes” Science 289, 602 (2000).
[2] G. Hummer, J. C. Rasaiah, and J. P. Noworyta, “Water conduction through the hydrophobic channel of a carbon nanotube.” Nature (London) 414, 188 (2001).
[3] Thomas Stöckli, Jean-Marc Bonard, André Châtelain, Zhong Lin Wang, and Pierre Stadelmann “Collective oscillations in a single-wall carbon nanotube excited by fast electrons” Phys. Rev. B, 64, 115424 (2001).
[4] You-Nian Wang and Z. L. Miškovic “Interactions of fast ions with carbon nanotubes: Self-energy and stopping power” Phys. Rev. A 69, 022901 (2004).
[5] Guiqiu Wanga and Yue Wang “Stopping Power and Self-Energy for a Fast Proton in Carbon Nanotubes” Applied Mechanics and Materials Vols. 488-489, 22-25(2014).
[6] Da-Peng Zhou, Yuan-Hong Song, You-Nian Wang, and Z. L. Miškovic, “Coulomb explosions and stopping of molecular ions channeled through carbon nanotubes” Phys. Rev. A 73, 033202(2006).
[7] Nestor R. Arista “Stopping of molecules and clusters” Nuclear Instruments and Methods in Physics Research B 164, 108(2000).
[8] Thiti Thitapura, Watchara Liewrian, Tula Jutrosaga, and S. Boonchui “Curvature effect on polarization of light emitted from chiral carbon nanotubes” Opt Express. 25, 25588(2017).
[9] S. Boonchui and J. Charoenpakdee “Chirality Effects on an Electron Transport in Single-walled Carbon Nanotube Coupling to a Quantum Dot” IOP Conf. Ser.: Mater. Sci. Eng. 526, 012017 (2019).
[10] Jia Wang,Wanrun Jiang, Bo Wang, Yang Gao, Zhigang Wang and Rui-Qin “Chirality dependent spin polarization of carbon nanotubes” New J. Phys. 18, 023029(2016).
[11] John R. Hiller, Ian D. Johnston, and Daniel F. Styer, Quantum Mechanics Simulations (Wiley, New York, 1995), pp. 96–101.

IOP Conf. Series: Materials Science and Engineering **773** (2020) 012031 doi:10.1088/1757-899X/773/1/012031

Electromagnetic interaction between talcum particle and topological insulator

C Summueang and S Boonchui*

Department of Physics, Kasetsart University, Bangkok, 10900, Thailand

*Corresponding author e-mail: fscistb@ku.ac.th

Abstract. We identify the electromagnetic interaction between talcum particle and surface of topological insulator which unsatisfied ordinary Maxwell's equations. For theoretical approach, we use the dyadic green's function including with the method of an image charge and the topological magneto-electric (TMEP). We describe the electromagnetic response taking into account the property of material in case of the dielectric constant. Our numerical approach demonstrates and discuss the interaction and characteristic of electromagnetic responses between talcum and topological insulator.

1. Introduction

Talcum particle or talcum powder is a particle in nanometer-scale that made from talc consisting of a mineral composed mainly of the element magnesium, silicon, and oxygen: $H_2Mg_3O_{12}Si_4$ [1] and has a property of anisotropic crystals. The structure of particle behaves a pole from different of the elements that cause the electrical interaction from itself [2].

Topological insulators (TIs) are a new state of quantum matter that presents quantum-Hall-like properties which are insulator in the bulk but their surface exhibit the conducing states that cause by time-reversal symmetry [3,4]. What makes TIs different from ordinary insulator is that their material has term of the topological magneto-electric polarizability (TMEP). This material has a special property about the interaction when a particle comes closer to the surface inducing the electromagnetic field in the system that occurs the interesting interaction [5].

The electromagnetic interaction between the TIs and a talc particle that exhibit the electric and magnetic field in the system [6,7]. The different types of materials including with dielectric permittivity and magnetic permeability of materials give different and interesting interaction [8,9]. The topological magneto-electric effect (TME) consists of the transmutation of electric and magnetic induction fields that result in term of the modified Maxwell's equations causing the characteristic property of this materials [10]. When a particle near the surface of TIs in the presence of an electric field, that has the molecular particle–surface interaction with a molecular–image interaction applying the method of images approach in part of the dyadic Green's function for the system [11,12].

In our work, we indicate the theoretical and numerical calculation in the presence of the dielectric constants which identify the different materials. Describing behavior of the electromagnetic interaction has affect from a property of the materials. We analyze the intensity of the electric and magnetic field in part of contour plot and demonstrate this field in part of quiver plot. Finally, we display results to compare with a property of talc particle.

2. Theoretical approach

A talc particle is brought close to the surface of topological insulator which characterized by dielectric constant ε_s and magnetic susceptibility μ. The lower-half space is occupied by a TI and the upper-half space is occupied by vacuum [8,13]. We determine a charged particle which is located at (0, 0, d) in XYZ-plane as shown in Figure 1.

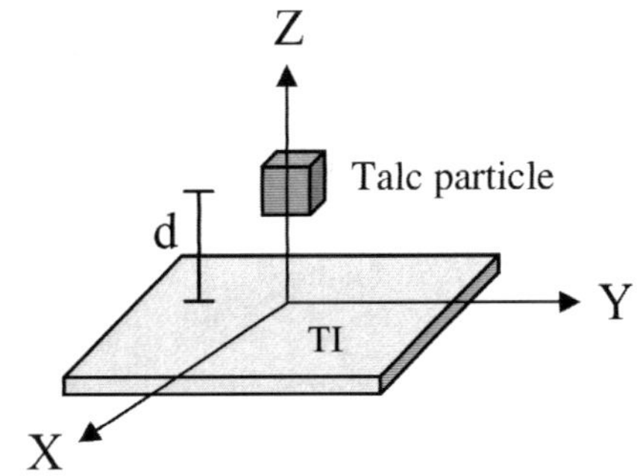

Figure 1. A charged particle or talc particle near a TI surface.

That arise the inducing electric polarization by electric field and magnetic polarization by magnetic field in the system. This effect can be written Maxwell's equations including the topological term as:

$$\nabla\cdot\mathbf{D} = 4\pi\rho, \qquad \nabla\times\mathbf{H} = \frac{\partial\mathbf{D}}{\partial t} + 4\pi\mathbf{J},$$

$$\nabla\cdot\mathbf{B} = 0, \qquad \nabla\times\mathbf{E} = -\frac{\partial B}{\partial t},$$

but with the modified constituent equations causing by the TME effect [11,13]:

$$\mathbf{D} = \varepsilon\mathbf{E} + \frac{\alpha}{\pi}\theta\mathbf{B}, \qquad \mathbf{H} = \frac{\mathbf{B}}{\mu} - \frac{\alpha}{\pi}\theta\mathbf{E},$$

where $\alpha = e^2/\hbar c$ is the fine structure constant and θ is TMEP.

We focus on the electromagnetic interaction which is convenient to use the Green's function method. The dyadic Green's function distinguishes between the direct and indirect parts including the method of image and that the indirect part must refer to a topological insulator term in the following form [6,14]:

$$G_{ij}(\mathbf{R},\mathbf{R}') = G_{ij}^{(Direct)}(\mathbf{R},\mathbf{R}') + G_{ij}^{(Indirect)}(\mathbf{R},\mathbf{R}'),$$

to calculate the electric field and magnetic field, indirect part in Eq. (3) have different coefficient following in refs. [6,13]. That can be expressed in two parts:

2.1. Electric field calculation

The indirect part of Green's function can be written in form:

$$G_{ij}^{(Indirect)}(\mathbf{R},\mathbf{R}') = G_{il}^{(Direct)}(\mathbf{R},\mathbf{R}'_M)M_{lj} = G_{il}^{(Direct)}(\mathbf{R},\mathbf{R}'_M)\frac{(\varepsilon_s-1)(\frac{1}{\mu}+1)+\tilde{\theta}^2}{(\varepsilon_s+1)(\frac{1}{\mu}+1)+\tilde{\theta}^2}\begin{pmatrix}-1 & 0 & 0\\ 0 & -1 & 0\\ 0 & 0 & 1\end{pmatrix},$$

to obtain the electric field, using the Lippmann-Schwinger equation in the following equation [7]:

$$E_i(\mathbf{R}) = E_i^{(0)}(\mathbf{R}) + \frac{k_0^2}{\varepsilon_0}\int_{V_p} d\mathbf{R_p}G_{ij}(\mathbf{R},\mathbf{R}'_\mathbf{p})\cdot P_j(\mathbf{R}'_\mathbf{p})$$

where $\tilde{\theta} = \alpha\theta/\pi$, $k_0 = \omega/c$, $\mathbf{R}'_M = (x', y', -z')$, P_j is particle polarization and c is the speed of light.

IOP Conf. Series: Materials Science and Engineering **773** (2020) 012031 doi:10.1088/1757-899X/773/1/012031

2.2. Magnetic field calculation

The indirect part of Green's function can be written in form:

$$G_{ij}^{(Indirect)}(\mathbf{R},\mathbf{R}') = G_{il}^{(Direct)}(\mathbf{R},\mathbf{R}'_M)M_{lj} = G_{il}^{(Direct)}(\mathbf{R},\mathbf{R}'_M)\frac{2\tilde{\theta}^2}{(\varepsilon_s+1)(\frac{1}{\mu}+1)+\tilde{\theta}^2}\begin{pmatrix}-1 & 0 & 0\\ 0 & -1 & 0\\ 0 & 0 & 1\end{pmatrix},$$

to obtain the magnetic field, using the Lippmann-Schwinger equation with magnetic polarization μ_j in the following equation [7]:

$$B_i(\mathbf{R}) = B_i^{(0)}(\mathbf{R}) + \frac{k_0^2}{\varepsilon_0}\int_{V_p} d\mathbf{R}_\mathbf{p} G_{ij}(\mathbf{R},\mathbf{R}'_\mathbf{p})\cdot\mu_j(\mathbf{R}'_\mathbf{p})$$

3. Numerical approach

Let us consider the distribution of electromagnetic field applying electric field in this system. We describe by setting same size of the square particle in this situation. For the system, it is convenient to explain the interaction that can be considerably simplified in YZ-plane as shown in Figure 2.

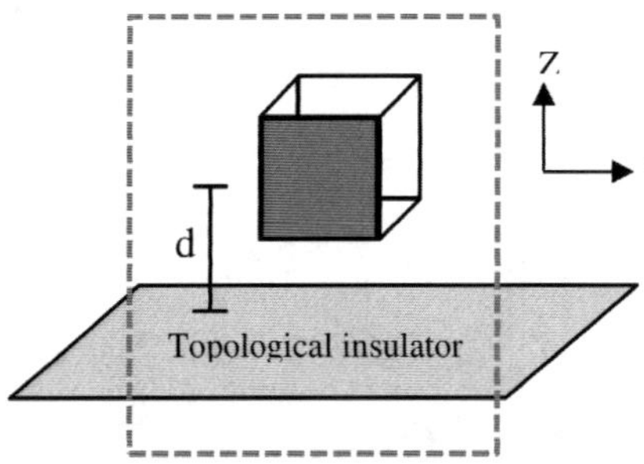

Figure 2. A square talc particle focuses on YZ-plane.

We illustrate and show calculation of the distribution with different dielectric constant.

3.1. Electric field distribution

We obtain the distribution and intensity of the field (insert) as shown in Figure 3 and Figure 4.

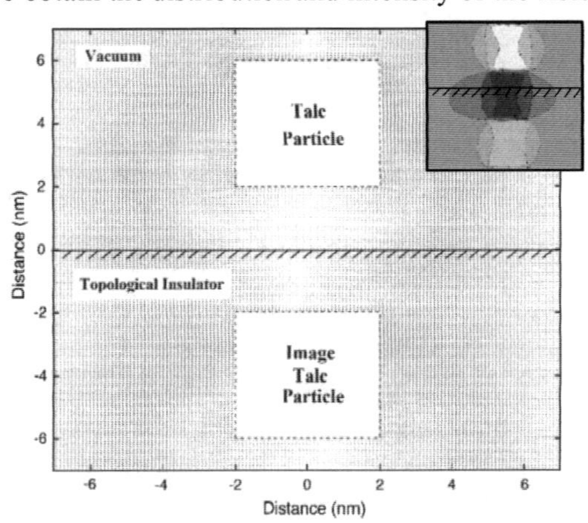

Figure 3. Illustration of electric field caused by TI with dielectric constant = 10.

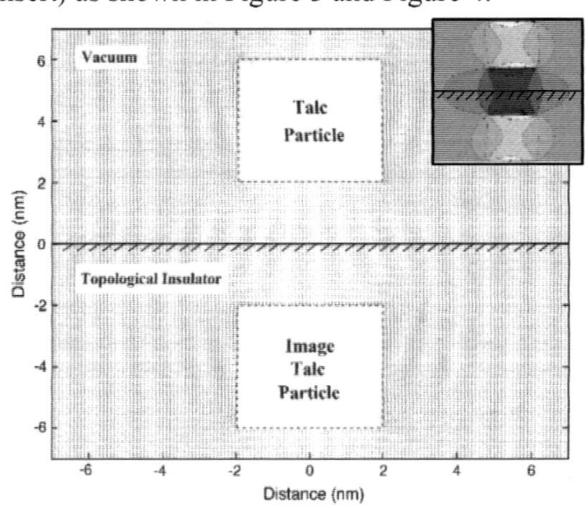

Figure 4. Illustration of electric field caused by TI with dielectric constant = 130.

3.2. Magnetic field distribution

We obtain the field distribution as shown in Figure 5 and Figure 6.

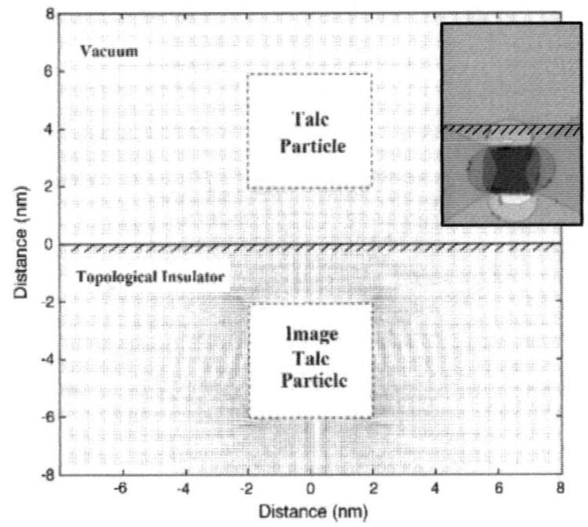

Figure 5. Illustration of magnetic field caused by TI with dielectric constant = 10.

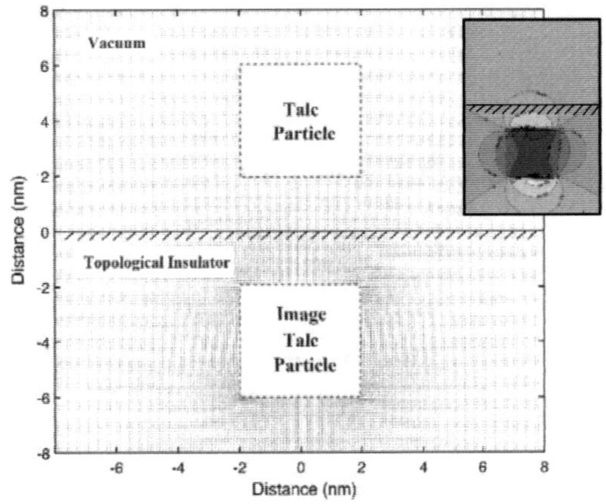

Figure 6. Illustration of magnetic field caused by TI with dielectric constant = 130.

4. Conclusion

In this work, we have shown the theoretical and numerical framework to estimate the electromagnetic response due to talc particle and different topological insulator. In electric distribution, As the results are shown that the density of field direction in high dielectric constant has more than low dielectric constant which corresponds to insert that $\varepsilon_s = 130$ has more intensity of image particle than $\varepsilon_s = 10$ since the higher dielectric constant is the great of the material's ability to store the energy, see ref. [8,14]. When we discuss on the magnetic distribution that give the similar results of field density to electric field but no variation of field above TI because the talc particle has a very weak magnetization that affect the magnetic field in vacuum about the intensity. That is an interesting property of TI, although the talc particle has weak magnetic property, but this material can induce magnetic field in the system.

5. References

[1] Flegmann A W and George Raymond A T 1975 *Soils and other growth media* (New York: The Macmillan Press)
[2] Burdukova E, Laskowski J S and Bradshaw D J 2006 *XIII international mineral processing congress* (Istanbul: The University of Belgrade) p 904-910
[3] Bernevig B A and Hughes T L 2013 *Topological Insulators and Topological Superconductors* (New Jersey: Princeton University Press)
[4] Zhu Z, Cheng Y and Schwingenschlögl U 2012 *Phys. Rev. B* **85** 235401
[5] Khudik B I and Lozovski V 1990 *Phys. Status Solidi B* **158** 511
[6] Ohtsu M 1998 *Near-Field Nano/Atom Optics and Technology* (Yokohama: Springer) p 273
[7] Novotny L and Hecht B 2006 *Principles of nano-optics* (New York: Cambridge University Press) p 25
[8] Kysylychyn D, Piatnytsia V and Lozovski V 2013 *Phys. Rev. E* **88** 052403
[9] Albanese A, Tang P S and Chan W C W 2012 *Annu. Rev. Biomed. Eng.* **14** 1
[10] Qi X L, Hughes T L, and Zhang S C 2008 *Phys. Rev. B* **78** 195424
[11] Qi X L, Li R, Zang J and Zhang S C 2009 *Science* **323** 1184
[12] Martín-Ruiz A, Cambiaso M and Urrutia L F 2015 *Phys. Rev. D* **92**, 125015
[13] Martín-Ruiz A, Cambiaso M and Urrutia L F 2016 *Phys. Rev. D* **94**, 085019
[14] Summueang C and Boonchui S 2019 *The International Conference on Materials Research and Innovation* (Bristol: IOP Publishing)

Application of polymer blend based on natural rubber latex and acrylic resin as a binder for wall paints

A Worlee[1],*, N Homdong[2] and N Hayeemasae[3]

[1] Research and Development of Halal Products, Faculty of Science and Technology, Fatoni University, Pattani, 94160, Thailand.
[2] Industrial Metrology and Testing Service Centre, Thailand Institute of Scientific and Technological Research, Patumthani, Bangkok 12120, Thailand.
[3] Department of Rubber Technology and Polymer Science, Faculty of Science and Technology, Prince of Songkla University, Pattani Campus, 94000, Thailand.

*Corresponding author: anuwatworlee@yahoo.com

Abstract. In this study, the polymer blends based on natural rubber latex and acrylic resin was prepared. It was then utilized as a binder for paint. The blending ratios between natural rubber latex and acrylic resin were varied from 30/70, 40/60, 50/50, 60/40, and 70/30 (w/w) where the natural rubber latex was pre-vulcanized prior to be blended with acrylic resin. The investigation was monitored through the physical and mechanical properties of the blends. Highest tensile strength and elongation at break were found at the blending ratio at 30/70 (w/w) of natural rubber latex and acrylic resin. This may be attributed to the compatibility between elastic phase of the natural rubber and hard phase of acrylic resin. When adding higher ratio of natural rubber latex, the tensile strength and elongation at break dropped continuously, indicating inhomogeneity of these two phases. To confirm the compatibility of the blend at 30/70 (w/w), the gel content was also observed. It is clear that such blend completely dissolved in non-polar solvent unlike other blending ratios. Therefore, this blending ratio is highly suggested and appropriate for applying as binder for paint.

1. Introduction

Wall paints are composite materials which consist of many compositions such as polymer, filler and pigment powder. They can be used for adding the colourful to the building and general equipment or materials [1-3]. There are many types of paints currently used either water-based or solvent-based depending on the usage, properties and surfaces of the substrate. Each type of paint has different properties and qualities. In general, the paint consists of two important functional components [2],[3]: to make different colouring according to customer's needs and to perform as an binder which is very important part indicating the quality of the paint. Generally, the binder commonly used in paint and paint is acrylic polymer or derivative of acrylic [2-4]. Due to it possesses several prominent properties that appropriate for using as binders to prepare the paint.

One of the important properties of paint is adhesion ability on the surface of the substrate, because of it is acknowledged that natural rubbers have a good adhesion property on many types of substrate [5, 6]. According to this advantage, the researchers are interested in using natural rubbers as the binder of water-based paint in order to increase the adhesion ability of paint. Therefore, the polymer blend was prepared by compounding the pre-vulcanized natural rubber latex with the acrylic resin at various blending ratios. Moreover, the physical and mechanical properties of these polymer blends also studied.

2. Experimental

The pre-vulcanized natural rubber latex [7-8], which has 90% of crosslink density degree (Moderately vulcanized), was firstly prepared using the formulation as show in Table 1 under condition of 70 oC for 2 h. The blending of pre-vulcanized natural rubber latex with acrylic resin at various blending ratios (using as the binder for wall paints) was prepared later. Some physical and mechanical properties as gel content and tensile property of pre-vulcanized natural rubber latex, acrylic resin and each polymer blends were also investigated. The paint was prepared by mixing the optimum ratio of polymer blend as a binder with $CaCO_3$, CMC, pigment, forming film agent and dispersing agent as shown in Table 2. The properties of paint were investigated to follow as the procedure according to the Thai industrial standard of weather-resistant emulsion paint (TIS. 2321-2549) [9].

Table 1. Formulas of pre-vulcanized natural rubber latex

Formula	phr
HA latex	100
Sulpur	3
ZDEC	1
ZDBC	0.5
Wing Stay L	1.5
ZnO	1
Teric N30	1.7

Table 2. The ingredients use to prepare paint using polymer blend between natural rubber latex and acrylic resin as a binder

Formula	(%)
Polymer blend	20
$CaCO_3$	40
CMC	0.5
Pigment	1
Forming film agent	2.5
Dispersing agent	1
Water	35

3. Results and discussion

Table 3 shows that the increasing of natural rubber content increased percentage of total solid content (%TSC) and pH value of polymer blends. The acrylic emulsion had %TSC 50.5% and pH 7.8 while pre-vulcanized natural rubber latex had %TSC 62% and pH 10.4. Therefore, it caused %TSC and pH value of polymer blends to be increased as a function of natural rubber content. In the other hand, the mechanical stability testing (MST) and viscosity value of polymer blend were increased as an increasing of acrylic resin content. The viscosity of acrylic resin was higher than pre-vulcanized natural rubber latex, and some surfactant were also added to remain the stability of acrylic resin during processing and storage. With these reasons, the viscosity and MST values of polymer blend increased upon the increasing of acrylic resin contents. Table 3 also shows that the pre-vulcanized natural rubber was not dissolved in Toluene but it swelled due to the crosslinking of rubber chain after pre-vulcanization [10, 11]. Acrylic resin was an un-polar as well as Toluene, thus it was completely dissolved. In the case of polymer blends, it can be dissolved 100% except the blending which contained natural rubber latex for 70% wt.; dissolved only for about 90%. This indicated that blending under 70% wt. of pre-vulcanized natural rubber latex resulted in good compatibility between rubber phase and acrylic resin. The completely dissolving of these polymer blends in Toluene showed that acrylic phase could be in the micro-scaled of the rubber phase, caused the separation of rubber molecules.

Table 3. The physical properties of pure acrylic resin, pre-vulcanized natural rubber latex and polymer blends

Properties	Pre-vulcanized rubber : Acrylic resin				
	30:70	**40:60**	**50:50**	**60:40**	**70:30**
TSC, %	52	53	55	56	57
pH	8.6	8.1	9.2	9.5	9.8
MST, sec.	1834	1754	1624	1542	1473
Viscosity, cps	78	70	64	59	56
Remaining get content, %	-	-	-	-	10.3

Figure1 (a) shows the Stress-strain curve of polymer blend compared with pure acrylic resin and pre-vulcanized natural rubber latex. The Stress-strain curves of each sample have different characteristics. Since natural rubber has excellent elasticity property, while the acrylic resin is a stiffness polymer due to it has high glass transition temperature (Tg) [12]. Therefore, the slope of each graph decreased according to the increasing of natural rubber latex content. Figure 1 (a) also shows that blending of 30% wt. pre-vulcanized natural rubber latex in polymer blends would make the most outstanding polymer blends, high tensile strength and elongation at break, which is the best ratio that gives water and rubber the best compatibility. This blend also shows good elasticity and hardness properties that complement each other. Therefore, the mixing of polymers at this ratio was used as a binder to prepare the wall paints in the study (Figure 1 (b) and (c)).

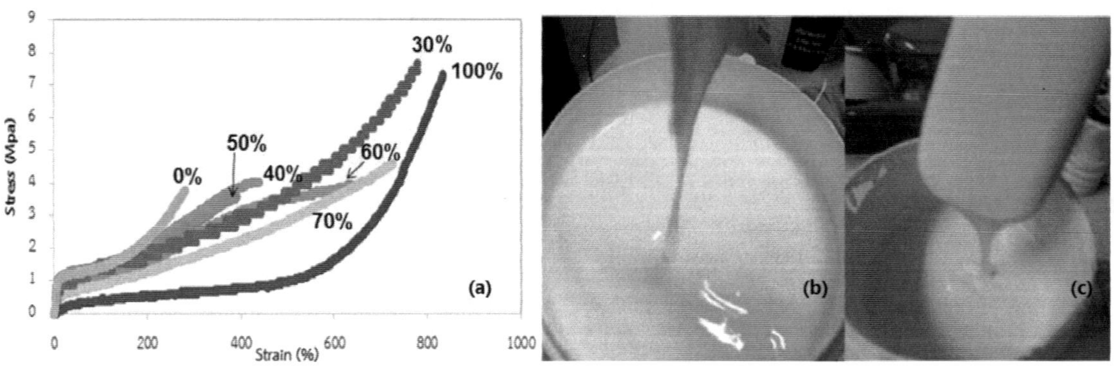

Figure 1. Stress-strain curve of each polymer and polymer blend (a) and the wall paint which used polymer blend as a binder before (b) and after add pigment (c)

The properties of wall paint was characterized as in Thai industrial standard of weather-resistant emulsion paint (TIS. 2321-2549). All properties of the paint which prepared in this study were passed Thai industrial standard of weather-resistant emulsion paint (TIS. 2321-2549). Especially, the determine of the resistance to erosion of wall paints (ASTM D 2486 method A)[13], which related to the adhesion properties of paint and the results were outstanding. In general, TIS. 2321-2549 required the scrub resistance of wall paint at least 3000 cycles. But, in the case of wall paint which prepared in this study had the scrub resistance at least 1,000 cycles. This indicated that the blending between natural rubber and acrylic resin at 30/70 ration was the excellent ration to use as a binder of wall paint and these results were consistent with the results of mechanical properties as mentioned before.

4. Conclusions

The binder is considered as an important functional component which indicating the paint quality. Using a high-performance binder will result in the paint with a high adhesion ability. In this study, the polymer blend at an appropriate ratio of the natural rubber and acrylic resin can be used as a binder of wall paints. The properties of wall paints that contain 30% wt. of natural rubber would pass Thai industrial standard of weather-resistant emulsion paint (TIS. 2321-2549). Moreover, it has excellence in scrub resistance.

5. References

[1] B. Emélie, U. Schuster, and S. Eckersley, 'Interaction between styrene/butylacrylate latex and water soluble associative thickener for coalescent free wall paints', *Prog. Org. Coat.*, vol. 34, no. 1, pp. 49–56, Jan. 1998.

[2] T. Yuzawa, C. Watanabe, S. Tsuge, N. Shimane, and H. Imai, 'Application of a pyrolysis-GC/MS system incorporating with micro-UV irradiation to rapid evaluation of the weatherability of acrylic coating paints for house exterior walls', *Polym. Degrad. Stab.*, vol. 96, no. 1, pp. 91–96, Jan. 2011.

[3] M. Eren and H. K. Can, 'Preparation of zinc methacrylate-methylmethacrylate-butyl acrylate emulsions and their application in exterior paints', *Prog. Org. Coat.*, vol. 135, pp. 424–437, Oct. 2019.

[4] T. K. Akchurin, A. V. Tukhareli, and T. F. Cherednichenko, 'Effective Concrete Modified by Complex Additive Based on Waste Products of Construction Acrylic Paints', *Procedia Eng.*, vol. 150, pp. 1468–1473, Jan. 2016.

[5] A. I. Abou-Kandil, A. Awad, N. Darwish, A. B. Shehata, and B. K. Saleh, 'Adhesion of brass plated steel cords to natural rubber: Dynamic and statistical study of adhesion failure', *Int. J. Adhes. Adhes.*, vol. 44, pp. 26–35, Jul. 2013.

[6] J. Sang *et al.*, 'Adhesion of carbon steel and natural rubber by functionalized silane coupling agents', *Int. J. Adhes. Adhes.*, vol. 72, pp. 70–74, Jan. 2017.

[7] W. D. S. Deniz, T. H. C. Lima, E. P. S. Arlindo, and G. C. F. Junior, 'Prevulcanized natural rubber and carbon black: High-deformation piezoresistive composites', *Mater. Lett.*, vol. 253, pp. 427–429, Oct. 2019.

[8] T. Pojanavaraphan and R. Magaraphan, 'Prevulcanized natural rubber latex/clay aerogel nanocomposites', *Eur. Polym. J.*, vol. 44, no. 7, pp. 1968–1977, Jul. 2008.

[9] TIS. 2321. 2549. Thai industrial standard of weather-resistant emulsion paint.

[10] R.F. Mausser. The vanderbilt latex handbook, vol. 3, 1987, R.T. Vanderbilt Company, inc.

[11] Blackley, D.C. 1997b. Ibid., 35 – 39.

[12] M. Klähn, R. Krishnan, J. M. Phang, F. C. H. Lim, A. M. van Herk, and S. Jana, 'Effect of external and internal plasticization on the glass transition temperature of (Meth)acrylate polymers studied with molecular dynamics simulations and calorimetry', *Polymer*, vol. 179, p. 121635, Sep. 2019.

[13] ASTM D 2486 Method A. 1998. Standard Test Methods for Scrub Resistance of Wall Paints.

Acknowledgement

The authors gratefully acknowledge the National Research Council of Thailand (NRCT), the Thailand Research Fund (TRF) and Thailand's Education Hub for Southern Region of ASEAN Countries for PhD Students (THE-AC) for financial support throughout this work.

Effect of sintering temperature on microstructure of TiO_2 scaffold

S Yanawarutwong[1], P Thunyakitpisal[2], D P Kashima[1,3,*]

[1] Research Unit of Advanced Ceramics, Department of Materials Science, Faculty of Science, Chulalongkorn Univer sity, Patumwan, Bangkok 10330, Thailand.
[2] Department of Anatomy, Faculty of Dentistry, Chulalongkorn University, Patumwan, Bangkok 10330, Thailand.
[3] Center of Excellence on Petrochemical and Materials Technology, 7th floor, Chulalongkorn University Research Building, Patumwan, Bangkok 10330, Thailand.

*Corresponding author: dujreutai@gmail.com

Abstract Due to an excellent biocompatibility of titanium dioxide (TiO_2), it is promising to use as scaffolds for inducing bone formation. The well-interconnected pore scaffolds were fabricated by sponge replication method in order to promote osteoblast ingrowth and vascularization. The viscosity of the slurry was adjusted by addition of calcium chloride ($CaCl_2$) which led to more uniformity and densification. The effects of sintering temperatures on microstructure were analyzed by scanning electron microscope (SEM) and microcomputed tomography (Micro-CT). Our experiments showed that the higher sintering temperature increased the grain size and more uniformity of the scaffold. Furthermore, the struts became more rounded and microcracks were not observed which probably correlated to the higher density of scaffold. Thus, it is suggested that optimization of sintering temperatures might allow better control of grain growth and result in more microstructural uniformity.

1. Introduction

Titanium dioxide (TiO_2) is well known as the promising material for biomedical applications due to its excellent biocompatibility and good osteoconductivity [1-2]. The most widely used technique for fabrication of well-interconnected pore TiO_2 scaffolds is sponge replication method. In order to promote osteoblast ingrowth and vascularization, the polymer that used as the template should have the desired pore size that must be open and large enough so that the cells can migrate into the scaffolds [3].

It is worth noting that the biocompatibility and bone cell proliferation are crucial factors for these custom-made scaffolds. Therefore, the interconnectivity of the scaffold will be more considerable than mechanical properties. However, since the scaffolds are highly porous, the mechanical properties are limited and may thus have insufficient strength to be machinable shaping for used [4].

The strength of the scaffolds is strongly dependent on the size and observation of cracks and flaws in the struts structure [4]. Nevertheless, the mechanical properties may be improved by optimizing the processing steps in the fabrication technique. The aim of this study was to investigate the effect of sintering temperatures on the microstructure of well-interconnected TiO_2 scaffolds.

2. Experimental

2.1. Polymer foams

The polyurethane ester foam templates with 60 pores per inch (Bulbren S, EurofoamGmbH, Germany) were used in this study. The foams were cut to cylindrical templates of 15 mm in diameter and 10 mm in length by punched them out with a metal stamp.

2.2. Scaffold fabrication

The slurry was controlled with the solid content of 65 wt% which prepared by adding 65 g of TiO_2 powder (Degussa, P25) into 35 ml of 0.1 M $CaCl_2$ solution. The TiO_2 slurry was stirred at rotation speed of 1,000 rpm for 30 minutes until the slurry mixture was homogeneous, but the higher amount of powder resulted in agglomerated slurry. The slurry viscosity was adjusted by small addition of 1 M HCl.

The polymer foams were then dipped into the TiO_2 slurry, excess slurry was squeezed out by a roller, make sure that only thin layer of the slurry left on the surface of polymer foams. According to the sponge replication method which is one of the template techniques, the surface charge of the replica and TiO_2 powder in slurry must be in opposite charge in order to promote the good attraction for both components. This is a main reason to select the polyurethane ester sponge used as the replica because the carbonyl group in sponge can act as electron donor. TiO_2 powder in the slurry at acidic pH displayed the positive charge. The sample were dried at room temperature for at least 12 hours. Before sintering the scaffolds, the polymer template was burned out at 450$_o$C at a heating rate of 1$_o$C/min. After 1 h soaking time at 450$_o$C, the temperature was raised to between 1350$_o$C and 1500$_o$C at a rate of 1$_o$C/ min and the soaking time was set to 3 h. After that the sintered scaffolds were then cooled down to room temperature.

2.3. Characterization

The crystal structure of the sintered TiO_2 scaffolds was studied by X-ray Diffractometer (XRD, D8-Advanced, Bruker AXS Model D8, German), using Cu Kα radiation (λ = 1.5406 Å; 40 kV and 40 mA). The microstructure was observed by scanning electron microscope (SEM, JEOL JSM-6480LV). The scaffolds were fixed on aluminium stubs and viewed with backscattered electrons at 15 kV accelerating voltage. Micro Computed Tomography (Micro-CT, Scanco Medical, Switzerland) was used to determined 3D scaffold structure. Scaffolds were placed in a plastic holder and using a voltage of 70 kV, current of 113 μA.

3. Results and discussion

The sintered TiO_2 scaffolds between 1350$_o$C and 1500$_o$C were analysed by XRD to determine their phase are presented in Figure 1. The data showed that the sintering process of the TiO_2 scaffolds produced TiO_2 with rutile crystalline structure.

After sintering of TiO_2 scaffolds at various temperatures between 1350$_o$C and 1500$_o$C with soaking time for 3 h, the microscopic appearances are shown in Figure 2. As-sintered scaffold at 1350$_o$C showed the scaffold struts with hollow structure (indicated by an arrow) due to elimination of the polymer foam template. One side of the walls of the struts had collapsed inward resulting in folded strut appearance which had a direct effect on mechanical strength of the scaffold. Furthermore, the longitude cracks along some of the struts were observed (indicated by a rectangle). At 1400$_o$C, the cracks along the edge of the struts still observed but decreased significantly. By the time, the grains size was increased. As the sintering temperature was increased to 1450$_o$C, the folded struts had developed into a solid structure with rounded corners and the earlier cracks along the struts were disappeared. Moreover, the small grains were reduced and resulted in more uniform grain size. At 1500$_o$C, the scaffold struts showed denser and among the grain boundaries became more three-dimensional. The struts showed the uniform small grain size and the grain boundaries were clearly observed. This is confirmed that the higher sintering temperature, the better strength of TiO_2 scaffolds.

IOP Conf. Series: Materials Science and Engineering **773** (2020) 012033 doi:10.1088/1757-899X/773/1/012033

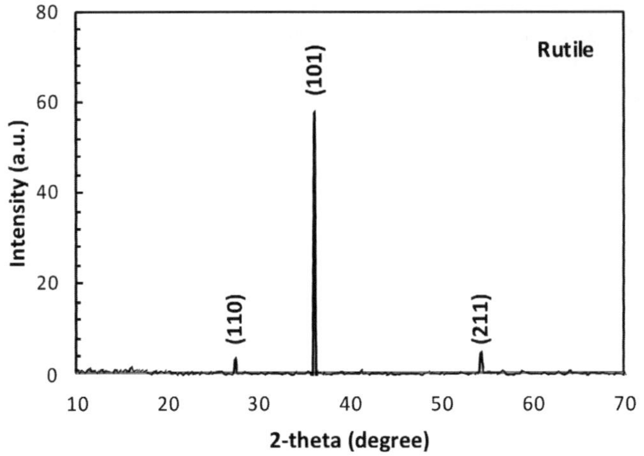

Figure 1. XRD pattern of sintered TiO_2 scaffold at selected sintering temperature, 1500oC.

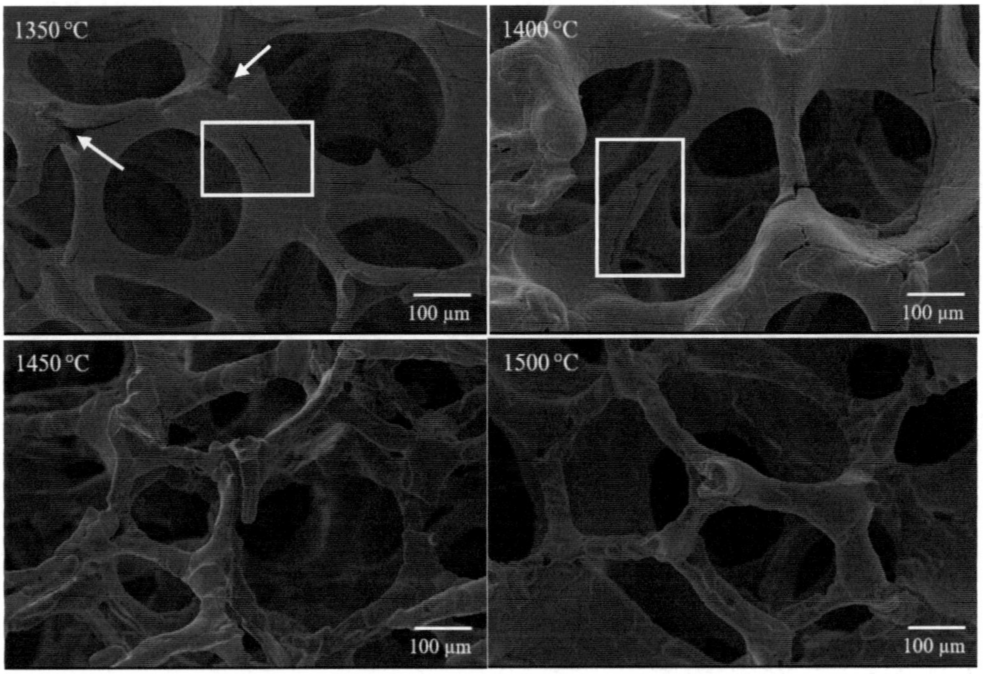

Figure 2. The effect of sintering temperatures on the microscopic appearance of TiO_2 scaffolds.

The pore structure of sintered TiO_2 scaffolds are showed in Figure 3. The sintered scaffolds maintained the original struts and pore shape of the template and formed highly interconnected pore structure as shown in Figure 4. For all four different scaffold sintering temperatures, the average porosity was above 92%.

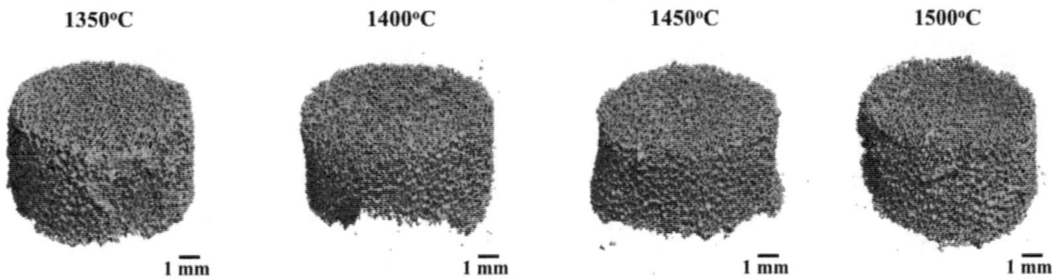

Figure 3. 3D micro-CT images of the sintered TiO_2 scaffolds at different sintering temperatures.

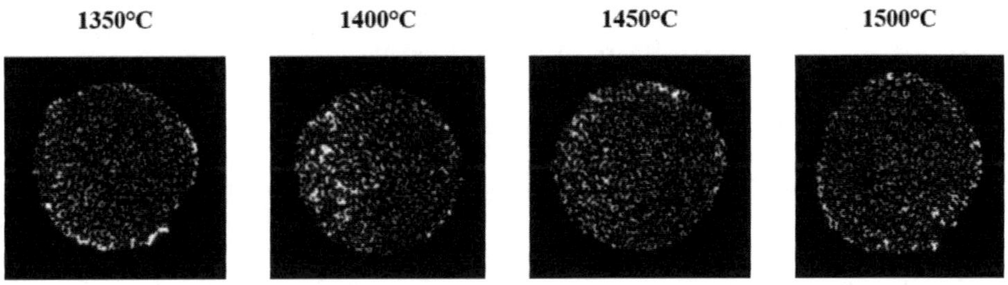

Figure 4. Micro-CT cross-section images of the sintered TiO_2 scaffolds at different sintering temperatures.

4. Conclusions

This study showed that increasing of the sintering temperatures could improve the microstructural uniformity of TiO_2 scaffolds which fabricated by sponge replication method. The desired pore architecture was also controlled as shown in well-interconnectivity and high porosity as high as 92%, which is good for promoting of bone cell proliferation. The crucial point that should take into account in the next step is cytotoxicity testing of the as-prepared scaffolds to bone cell. Moreover, the strength of the scaffolds might be improved by the optimization of the sintering temperatures, soaking time, multi-step dipping in the slurry. However, improvement in other processing parameters may also allow further enhancement in mechanical strength in order to produce machinable custom-made scaffolds.

5. References

[1] Klemm A and Tiainen H 2018 *Cer. Int.* 44 6265.
[2] Narkevica I, Stradina L, Yakushin V and Ozolins J 2015 *Mat. Sci. and App. Chem.* 31.
[3] Haugen H, Will J, Kohler A, Hopfner U, Aigner J and Wintermantel E 2004 *J. of the Euro. Cer. Society* 24 661.
[4] Tiainen H, Wiedmer D and Haugen H 2013 *J. of the Euro. Cer. Society* 33 15.

IOP Conf. Series: Materials Science and Engineering **773** (2020) 012034 doi:10.1088/1757-899X/773/1/012034

A development of automatic builder machine for retreaded tire manufacturing

C Kamyat[1], C Raksiri[1] and R Masakasin[1]

[1]Department of Industrial Engineering, Faculty of Engineering, Kasetsart University, Bangkok 10900, Thailand

*Corresponding author: chana_raksiri@yahoo.com

Abstract. Tread building is one of tire retreading process that used tire from buffing process to formed a retreaded tire. Building process was required high expertise worker because most builder machine required worker to work with machine in any point of process. For this reason, it is difficult to control processing time and quality of products. The objective of this research is to develop an automatic builder machine in retreaded tire manufacturing which is available with a variety of size of rubber tires and reduce the workforce. The results of this research are composed of the design and installation of automatic tire loading system, tread loading system, tire rotating and tire pressing system. The results showed that building cycle time of proposed automatic builder machine is average 8 min per tire and improved the repeatability. Another advantage of proposed system was related with consistent quality. It also affects to the number of workers and make fewer operated machines. Since it is produced by automation system, it can transmit the production information to next operation and can support the strategy of Thailand 4.0.

1. Introduction

Tread building is one operation of tire retreaded process [1-2] for removing old tread from buffing process and then take new tread pattern to build it back on the used tire. This process requires more expertise of worker because the existing machine is a manual machine. The operation requires people to help almost in every steps such as attach new tread on used tire, align the new tread to center while rolling and press the new tread on used tire (see in Figure 1).

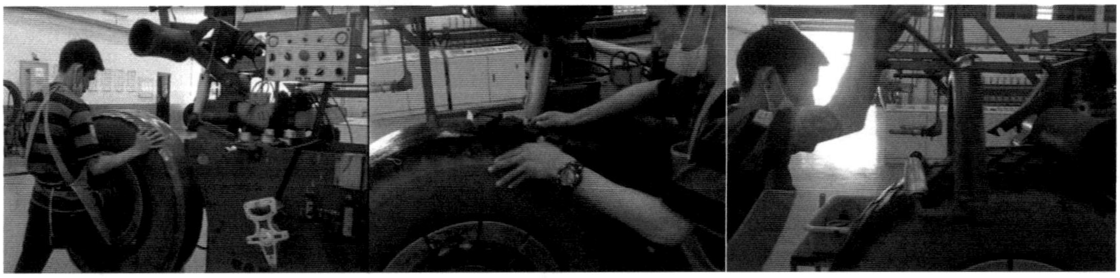

Figure 1. Traditional building process in case study.

Due to traditional operations, it is difficult to control production time and quality of the tire. This operation takes about 10-30 min per piece. In addition, force that used to press on tire cannot be known or controlled as well because the machine does not have a sensor to check or control it. These problems lead to concept of the development of automatic builder machines. The development of automatic tread builder machine consists of tire loading/unloading system, tread pattern adjusting and mounting system and tread pressing system on the tire. The control system (PC Base Control) allows us to easier use and can also link data to various parts. In addition, the system can be developed to produce, inspect and track in real time. It can be linked to the previous process or the next process as well.

2. Experimental

Machine design as following steps: (1) mechanical design, (2) electrical design, (3) building and assembling mechanical parts, (4) building and assembling electrical part, (5) developing program and control system and (6) testing and adjusting system. Overall mechanical design of builder machine is shown in Figure 2.

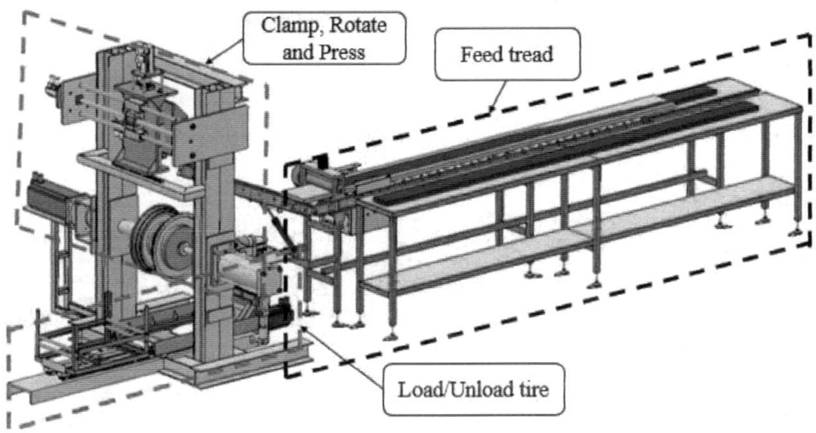

Figure 2. Mechanical design of builder machine.

From figure 2, there are three main sections of this concept design. First section is load/unload section for loading/unloading tire in/out machine. Next section is clamp, rotate and press tire for building tread on tire. Last section is load tread for feeding tread pattern to attach on tire in building section. All sections have axis motion to adjust position of tire, tread and pressing system for working with various sizes. Electrical design and machine control system include axis motion control for each section, sensor and wiring system for connecting all electrical devices (Figure 3).

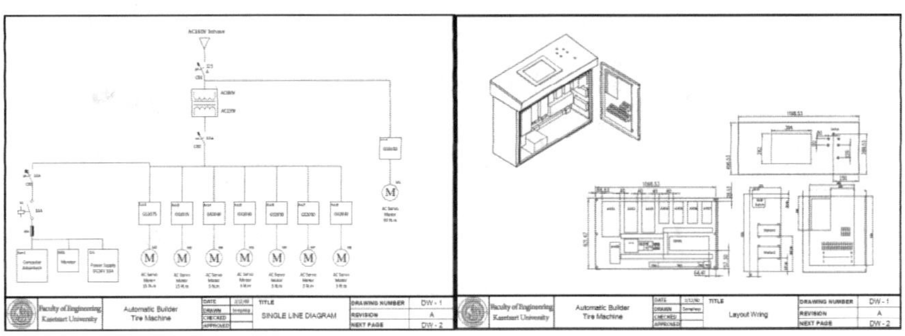

Figure 3. Electrical design of builder machine.

The development of control program design for human interface consists of two main windows. First window is manual mode that operator can control or move each mechanical motion independent such as servo motor and air cylinder. Second window is auto mode that operator can select model of tire and start/stop machine operation (Figure 4). The auto mode is separated into semi-automation and automation. Semi-automation: machine will be working step by step. Then after finish each step, machine will be waiting operator to start run for next step. For Automation: machine will be working all step.

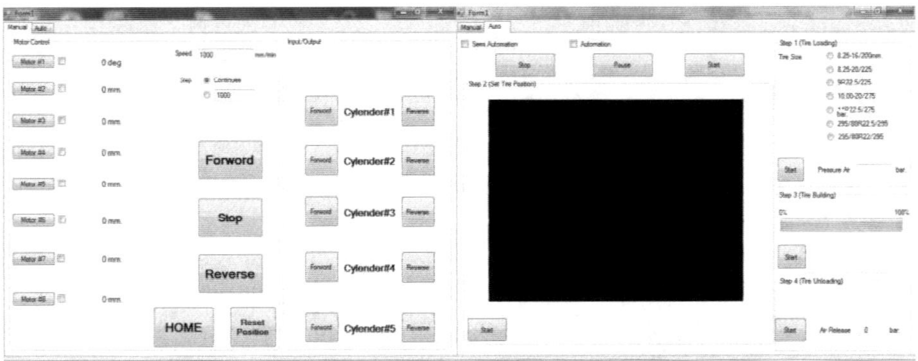

Figure 4. Program windows for builder machine.

3. Results and discussion

All machine has been tested with the sample truck tire model 10R20. The operation of machine was working as following steps: (1) Load tire to clamp and inflate, (2) Check position to adjust tread and press head to tire center, (3) Feed tread to attach on tire, (4) Press and roll across the surface of tread and (5) Deflate and unload tire off. Each tire model required the parameters which include tire rotation speed, tread feed and press load. The result of functional testing is shown in Figure 5. The operating time from start to end process was around 8 minutes.

Figure 5. Functional testing of builder machine.

4. Conclusions

The objectives of this research are to increase number of tires retreaded per day, to reduce time and cost in process and to improve quality of tire. The result of machine testing with truck tire model 10R20 showed that the processing time was around 8 minute per piece. In addition, the machine can attach tread pattern in the center of wheel and can provide consistent pressure throughout the tire. In term of manufacturing process, the machine can reduce workforce in process which a worker can operate more than one machine. Also, there is information that obtained from machine which can be used to link with other machine in processing line or can be used to improve processing plan.

5. References

[1] https://www.thailandautocars.com/ยางหล่อดอกคืออะไร; Jun 2017.

[2] https://www.piedmonttrucktires.com/retreads.html;

Acknowledgement

This research was funded from The Thailand Research Fund (TRF). For data collection, material and knowledge was supported by Bumrung Yang LTD., Partnership. For facilities and staffs to machine testing was supported by Thai R&D Solutions Co., Ltd. And finally, thank you to those involved persons that not mentioned for together developed successfully in this research.

Application of Kawabata evaluation system for the tactile properties of woven silk fabrics in textile industry

P Phoophat[1,*], P Kumphai[1], S Suwonsichon[2], J Boonyarit[3], C Plangmon[1], and R Chollakup[3]

[1] Department of Textile Science, Faculty of Agro-Industry, Kasetsart University, Bangkok, 10900 Thailand.
[2] Kasetsart University Sensory and Consumer Research Center,
Department of Product Development, Faculty of Agro-Industry, Kasetsart University, Bangkok, 10900 Thailand.
[3] Kasetsart Agricultural and Agro-Industrial Product Improvement Institute, Kasetsart University, Bangkok, 10900 Thailand.

*Corresponding author: pithalai.p@ku.th

Abstract. Thailand is one of well-known countries for the silk fabric production. However, there are just few studies focusing on tactile properties of silk fabrics. This study collected forty-one machined-woven silk fabrics manufactured in Thailand, Japan, China, Korea, Italy, France, India and USA. The fabrics were classified into 4 groups categorized by the fabric weight according to the "silk fabric" industrial product standard of the Thai Industrial Standards Institute (TISI) 176-2540 ($\leq$ 90 g/m^2, 91-120 g/m^2, 121-160 g/m^2, >160 g/m^2). Mechanical properties of the silk fabrics were measured by the Kawabata evaluation system for fabric (KES-F). Correlations between fabric weight groups were analyzed to investigate the relationship among properties. The results showed significant effect of the fabric weight (per unit area) on only compression resilience (RC) and thickness. The thickness was also correlated with the compression properties (WC, RC), the bending rigidity (B) of the warp direction and the surface roughness (SMD). Moreover, the air permeability was associated with the coefficient of friction (MIU), the shearing properties (G, 2HG, 2HG5) and the tensile properties of the weft direction (EM, LT, WT). The total hand value (THV) was calculated by using the Kawabata hand evaluations KN-201-LDY for women's thin dress fabric. The fabrics with the THV scores above 4 showed high quality for women's dresses. The principal component analysis (PCA) was conducted to examine the relationship between the mechanical properties and the fabrics. The results showed that the high-quality fabrics were plotted near each other as a group and related to some properties such as the coefficient of friction (MIU), the extensibility (EM), the tensile energy (WT) of the weft direction and the linearity of compression (LC). Finally, the biplot graph of the fabric distribution can be a guideline for a variety of applications. The fabrics for apparels mainly located in the lower-left quadrant while those located in the upper quadrant were appropriate for home textiles.

1. Introduction

Thailand has been famous for producing silk with good quality. Thai silk is recognized as one of the symbols of Thailand by people from many nations especially USA, Japan, European countries and etc. The sensory attraction both visual and tactile influences consumers' purchase decision. The sensory analysis is one of the methods suitable for assessing the tactile properties of fabrics [1-5]. Kawabata and Niwa developed a reliable instrument named Kawabata's Hand Evaluation System for Fabric (KES-F) [6,7] with the aim of consuming less time and expenses than when using the sensory analysis. KES-F would conduct the fabric texture measurement, and, the results would be the substitute for human perception. This substitution reliability was confirmed by some research which examined the correlation between the textile perception measured by subjective and objective measurements [3, 8-10]. Since the KES-F had been generally utilized as a measurement in textile, this research studied the relation between the physical and the mechanical properties of silk fabrics by using KES-F and calculating Total hand value [6] in order to identify certain silk fabrics with qualities suitable for apparels or for other applications.

2. Experimental

This study collected 41 machine-woven silk fabrics from various leading sources of production (23 samples from Thailand, 5 samples from Japan, 5 samples from China, 3 samples from Korea, 2 samples from Italy, 1 sample from France, 1 sample from India and 1 sample from USA). The 41 silk fabrics were classified into 4 groups, according to the silk fabric industrial product standard of the Thai Industrial Standards Institute (TISI) 179-2540, which were categorized by the fabric weight per area unit. As shown in Table 1, Group A was consisted of 19 very thin samples (weight per area unit ≤ 90 g/m_2). Group B contained 9 thin samples (weight per area unit = 91-120 g/m_2). Group C stood for 7 thick samples (weight per area unit = 121-160 g/m_2). Group D was of 6 very thick samples (weight per area unit >160 g/m_2). The specification of samples is shown in Table 1.

The mechanical properties of fabrics were measured at 20°C and 65% relative humidity by using KES-F (Kato Tech Co., Ltd.). The mechanical properties were as follows. First, the tensile properties were consisted of the fabric extension at 500 N/m width (EM, %), the linearity of load-extension curve (LC), the tensile energy (WT, J/m_2) and the tensile resilience (RT, %), which were tested by KES-FB1. Second, the bending properties were consisted of the bending rigidity (B, μNm) and the hysteresis of bending moment (2HB, mN), which were tested by KES-FB2. Third, the shear properties were consisted of the shearing stiffness (G, N/m), the hysteresis of shear force at a shear angle of 0.5° (2HG, N/m) and the hysteresis of shear force at a shear angle of 5° (2HG5, N/m), which were tested by KES-FB1. Fourth, the compression properties were consisted of the linearity of compression-thickness curve (LC), the compression energy (WC, J/m_2) and the compression resilience (RC, %), which were tested by KES-FB3. The Last conditions were the surface properties consisted of the coefficient of friction (MIU), the mean deviation of MIU (MMD) and the geometrical roughness (SMD, μm), and tested by KES-FB4.The air permeability was measured by the air permeability tester following the method of ASTM D737. The q_{max} was measured by KES-F7.

The results were analyzed by using the correlation analysis with SPSS version 12.0 to see the correlations between the silk fabric weight, the thickness, the air permeability and all the mechanical property values from KES-F. To plot a graph exhibiting the relations between the silk fabric samples and the physical and the mechanical properties, Principal Component Analysis (PCA) was conducted by utilizing the XLSTAT-Sensory statistical software 2017 (Add in soft, New York, USA). After that, Total Hand Value (THV) and the Kawabata hand evaluation KN-201-LDY for women's thin dress fabric played the role to evaluate propriety of certain silk fabrics which should be made thin dresses for women.

3. Results and discussion

According to the 41 silk fabrics, the q_{max} and the 5 mechanical properties (the tensile properties, the bending properties, the shear properties, the compression properties and the surface properties) were measured and analyzed by using the correlation analysis to find the correlations between the silk fabric weight, the thickness and the air permeability. The results revealed that the silk fabric weight was significantly related only to the compression resilience (RC) and the thickness ($P<0.05$) (with Pearson's correlation coefficients: r = -0.424 and -0.386). Moreover, the silk fabric thickness was significantly related to various mechanical properties: WC (0.708), RC (0.613), B-warp (0.600), HB-warp (0.540), SMD-warp (0.705) and SMD-weft (0.580) ($P< 0.01$). Meanwhile, the air permeability was significantly related to various mechanical properties: LC (0.558), MIU-warp (0.635), G-warp (-0.664), G-weft (-0.659), 2HG-warp (-0.553), 2HG-weft (-0.487), 2HG5-warp (-0.599), 2HG5-weft (-0.572), EM-weft (0.680), LT-warp (-0.604), LT-weft (-0.765), WT-weft (0.670) ($P< 0.01$).

Table 1. Sample specifications of four categories of fabric weight

Fabric Sample	Structure	Made in	Export to	Weight (g/m_2)	Thick-ness (mm)	Air permeability ($cm_3/cm_2/s$)	THV
A01	Plain	Japan	Japan	18.71	0.20	-	3.261
A02	Plain	Japan	Japan	25.10	0.22	-	4.021
A03	Plain	Japan	Japan	37.41	0.11	88.73	1.228
A04	Plain	China	USA	63.95	0.14	71.35	3.698
A05	Plain	Japan	Japan	65.34	0.09	8.07	2.709
A06	Twill 2x1	Thailand	Japan	68.63	0.15	60.47	4.486
A07	Plain	Korea	USA	69.76	0.17	56.67	4.344
A08	Satin	Japan	Japan	69.80	0.09	56.67	4.529
A09	Plain	Thailand	Eastern Europe	70.82	0.13	8.37	3.991
A10	Plain	China	USA	73.37	0.14	25.77	3.214
A11	Plain	China	USA	74.81	0.15	34.07	3.452
A12	Plain	China	USA	75.07	0.16	8.03	4.514
A13	Plain	Thailand	Eastern Europe	76.37	0.13	19.70	3.577
A14	Plain	Thailand	Eastern Europe	78.53	0.15	7.44	3.719
A15	Herringbone Twill	France	USA	79.84	0.15	8.62	4.406
A16	Jacguard	Thailand	Britain	80.03	0.13	49.20	2.592
A17	Plain	Korea	USA	82.01	0.16	2.62	3.594
A18	Plain	Thailand	Eastern Europe	88.16	0.14	0.93	3.632
A19	Broken Twill	Thailand	France	90.41	0.19	7.45	4.147
B01	Plain	Thailand	USA	94.93	0.21	8.00	3.679
B02	Plain	Thailand	Eastern Europe	101.71	0.17	1.97	4.169
B03	Twill 2x1	Thailand	France	101.87	0.20	8.46	3.805
B04	Plain	Thailand	Eastern Europe	103.47	0.17	1.46	4.119
B05	Plain	Italy	USA	104.68	0.20	2.07	2.592
B06	Plain	China	USA	104.78	0.27	66.17	4.145
B07	Plain	India	USA	105.58	0.20	1.94	3.731
B08	Plain	Thailand	Eastern Europe	106.65	0.23	6.24	3.285
B09	Plain	USA	USA	109.18	0.20	2.24	3.993
C01	Plain	Thailand	Britain	127.27	0.17	1.70	2.750
C02	Twill 2x1	Thailand	Norway	129.37	0.32	35.97	4.353
C03	Plain	Thailand	Eastern Europe	131.79	0.25	6.33	2.376
C04	Twill 2x1	Thailand	Eastern Europe	135.30	0.23	2.00	4.448
C05	Plain	Thailand	Eastern Europe	139.33	0.26	2.69	2.979
C06	Twill 2x1	Thailand	Eastern Europe	139.80	0.24	7.76	4.155
C07	Twill 2x1	Thailand	Eastern Europe	142.66	0.21	1.03	4.580
D01	Plain	Korea	USA	173.41	0.39	57.34	4.135
D02	Dobby	Thailand	Hongkong	191.44	0.38	1.58	3.654
D03	Twill 2x1	Italy	USA	229.84	0.42	55.90	4.511
D04	Dobby	Thailand	France	231.65	0.50	6.52	3.829
D05	Plain	Thailand	Canada	267.43	0.63	7.28	3.993
D06	Plain	Thailand	Eastern Europe	366.95	0.79	9.07	3.170

After analyzing by using Principal Component Analysis (PCA), it was shown in the biplot graph (Figure 1.) that the relations between the silk fabric weight and the mechanical properties were not strong. According to Dimension 2, most of the heaviest silk fabrics in Group C and Group D were plotted on the upper half of the graph, which showed the relations to the thickness, the compression properties, the bending properties and the surface roughness. Dimension 1 represented the silk fabrics with different the tensile properties, the shear possessions and the surface friction. Dimension 3 linked the tensile properties in warp direction and the surface roughness

After calculating the mechanical property values to find Total Hand Value (THV) for thin dresses for women by using the Kawabata hand evaluation KN-201-LDY for women's thin dress fabric, it was found that the silk fabrics with good quality (scored above 4) appropriate for being made dresses for women were the silk fabric A02, A06, A07, A08, A12, A15, A19, B02, B04, B06, C02, C04, C06, C07, D01 and D03. These fabrics with good quality were related to the MIU, EM, WT and LC values. Meanwhile, the thick and heavy silk fabrics plotted on the upper part or the right axis were related to the shear properties and suitable for home textiles.

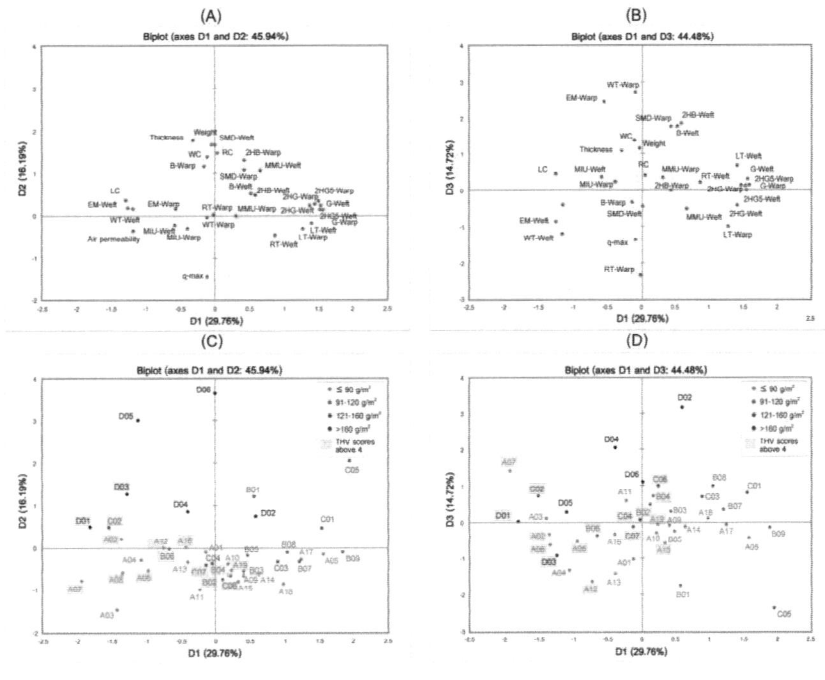

Figure 1. Biplot graph by using Principal Component Analysis indicating correlations between (A,B) the physical and the mechanical properties and (C,D) silk fabric samples categorized by silk weight.

4. Conclusions

The association between the silk weight and the mechanical properties was not explicit. The thickness and the air permeability correlated with the mechanical properties. The thickness correlated with the compression properties, the bending properties of the warp direction and the surface roughness. The air permeability was associated with the surface friction, the shearing properties and the tensile properties of the weft direction. The mechanical properties can predict the quality of silk fabrics suitable for different usages.

5. References

[1] Howorth WS and Oliver PH 1958 *J Text Inst*, 49 (11): 540-553.

[2] Brand RH 1994 *Text Res J*, 34: 791-804.

[3] Cardello VC, Winterhalter A and Schutz HG 2003 *Text Res J*,73(3): 221-237.

[4] Hui CL, Lau TW, Ng SF and Chan KC 2004 *Text Res J,* 74(5): 375-383.
[5] Chollakup R, Sinoimeri A, Philippe F, Schacher L and Adolphe D 2004 *Int J Cloth Sci Tech*, 16: 132-140.
[6] Kawabata S 1980 *The Textile Machinery Society of Japan*.
[7] Kawabata S and Niwa M 1991 *Int J Cloth Sci Tech*, 3(1), 7-18.
[8] Postle R and Dhingra RC 1989 *Text Res J*, 59: 448-459.
[9] Raheel M and Liu J 1991 *Text Res J*, 61: 31-38
[10] Alimma D, Matsuo T, Nakajima M and Takahashi T 2000 *Text Res J*, 70 (11): 985-990.

Acknowledgement
The authors are grateful to Kasetsart University Research and Development Institute (KURDI) for the financial support and Specialized Center of Rubber and Polymer Materials in Agriculture and Industry (RPM), Faculty of Science, Kasetsart University for the presentation support.

IOP Conf. Series: Materials Science and Engineering **773** (2020) 012036 doi:10.1088/1757-899X/773/1/012036

A wide energy range neutron shielding material based on natural rubber and boron

Chayanit Jumpee[1,*], Tithinun Rattanaplome[2] and Natthaporn Kumwang[3]

[1]Department of Radiologic Technology, Faculty of Associated Medical Sciences, Chiang Mai University, Chiang mai 50200, Thailand
[2]Faculty of Engineering and Agro-Industry, Maejo University, Chiang Mai, 50290 Thailand
[3]Nuclear Technology Service Center,Thailand Institute of Nuclear Technology (Public Organization), Ongkharak, Nakorn Nayok, 26120 Thailand

*Corresponding author: chayanit.j@cmu.ac.th

Abstract. Neutron have been used in many applications. Neutrons have no electric charge and have high penetrating power. Therefore, the use of neutron shielding material is one of the methods to effectively prevent neutron radiation hazard. In this work, neutron shielding materials were designed using the Monte Carlo N-Particle (MCNP) transport code in order to select proper materials for a wide energy range neutron shielding. Thicknesses of 1 cm and 4 cm were chosen for shielding neutron in the wide energy spectrum (10E-8 - 100 MeV). MCNP simulation results indicated that the best shielding material was 2 layers of 2 cm natural rubber (NR) and 2 cm NR with 20 part per hundred rubber (phr) boron powder. This shielding material can reduce the neutron dose by 17.91 $\pm$ 0.09 % with the total macroscopic cross section of 0.042 cm^{-1}. The optimal neutron shielding materials were fabricated. The mechanical properties and aging properties of shielding materials was found to meet the requirement of Thai Industrial Standard (TIS 2377-2551 Rubber flooring). Therefore, the designed shielding material can be utilized to shield against wide energy range neutrons.

1. Introduction

Neutron is widely used in many applications. As neutrons are high penetrative, they can directly interact with atomic nucleus of living cells. Therefore, exposure to free neutrons could be increased the risk of health problems. Neutron shielding requires materials with an atomic mass close to that of the neutron, for example, hydrogen atom since hydrogen both slows down and scatters neutron. Therefore, hydrogenous materials including water, paraffin and hydrocarbon polymer are usually used for neutron shielding. Natural rubber (NR) is a hydrocarbon polymer which contains a lot of hydrogen atoms. Besides, NR can be easily mixed with high neutron cross-section materials. The combination of NR and high neutron cross-section materials could improve neutron shielding's performance.

There were several studies to investigate the shielding performance of hydrocarbon polymer compounds. Materials composed of epoxy resin and borates mineral, for example, colemanite and ilmenite are used to shield against neutrons.(1; 2) Polyethylene/ boron compound composites are also wildly used for neutron shielding.(3; 4) There are many applications require

IOP Conf. Series: Materials Science and Engineering **773** (2020) 012036 doi:10.1088/1757-899X/773/1/012036

flexibility materials for radiation shielding, for example wrapping a neutron shielding layer around irregular shape surface of the neutron source container, NR based material which combined with boron compound materials could be applied (5; 6), however, they were competent to shield only thermal or low energy neutron and, the neutrons emitted from the sources are fast neutrons or high energy neutron which may gradually slow down by collision with the hydrogen atom. Hence, the effective neutron shielding requires a wide energy range neutron shielding materials.

2. Experimental

2.1. Material selection and Design by MCNP simulation

The Monte Carlo method was performed with Monte Carlo N Particle Version 5 (MCNP5) transport code to optimize the designs and compositions of neutron shielding materials. The shielding materials 4 cm thickness were simulated in a spherical shape because preliminary simulation exhibited that fast neutron dose rate initially decreasing. A simulated source of isotropic wide energy range neutron (1E-8 – 100 MeV) was placed at the center of the spherical shielding model. Surface flux tally (F2) with flux to dose rate conversion factors based on the International Commission on Radiological Protection (ICRP-21) was used to determine the neutron dose equivalent rate [(rem/h)/neutron] on the outer surface of the spherical model. The main compositions of each neutron shielding material were listed in Table 1. The amount of 50 phr Fe_2O_3 and/or boron were assigned for each neutron shielding sample beneficial to compare their shielding performance. Then determined the optimum amount of Fe_2O_3 and/or boron additions by simulation shielding material with varying amounts of them

Table 1. Main compositions of neutron shielding materials (50 phr each of fillers)

Sample No.	Material composition	Density (g/cm^3)	Mass ratio of element content					
			H	C	^{10}B	^{11}B	O	Fe
0	No material (Air)	0.00120	Air					
1	NR	0.9200	10.378	89.622	-	-	-	-
2	NR+Boron	1.1557	10.378	89.622	2.460	9.963	27.577	-
3	NR+Fe_2O_3	1.2678	0.789	0.5878	-	-	0.1002	0.2331
4	NR+Boron+Fe_2O_3	1.4346	0.0592	0.4408	0.0498	0.2002	0.0751	0.1748
			Two-layers sample					
	Layer1*					Layer2		
5	Sample 2 (NR+Boron)					Sample 3 (NR+Fe_2O_3)		
5.1	Sample 3 (NR+Fe_2O_3)					Sample 2 (NR+Boron)		
6	Sample 1 (NR)					Sample 2 (NR+Boron)		
6.1	Sample 2 (NR+Boron)					Sample 1(NR)		
7	Sample 1 (NR)					Sample 3 (NR+ Fe_2O_3)		
7.1	Sample 3 (NR+ Fe_2O_3)					Sample 1 (NR)		

* Layer1 is adjacent to neutron source.

2.2. Fabrication of neutron shielding

Material compositions which were the optimum compounds, that got from MCNP simulation were fabricated by semi-efficient vulcanization system including with appropriate rubber chemicals. All compositions were mixed in a two-roll mill, then an electrically heated hydraulic mold at 150±2 °C was used to vulcanize and fabricate the rubber compound into slab shape.

2.3. Mechanical properties test

All vulcanizing materials both before and after aging at 70 °C for 72 h were investigated the mechanical properties consisting of hardness (Durometer type A: ASTMD2240) and tensile properties (ISO37) that consist of Tensile strength and Elongation at the break.

3. Results and discussion

3.1. MCNP simulation

The MCNP simulation results showed that the sample No.6 (2 layers) exhibited the best neutron shielding efficiency as shown in **Figure 1.** The initial neutron intensity remining after being attenuated with shielding material was represented by % neutron transmission which its small value is considered as excellent shielding performance. Although both of sample No.6 and No.6.1 were made by the same compositions, their layers arrangement was difference. **Figure 2.** Illustrated neutron dose equivalent rate spectra. In fast neutron region (above 1E-2 MeV), neutron dose rate of all samples was similarly attenuated. This was due to the fast neutron slowing down by the elastic scattering with hydrogen atoms which contained in NR, then thermal neutron was produced. For sample No.6 as shown by pink dash-dotted line (**Figure 2.**), the 2 cm layer1 of NR which was adjacent to neutron source works as both fast neutron moderator and thermal neutron attenuator because there are the highest hydrogen density, then lower neutron energy as thermal neutron (1E-8 – 1E-2 MeV) was effectively absorbed by boron content which contained in another 2 cm layer2 of NR+Boron. On the other hand, when material of NR+boron become to layer1 as the sample No. 6.1, boron cannot properly absorb neutrons as exhibited by the blue dashed line in **Figure 2.**

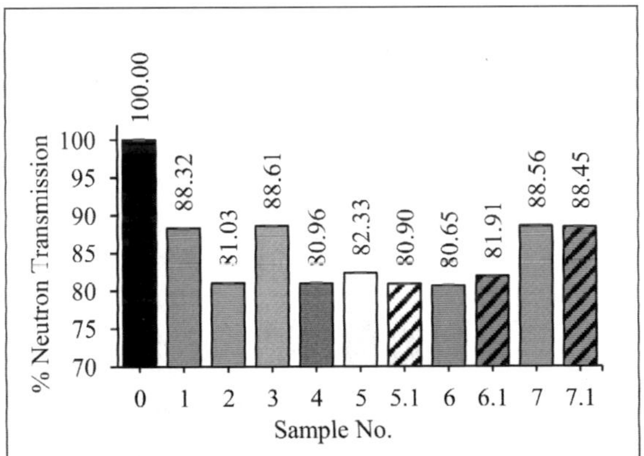

Figure 1. % Neutron transmission (I/I_0) of neutron dose equivalent rate.

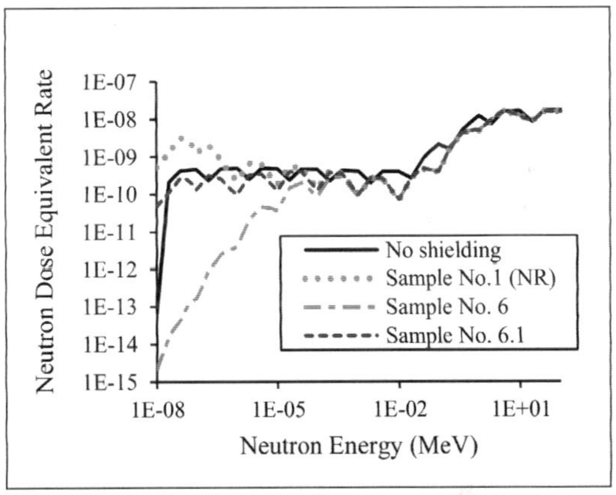

Figure 2. Neutron dose equivalent rate spectra for various materials.

A simulation of sample No.6 with varying boron amounts of layer2 was carried out to determine the optimized amount of boron. Result reveals that the most appropriate boron amounts were 20 phr as illustrated in **Figure 3.**, it was exhibited that as the amount of boron added to layer2 of sample No. 6 increased, the neutron shielding performance was not increase due to hydrogen atom plays a key role in a wide energy neutron shielding more than that of boron that can only shield against thermal neutron. From MCNP calculation of neutron attenuation, the 4 cm shielding material No.6 having 2 layers of NR and NR with 20 phr boron can reduce the neutron dose by 17.91 ± 0.09 % with the total macroscopic cross section of 0.042 cm^{-1}.

3.2. Mechanical properties test

From MCNP simulation, the optimized neutron shielding designs were fabricated into 2 mm slab sheets. Because of the neutron shielding material unnecessary to stretch and impact, the standard mechanical properties of rubber flooring was considered. The results of mechanical properties are shown in **Table 2**, Both layers, the elongation at break value were more than 150% and the difference of tensile strength and hardness values of shielding materials between before and after heating (aging) tests were less than 25% and 5, respectively which found to meet the requirement of Thai Industrial Standard (TIS 2377- 2551 Rubber flooring)

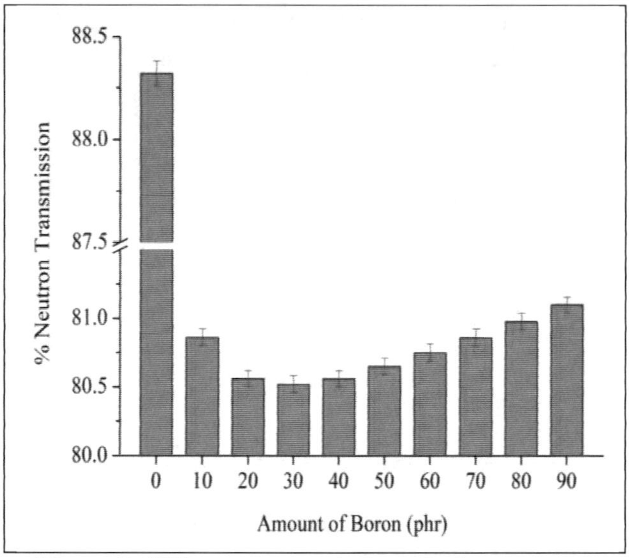

Figure 3. %Neutron transmission of sphere neutron shielding at different amount of boron contents (phr)

Table 2. Mechanical properties of NR and NR+20 phr boron before and after aging at 70 °C for 72 h.

Properties	Material					
	NR of Layer1			NR + 20 phr boron of Layer2		
	Before Aging	After aging	Difference	Before Aging	After aging	Difference
Tensile strength (MPa)	22.9±1.6	25.3±1.1	+10.5%	24.5±2.6	25.2±1.4	+2.8%
Elongation at break (%)	645±14	641.0±12	-0.6%	727±11	712.0±16	-2.1%
Hardness (Shore A)	36.8±0.5	39.4±0.3	+2.6	33.3±1.3	35.8±0.5	+2.5

4. Conclusions

The optimized neutron shielding materials were designed using the MCNP transport code. For 4 cm thickness of shielding material, 2 layers of NR and NR with 20 phr boron powder (2 cm each) was the most appropriate for a wide energy range of neutron shielding. The designed shielding materials were well fabricated, and their mechanical properties meet the requirement of TIS 2377-2551 Rubber flooring. The thickness of shielding material should be considered to achieve a desired attenuation level.

Acknowledgement

Authors would like to acknowledge financial supports from the Thailand Research Fund under Grant Number RDG6T0126.

References

[1] Okuno K. 2005. *Radiation protection dosimetry* 115:258-61
[2] A. El-Sayed Abdo MAE-S, F.A. Gaber. 2003. *Annals of Nuclear Energy* 30:175–87
[3] Harrison C, Weaver S, Bertelsen C, Burgett E, Hertel N, Grulke E. 2008. *Journal of Applied Polymer Science* 109:2529-38
[4] Shin JW, Lee J-W, Yu S, Baek BK, Hong JP, et al. 2014. *Thermochimica Acta* 585:5-9
[5] Gwaily S, Badawy M, Hassan H, Madani M. 2002. *Polymer testing* 21:129-33
[6] Gwaily S, Hassan H, Badawy M, Madani M. 2002. *Polymer testing* 21:513-7

IOP Conf. Series: Materials Science and Engineering **773** (2020) 012037 doi:10.1088/1757-899X/773/1/012037

Mechanical characteristics of airless tyre by laboratory testing

R Rugsaj[1,2], and C Suvanjumrat[1,2,*]

[1] Department of Mechanical Engineering, Faculty of Engineering, Mahidol University, Salaya, Nakhon Pathom, 73170, Thailand.
[2] Laboratory of Computer Mechanics for Design (LCMD), Department of Mechanical Engineering, Faculty of Engineering, Mahidol University, Salaya, Nakhon Pathom, 73170, Thailand.

*Corresponding author: chakrit.suv@mahidol.ac.th.

Abstract. This paper aimed to study the mechanical characteristics of non-pneumatic tyre (NPT) or airless tyre upon usage and testing. The NPT was recently invented to replace traditional pneumatic tyre subjected to specific applications while eliminate the risk of flatten while using or periodically air pressure maintenance. For that purpose, the NPT must exhibit desired tyre characteristics such as load carrying capacity, vertical stiffness, contact pressure, and rolling. The tyre mechanical characteristics can be observed by mean of experimental tests in laboratory. In this article, the commercial NPT Tweel by Michelin was tested using vertical tyre testing machine and drum testing machine. The vertical stiffness and footprint were obtained from tyre testing machine, which the latter was obtained using the pressure measurement film. The results were then observed and discussed. The vertical stiffness of tested NPT was compared with the estimated value of traditional pneumatic tyre, in which the same size and appropriate inflation pressure were used. The rolling characteristic of NPT was observed from drum testing machine. The rolling resistance at various load and rolling speeds and was recorded. In addition, the high speed video camera was set up to capture NPT spoke deformation during rolling, while the image processing technique was used to collect the coordinate of spoke position at various time. The results were discussed and can be useful in development of NPT tyre in the near future.

1. Introduction

The tyres are an important component of the vehicle. However, the conventional pneumatic tyres have several disadvantages since the tyre possesses the risk to undergo severed damage or flatted if the tyre traversing across various types of critical obstacles [1]. The non-pneumatic tyre (NPT) or airless tyre was recently developed to overcome these disadvantages. The NPT was designed to exhibit desired characteristics of the pneumatic tyres without requirement of the inflation pressure [2]. The first available commercial NPT is TWEEL developed by Michelin. The Tweel consists of two main components which are 1) circular shear beam and tread 2) elastic spokes which connected to a hub. The NPT can be designed to have required vertical stiffness and contact pressure upon spoke geometry and material without limit to size and pressure, unlike traditional pneumatic tyre. The NPT's lateral stiffness which affects handling and cornering can also be independently optimized with vertical stiffness which affects the riding comfort [3]. The pneumatic tyres are required to exhibit some performance parameters to meet desired characteristic. To name a few, the interested tyre parameters are including load carrying capacity, vertical stiffness, contact pressure, and rolling resistance. Design of NPT is needed to match the pneumatic tyre performances, which emphasised on mechanical behaviour and characteristics, such as load carrying capacity, vertical stiffness, contact pressure, and rolling resistance [4]. In order to achieve those objectives, the in-depth study and discussion regards testing and analysis on NPT performances is still required. The tyre mechanical characteristics can be observed by mean of experimental tests in laboratory. This paper aimed to study the mechanical characteristics of non-pneumatic tyre (NPT) or airless tyre upon usage and testing. The commercial NPT Tweel developed by Michelin was tested using vertical tyre testing machine and drum testing machine. The results were discussed and can be useful in development of NPT tyre in the near future.

2. Experimental

The commercial NPT Tweel 12N16.5 SSL ALL TERRAIN airless radial tyre developed by Michelin (Fig. 1) is selected to study the mechanical behaviour and characteristic of NPT. The vertical stiffness of NPT is tested by tyre testing machine Ektron PL-2003 as shown in Fig. 2 (a). The footprint is also collected using the same machine and the pressure measurement films, Prescale LLLW PS 270x220 5S-E, as shown in Fig. 2 (b). The NPT was then tested at 5 positions along the circumference, the results were then averaged. The vertical load of 14, 16, 18, 20 and 22 kN were used to test the vertical stiffness of NPT along with the footprint. The drum testing machine, Kayton DTM-350MS, developed by Kayton Industry Co., Ltd. (Fig. 3 (a)) is used to test rolling characteristic and rolling resistance of NPT. In addition, the high-speed video camera is set up to capture the deformation of spokes while the NPT is rolling (Fig. 3 (b)). The NPT was rotated at various speed including of 11, 12, 13, 14 and 15 km/h while the load was fixed at 14 kN. On other testing condition, the load was varied at 14, 16, 18, 20 and 22 kN while the speed was fixed at 11 km/h. The rolling resistance value and spoke deformation behaviour of rolling tyre were collected.

Figure 1. NPT Tweel 12N16.5 SSL ALL TERRAIN model by Michelin.

(a) (b)

Figure 2. (a) Tyre testing machine Ektron PL-2003 (b) The mounting of NPT onto the tyre testing machine.

(a) (b)

Figure 3. (a) The drum testing machine Kayton DTM-350MS developed by Kayton Industry Co., Ltd. (b) The instalment of the high-speed video camera onto drum testing machine.

3. Results and discussion

The force-displacement relationships obtain from vertical stiffness testing results at different loads is shown in Fig 4. The vertical stiffness of NPT could be calculated from using force-displacement relationship is equal to 869.93 N/mm. The expected vertical stiffness value of pneumatic tyre can be evaluated using simplified Rhyne's equation [1] as expressed by Eq. 1:

$$K_z = 0.00028P\sqrt{W \times OD} + 3.45 \tag{1}$$

where K_z is vertical stiffness, P is inflation pressure, W is tread or footprint width and OD is overall diameter. The vertical stiffness of NPT is compared to vertical stiffness of pneumatic tyre, which is calculated using Eq. 1 based on the same overall dimension under a range of inflation pressure as shown in Fig. 4 (b). From the results, it could be seen that NPT have higher vertical stiffness than conventional pneumatic tyre. The rolling resistance coefficient at various testing speed and load is shown in Fig. 5 (a) and Fig. 5 (b), respectively. In addition, the deformation of spoke at various position from the hub centre is digitized by image processing as shown in Fig. 6 (a) according to rolling angle positions as shown by schematic diagram in Fig. 6 (b). It should be noted that the spoke deformation are found to be in tension and compression at the upper and lower parts of the spoke structure, respectively. The testing results could explain the function of spokes to distribute the load along all portion of spoke. Therefore every spokes are deform, not just only the lower portion of spoke, which is directly in contact with the ground.

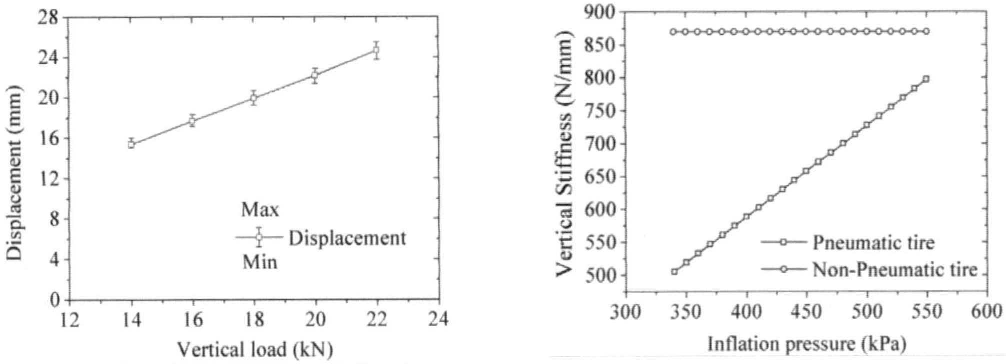

(a) (b)

Figure 4. (a) The force-displacement relationships of NPT obtained from vertical stiffness testing (b) The mounting of NPT onto the tyre testing machine.

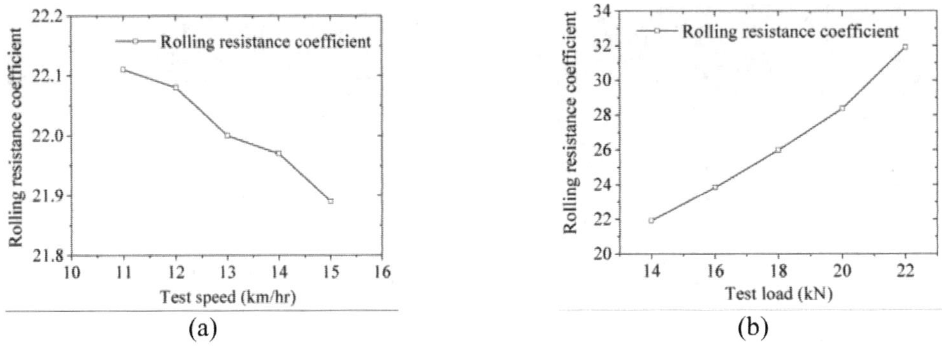

Figure 5. Rolling resistance coefficient (a) at various test speed (b) at various test load.

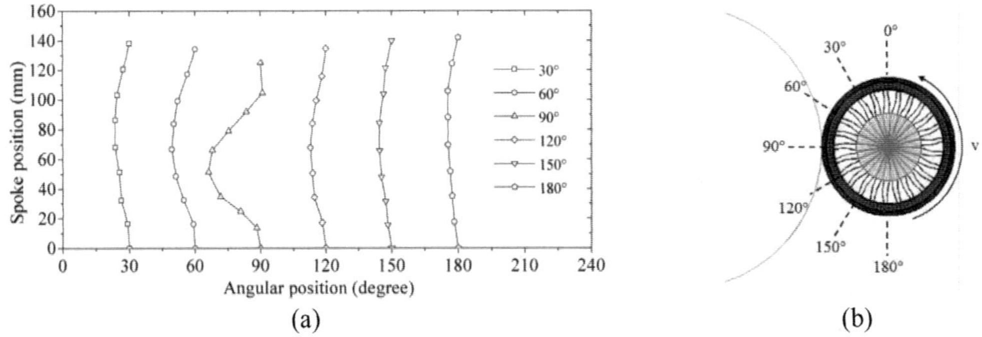

Figure 6. (a) The spoke deformation (b) Schematic diagram of angular recording position.

4. Conclusions

In this research, the mechanical characteristics of NPT or airless tyre were studied using the tyre testing machines. The force-displacement was obtained by a tyre testing machine. The vertical stiffness was achieved at 869.93 N/mm. It can be concluded that NPT has higher vertical stiffness, which is directly related to load carrying capacity, than conventional pneumatic tyre based on the same size. The rolling resistance value and spoke deformation of rolling NPT on the drum testing machine were collected, which the latter was recorded by a high speed camera. The spoke deformation are found to be in tension and compression at the upper and lower parts of the spoke structure, respectively. This emphasised the function of spokes to share the force distribution on all spokes. This knowledge is very useful in designing the NPT structure in the near future. It can concluded that a good NPT design should exhibit the function of spoke geometry to distribute the load on all portion of the spoke structure.

5. References

[1] Gent AN, Watler JD. The pneumatic tyre. Washington DC: National highway traffic safety administration; 2006.
[2] Rhyne TB, Cron SM. Development of a Non-Pneumatic Wheel. Tyre Sci Technol 2006; 34(3):150-69.
[3] Gasmi A, Joseph PF, Rhyne TB, and Cron SM. Development of a two-dimensional model of a compliant non-pneumatic tyre. Int J Solids Struct 2012; 49(13):1723-40..
[4] Jang IG, Sung YH, Yoo EJ, Kwak BM. Pattern design of a non-pneumatic tyre for stiffness using topology optimization. Eng Optimiz 2012; 44(2):119-132

Acknowledgement

This work was financial supported by Rubber Technology Research Center (RTEC), Mahidol University and the Thailand Research Fund (TRF) under the TRF Research Grant No. RDG60T0140.

IOP Conf. Series: Materials Science and Engineering **773** (2020) 012038 doi:10.1088/1757-899X/773/1/012038

Assessment of human health impact based on life cycle assessment: A case study of Thai retread tire

W Ounsaneha[1], T Buadit[2] and C Rattanapan[3]*

[1] Faculty of Science and Technology, Valaya Alongkorn Rajabhat University under the Royal Patronage, Klong Nuang, Klong Luang, Pathumthani, 13180, Thailand.
[2] Faculty of Environment and Resource Studies, Mahidol University, Salaya Phutthamonthon, Nakhonpathom, 73710, Thailand.
[3] ASEAN Institute for Health Development, Mahidol University, Salaya Phutthamonthon, Nakhonpathom, 73710, Thailand.

*Corresponding author: cheerawit.rat@mahidol.ac.th

Abstract. The sustainable production of tire supply chain has been recently highlighted to reduce the environmental and human health impacts. The main environment problem of tire supply chain is waste treatment. When the tires reach the end of their working life, they are recycled (obtaining raw materials and other components for certain manufacturing processes) or burned (obtaining energy). During the exploitation, a tire with worn tread can be processed in a certain way to allow its reuse. The retreated tire should be accepted for further treatment. Hence, this study conducted the human health impacts of retread tire using life cycle assessment. The recipe2016 technique and Sigmaoro 9.0 software were used as the method of human health assessment. The result showed that the total value of human health impact assessment for retreated tire production were 4.3007E-06 Daly. The hotspot of human health impact for retreated tire production was tire material and electricity consumption. Form this finding will help designers or managers recommend the installation scenario of retreated tire production by considering human health impacts.

1. Introduction

Due to the phenomenal growth in the transport sector as a part of rapid urbanization, especially in the countries where transportation of goods and people are essentially done through the existing road associated with production of tires and to identify the grey areas in which enough networks, tire industries have seen a rapid growth [1]. When the tires reach the end of their working life, they are recycled (obtaining raw materials and other components for certain manufacturing processes) or burned (obtaining energy). During the exploitation, a tire with worn tread can be processed in a certain way to allow its reuse. In the past few decades, the emphasis has been on its further use. If during the exploitation, a tire suffers the damages which can be fixed, it is planned for retreading [2]. In 2018, over six thousand retreading tires are manufactured annually in Thailand [3].

In order to be retreaded a tire must undergo a control realized in two basic steps before being accepted for further treatment. The first step consists of a visual control of tire treads or other damages. The second step consists of an X-ray of the interior of the tire. After retreading, within a final control, each retreaded tire is once again subject of a detailed visual check, before being returned in service [2]. Shanbag and Manjare [1] reported that there is a significant impact on the environment due to emissions from the generation of electrical energy and steam in the plant under study along with emissions due to various operations in the tire production process.

One of the effective environmental assessment tools for products is based on life cycle perspective. Called life cycle assessment (LCA), it is employed to assess environmental impacts of product throughout its life cycle, from raw material extraction to production processes, transportation, use, and disposal [4]. The information on LCA studies of tire industry is very scanty. A sustainable issue of natural rubber industry has been studied thorough LCA approach [5].

Generally, many LCA methods quantify single environmental consequences (e.g. IPCC Global Warming Potentials (GWP) and other LCIA methods aggregate many environmental consequences into one or more endpoint value(s). One of these endpoint values is measured as Disability Adjusted Life Years (DALYs), which quantifies the damage to human health as a consequence of aggregated environmental impacts. In addition to human health damage also damages ecosystems through, for example, adverse effects on climate change, water use, eco-toxicity, land use, eutrophication, and biodiversity loss [6]. Hence, the objectives of this study were to assess the human health impacts of Thai retread tire industry. A cradle-to-grave was used to compare the influence of different the resources and material consumption on its life cycle environmental impacts.

2. Material and Methods

The International Standards, ISO 14040 and ISO 14044 [7-8] were used to conduct the LCA study of retread tire industry for analyzing the human impact. The more detail of methodology, data and the assumptions are the following:

2.1. Goal and scope definition

2.1.1. The is the reprocess of the use tire with the good of casing. The retreading rubber can be the reduction the process on 30 – 50 percent of the new tire. Hence, the health impact assessment of the retreading rubber process consisted of the cool and hot processes including tire checker, repairing, buffing, rubber cementing, building, molding or envelope and curing (Shown in Figure 1). The cradle-to-gate approach was used in this study, was from raw material extraction through product manufacturing only and only related to product manufacturing [1]. The human health impact of the retreading rubber process was calculated by life cycle assessment approach.

2.1.2. The data from a one company of the retreading rubber process in the central parts of Thailand was collected in this study. The primary data of all process use was conducted by the input and output of mass balance.

2.1.3. The functional unit of this study was the 1 kg of retreading rubber product. The data inventory analysis was studied and calculated by the resources consumption, energy consumption and waste production or pollution generation among retreading rubber process. The input and output of process was collected according to the boundary scope of each unit. Secondary data for environmental impact assessment was obtained from the related study or report with strong in the academic reference. The database used for impact assessment was the ReCiPe2016.

2.2. Life cycle impact assessment

ISO 14044 [8] was used to the impact assessment includes the selection of impact categories, classification, and characterization. Two main orientations or classifications, namely midpoint methods and endpoint methods. In this study, endpoint is supportive for the decision-makers, while midpoint, is helpful when specific environmental concerns are targeted [9]. The allocation methods were chosen such that, where a market is already established, economic allocation was used, whereas in other cases system expansion was applied. Data storage and modelling were performed using the SimaPro 8.5.2.0.

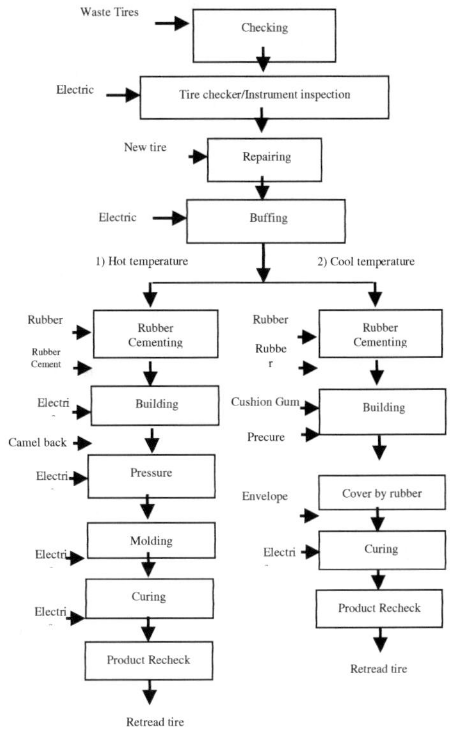

Figure 1. Retread Tire Process.

3. Result and Discussion

3.1 Life cycle inventory of retreading tire

From the data collection from a case study of Thai retreading rubber tire process, the Input of the production process consisted of used tire, repair tire, electricity, chemical and water consumptions. The result shown in Table 1, the average consumptions from 2014-2016 of used tire, waste tire, electricity, chemical and water consumptions per functional unit (1 kg of retreading rubber tire) were 0.850 kg, 0.250 kg, 3.957 kWh, 0.230 kg and 0.001 kg, respectively.

Table. 1 Life cycle inventory of retreading tire

Items	Value (kg) /1 kg tire/year			Average Value
	2014	2015	2016	
Used tire	0.850	0.900	0.800	0.850
Raw tire material	0.250	0.300	0.200	0.250
Electric consumption	3.750	4.320	3.800	3.957
Chemical consumption	0.2083	0.2600	0.2207	0.230
Water consumption	0.0009	0.0014	0.0011	0.001

3.2 Human health assessment

Figure 2 showed that the caparison of human impact based on the characterization for retreading rubber tire. The result indentified that electricity consumption was the majority impact on human health in all categories. Over 58 percents of human impact for retreading rubber tire were found in the human carcinogenic and non- carcinogenic toxicities.

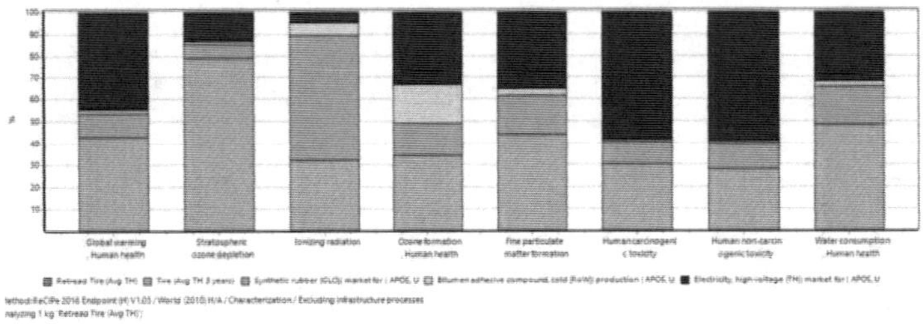

Figure 2. Identification of impact categories on human health for retreading rubber tire process.

For damage and normalization impact found that the main human impacts of retreading rubber tire were raw material (used tire) and electricity consumption (Figure 3 and 4). Then, the reduction approach of electricity consumption for retreading rubber tire industry should be proposed for decreasing the human health for the end point impact. Shanbag and Manjare [1] indentified that the significant reduction in environmental impact due to tyre production is possible if the source of electricity is non-conventional.

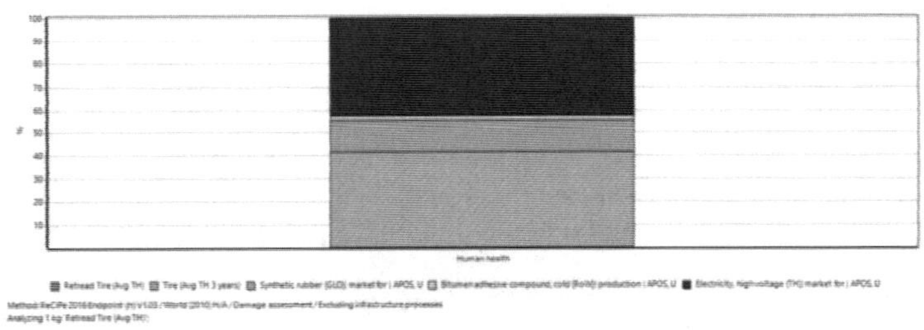

Figure 3. Identification of human health damage for retreading rubber tire process.

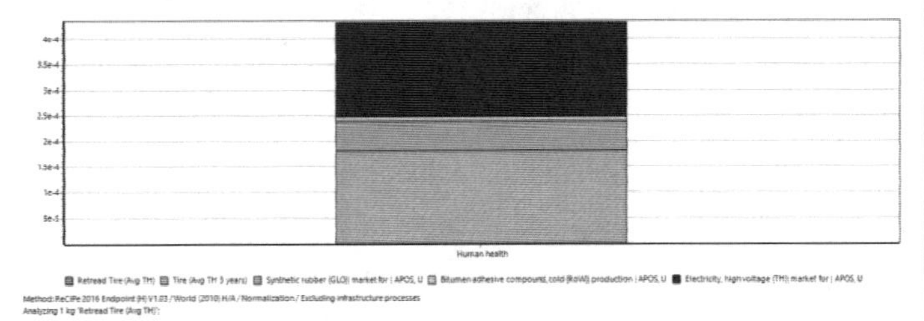

Figure 4. Identification of human health normalization for retreading rubber tire process.

4. Conclusion

This objective of this study was to identify the human health impacts of retread tire using life cycle assessment. The recipe2016 technique and Sigmaoro 9.0 software were used as the method of human health assessment. The result indentified that electricity consumption was the majority impact on human health in all categories. Over 58 percents of human impact for retreading rubber tire were found in the human carcinogenic and non- carcinogenic toxicities. The hotspot of human health impact for retreated tire production was electricity consumption. Then, the reduction approach of electricity consumption for retreading rubber tire industry should be proposed for decreasing the human health for the end point impact

5. References

[1] Shanbag A and Manjare S 2020 J Sustain Develop Energ Water Envi Sys. 8(1) 22-34.
[2] Dabic′-Ostojic′ S, Miljus′ M, Bojovic´S, Glisovic´ N, Milenkovic´M 2014. Applying a Resour., Conserv. Recycl. 86 107–117.
[3] https://www.diw.go.th/hawk/content.php?mode=spss61: November 2019.
[4] Phrommarat, B 2019. Environ. Nat. Resour. J 17(2) 96-108.
[5] Mohd Nor, Z 2012. Proceedings of the International Rubber Conference, Kerala, India.

[6] Walker C, Gibney ER, Mathers JC, Hellweg S. 2019. Sci. Total Environ. 685 609–620.
[7] ISO, ISO 14040 2006. Environmental management–lifecycle assessment–principles and framework. Geneva.
[8] ISO, ISO 14044: 2006 Environmental management–lifecycle assessment–principles and framework. Geneva.
[9] Obaideen, K, Tan Y.C, Yap, PW, Awang, M, Ghani AAA, Subramaniam V and Tai VC. 2018. IOP Conf. Ser.: Earth Environ. Sci. 268 012036

Acknowledgement
This research was supported by National Research Cousin of Thailand and Thailand Research Fund (contract reference number RDG60T0147).

IOP Conf. Series: Materials Science and Engineering **773** (2020) 012039 doi:10.1088/1757-899X/773/1/012039

The utilization of wastewater of Thai fermented rice noodle (Kanom-jeen) manufacturing process for the production of bacterial cellulose by *Acetobacter xylinim* TISTR 975

T Pongjinapeth, P Sudying, and P Jaturapiree*

Department of Biotechnology, Faculty of Engineering and Industrial and Technology, Silpakorn University, Nakhon Pathom 73000, Thailand

*Corresponding author: phimchanok@hotmail.com

Abstract. Bacterial cellulose (BC) is a naturally produced as an exopolysaccharide from some bacteria. It has excellent properties over the plant cellulose and has numerous applications in many fields including food, pharmaceutical, textile, paper manufacturing and other industries. However, a major limitation of bacterial cellulose production is the high cost of carbon substrate. The study aims to reduce the cost of bacterial cellulose using a cheap carbon source. This study presents feasibility in the production of bacterial cellulose using the starchy effluent waste from the Thai fermented rice noodle manufacturing process as a low-cost substrate by *Acetobacter xylinum* TISTR 975. The optimizations of culture conditions for bacterial cellulose production were also investigated under static culture. The results indicated that starchy effluent waste from the Thai fermented rice noodle manufacturing process performs well for the production of bacterial cellulose by supplementing with 50 g/L sucrose and 2% olive oil under the static condition. The structure and physical properties of bacterial cellulose were characterized using SEM, FTIR and XRD. In summary, the starchy effluent waste from the Thai fermented rice noodle manufacturing process can be used to produce bacterial cellulose which is a high value-added, sustainable and green product.

1. Introduction

Bacterial cellulose (BC), an exopolysaccharide, is produced by various species of bacteria, such as those of the genera *Acetobacter strains*, *Gluconacetobacter*, *Agrobacter*, *Sarcina*, and among others. They are hence called bacterial cellulose. BC demonstrates unique properties superior to cellulose from plants. These include high purity, good shape retention, high water binding capacity, high degree of crystallinity, good water retention, excellent tensile strength and moldability [1, 2, 3]. However, a limitation of the commercial uses of BC is its high cost of substrate (mainly carbon source). The carbon source may determine the cost of a production process, and in the case of cellulose production, it is up to 65% of the entire cost of biotechnological process [4]. More recently, some cheap and renewable substrates based on agro-industrial wastes and by-products can be used as the culture medium for BC production [5,6,7]. The wastewater derived from Khun Boonruen's Thai fermented rice noodles (TFRN) factory in Prong Madua Community, Nakhon Pathom, Thailand, was discharged at approximately 80-100 L per day. The study aimed to investigate the feasibility of the production of BC using wastewater from Thai fermented rice noodles (TFRN) without any pre-treatment in order to decrease cost of BC production. This would be a potential renewable carbon source differing from the traditional one. In addition, this study determined the structure and properties of these BC. So this study would benefit the provision of a potential low-cost and environmentally-friendly.

2. Experimental

Bacterial strain: *Acetobacter xylinum* TISTR 975 was obtained from the Thailand Institute of Scientific and Technological Research (TISTR) Culture Collection, Thailand.

The wastewater of Thai fermented rice noodles was obtained from Khun Boonruen Knom-jeen's factory in Prong Madua Community, Nakhon Pathom, Thailand. The pH, BOD and COD of the wastewater were 6.30, 0.30-1.26 mg/L and 17,000 - 17,600 mg/L, respectively [8]. For the preparation of wastewater from Thai fermented rice noodles (TFRN) medium, the wastewater was centrifuged at 12,000 rpm for 10 min to remove particles. The supernatant was then filtered and kept as a stock solution and sterilized in an autoclave at 121 $_{o}$C for 15 min The TFRN medium containing (g/L): glucose 50, ammonium sulfate 5 and 1% (v/v) acetic acid. The seed medium containing (g/L): glucose 4, peptone 5, yeast extract 5, disodium hydrogen phosphate 2.7, citric acid 1.15, and pH 6.0 [9].

Acetobacter xylinum TISTR 975 was cultivated in 500 mL flasks with 100 mL of seed medium and then incubated at 30 $_{o}$C under static culture conditions for 3 days. The seed inoculum (10% v/v) was then transferred into the TFRN medium and then incubated at 30 $_{o}$C under static culture conditions for 15 days. The thickness of BC was determined using a Vernier caliper. The experiments were done in triplicate.

To study the effects of olive oil for BC production, the olive oil (Bertolli®, Sino-Pacific Trading Thailand) concentration of 0.5%, 1% and 2% (v/v) was added to the TFRN medium and then incubated at 30 $_{o}$C under static culture conditions for 15 days. The thickness of BC was determined using a Vernier caliper. The experiments were done in triplicate.

After cultivation, the BC was picked from the surface and treated with 0.1 N NaOH at 90 °C for 10 min to remove the microorganisms. The cellulose was immersed overnight in distilled water or until the pH of the water became neutral. Purified BC was washed twice with distilled water. After that, the purified BC was freeze-dried for the characterization of the structure and physical properties.

Scanning Electron Microscopy (SEM). The morphology and microstructure of BC were characterized by SEM. After freeze drying, the BC was mounted on SEM grids, and gold particles were sputter coated with Rotary-pumped sputter (Q150R S model, Quorum, England) to the BC surfaces. The specimens were observed with a scanning electron microscopy at an accelerating voltage of 3 kV (MIRA3 model, TESCAN, Czech Republic).

Fourier Transform Infrared Spectroscopy (FTIR). The surface properties of the BC were analysed by a Nicolet iS5 FTIR Spectrometer (Thermo Fisher Scientific, USA). For each analysis, the scan range was 4,000 - 400 cm$_{-1}$with a resolution of 4 cm$_{-1}$ [10].

X-ray diffraction (XRD). Dried BC was investigated using a MiniFlex II X-ray diffractometer. The samples were scanned from $2\theta = 3°$ to $2\theta = 70°$ at a scanning rate of 2° 2θ min$_{-1}$. The operating voltage and current were 40 kV and 30 mA, respectively [10].

3. Results and discussion

This result shows feasibility in the production of bacterial cellulose by *Acetobacter xylinum* TISTR 975 using the starchy effluent waste from the Thai fermented rice noodle manufacturing process without any pre-treatment in order to save the costs of energy, chemicals and time for BC production. The thickness of the BC film obtained from TFRN and TFRN medium with sucrose concentration of 50 g/l that were significantly different in the static cultures. After 15 days of incubation under static culture, the thickness of the BC film obtained from TFRN and TFRN medium with sucrose concentration of 50 g/l were 5.33 and 13.67 mm, respectively. These results show that the TFRN medium could be used for BC production. It indicated that starchy effluent waste from the Thai fermented rice noodle manufacturing process performs well for the production of bacterial cellulose by supplementing with 5 g/L sucrose and

2% olive oil under the static condition as shown in Figure 1. The thickness of the BC film significantly increased by adding 2.0 % (v/v) olive oil. These condition gave the maximum thickness of 18.3 mm compared with 14.0 mm without olive oil (Table 1). Similarly, Żywicka et. al. (2018) report that the increased production of BC was achieved by supplementing culture medium with vegetable oil [11]. SEM displayed the morphology of the surface structure of the BC at a magnification of 25,000 times as shown in Figure 2. The width of BC fibrils was less than approximately 100 nm in size .The network of thin layers of BC structures composed of very neat fibrils. The two conditions of BC had similar microstructures.

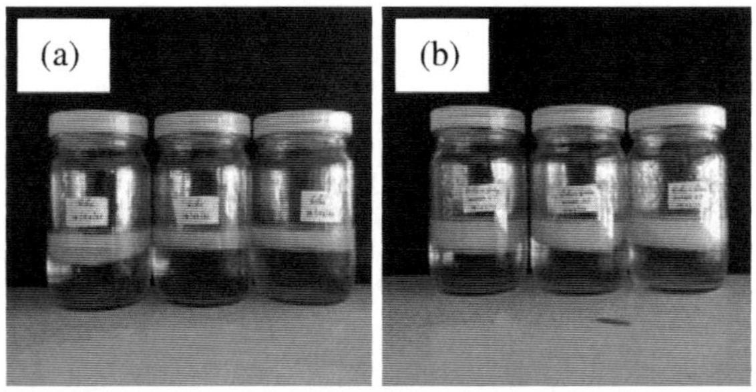

Figure 1. The production of bacterial cellulose using starchy effluent waste from the Thai fermented rice noodle manufacturing process (a) without and (b) with 2% olive oil.

Table 1. Effects of olive oil on BC production by *Acetobacter xylinum* TISTR 975 in TFRN medium with sucrose concentration of 50 g/l under static condition for 15 days.

olive oil concentration (v/v)	Thickness (mm)
0	14.0 ± 1.73
0.5	11.7 ± 1.15
1.0	17.0 ± 1.00
2.0	18.3 ± 1.52

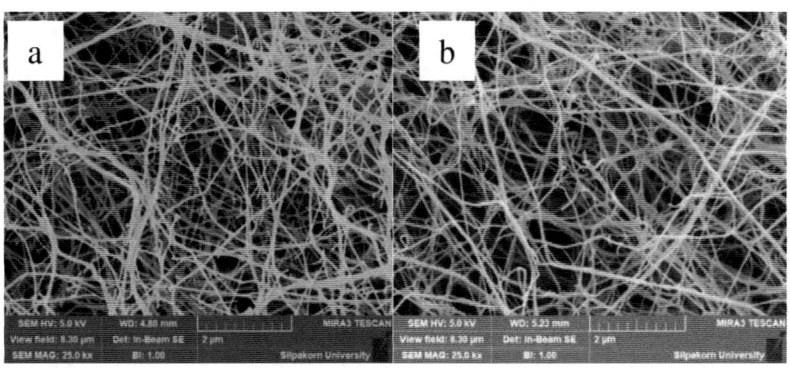

Figure 2. SEM images of the surface morphological structure of freeze-dried BC in TFRN medium with sucrose concentration of 50 g/l (a), with sucrose and olive oil concentration of 50 g/l and 2% (v/v) (b).

FTIR was used to obtain the specific functional groups or chemical bonds of the BC samples. The FTIR spectra of the BC from TFRN medium with sucrose concentration of 50 g/l (blue curve) and TFRN medium with sucrose and olive oil concentration of 50 g/l and 2% (v/v) (red curve) are shown in Figure 3. The two conditions of BC showed several important characteristic peaks at 3344 cm^{-1} (stretching vibration of hydroxyl groups (OH)), at 2895 cm^{-1} (asymmetric stretching vibration of $-CH_2CH$), at 1427 cm-1 (asymmetric deformation vibration of methyl and methylene), at 1161 cm-1 (asymmetric stretching of C-O at ring) and at 1056 cm-1 (stretching of C-O) [10,12]. The results confirmed that no obvious difference is observed for functional groups of BC samples obtained from both media. The X-Ray Diffraction patterns of BC from TFRN medium with sucrose concentration of 50 g/l (blue curve) and TFRN medium with sucrose and olive oil concentration of 50 g/l and 2% (v/v) (red curve) are shown in Figure 4. The blue curve, it had two main characteristic peaks at 14. 64° and 22.95°. The red curve, it showed two main characteristic peaks at 14.89° and 23.23°.

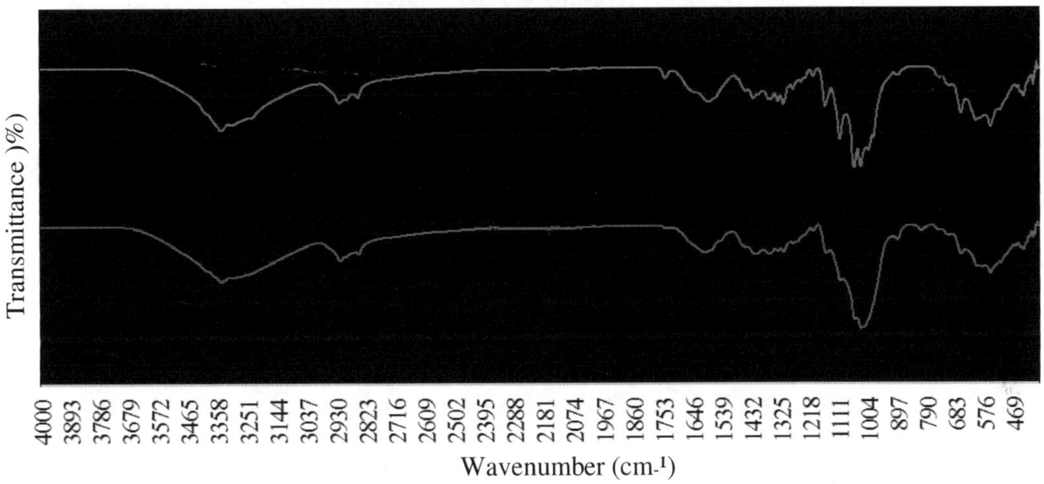

Figure 3. FTIR spectra of BC produced by TFRN medium with sucrose concentration of 50 g/l (blue curve), with sucrose and olive oil concentration of 50 g/l and 2% (v/v) (red curve).

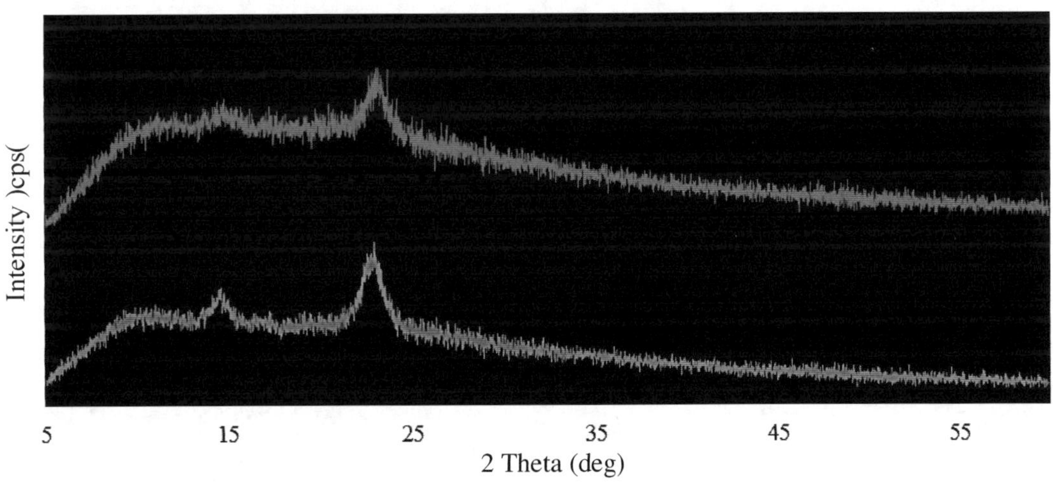

Figure 4. X-ray diffraction diagrams of BC produced by TFRN medium with sucrose concentration of 50 g/l (blue curve), with sucrose and olive oil concentration of 50 g/l and 2% (v/v) (red curve).

4. Conclusions

In this study, the TFRN without any pretreatment could be used directly as substrate for BC production by *Acetobacter xylinum* TISTR 975. The thickness of the BC film was significantly increased by adding sucrose and olive oil concentration of 50 g/l and 2.0 % (v/v) respectively. Overall, the fermentation of TFRN showed a small influence on BC structure as indicated by DSC, FTIR, and XRD technique. The BC from both media had similar properties. Using inexpensive waste materials as substrates in fermentation media can significantly reduce the production cost of BC. It also has an environmentally friendly effect by the removal of such waste from the environment.

5. References

[1] Lin S-P, Loira Calvar I, Catchmark J M, Liu J-R, Demirci A and Cheng K-C 2013 Cellulose. 20(5) 2191-2219.
[2] Tanskul S, Amornthatree K and Jaturonlak N 2013 Carbohyd. Polym. 92(1) 421-428.
[3] Zeng X, Small D P and Wan W, 2011 Carbohyd. Polym. 85(3) 506-513.
[4] Jedrzejczak-Krzepkowska M, Kubiak K, Ludwicka K and Bielecki S. 2016 (Eds.) In Bacterial Nanocellulose, Elsevier, 260 pages.
[5] Lin D, Lopez-Sanchez P, Li R and Li Z 2014 Bioresource Technol. 151 113-119.
[6] Castro C, Zuluaga R, Putaux J L, Caro G, Mondragon I and Gañán P 2011 Carbohyd. Polym. 84(1) 96-102.
[7] Kim S Y, Kim J N, Wee Y J, Park D H and Ryu H W 2006 Appl. Biochem. Biotech. 131(1) 705.
[8] Yeesang J, Yamsuan S and Kaewpuk W 2015 ABDR Journal. 7 6-15.
[8] Hong F, Guo X, Zhang S, Han S f, Yang G and Jönsson L J 2012 Bioresource Technol. 104 503-508.
[9] Fan X, Gao Y, He W, Hu H, Tian M, Wang K and Pan S 2016 Carbohyd. Polym. 151 1068-1072.
[10] Huang C, Guo H J, Xiong L, Wang B, Shi S L, Chen X F, Lin X Q, Wang C, Luo J and Chen X D, 2016 Carbohyd. Polym. 136 198-202.
[11] Żywicka A, Junka A F, Szymczyk P, Chodaczek G, Grzesiak J, Sedghizadeh P P and Fijałkowski K 2018 Carbohyd. Polym. 199 294-303.

Acknowledgement
The authors wish to thank the Department of Biotechnology, Faculty of Engineering and Industry Technology, Silpakorn University for financial support.

IOP Conf. Series: Materials Science and Engineering **773** (2020) 012040 doi:10.1088/1757-899X/773/1/012040

Confirmation molecular structure of the *Hevea* rubber molecule and its effects in storage hardening

P Theamsawade [1], S Kumarn [2] and J Sakdapipanich [1,2,*]

[1] Department of Chemistry and Center of Excellence for Innovation in Chemistry, Faculty of Science, Mahidol University, Bangkok 10400, Thailand
[2] Institute of Molecular Biosciences, Mahidol University, Nakhon Pathom 73170, Thailand

*Corresponding author: jitladda.sak@mahidol.ac.th

Abstract. Lipids are the one of major non-rubber components discovered in natural rubber (NR) latex, which reported to be the significant parameter affecting the properties of NR [1,2]. Our previous work suggested that lipids are the major factor dominate branching formation in NR [3], which could bring about the change in physical properties of NR during long term storage as called storage hardening (SH) phenomenon. However, lipids found in NR consist of neutral and polar lipids, *i.e.*, phospholipids and glycolipids [3]. The information on types of lipids affecting the storage hardening behaviour in NR have not yet been investigated. Therefore, the aim of this work was to study the relationship between types of lipids on the gel formation in NR under accelerated storage. In the present work, natural rubber was treated by deproteinization, acetone extraction, chloroform:methanol extraction and transesterification for eliminating proteins, neutral lipids, polar lipids and linked lipids, respectively. The molecular structure of the rubber samples and the remaining non-rubber components were analyzed using nuclear magnetic resonance (NMR) and Fourier-transform infrared spectroscopy (FTIR). Moreover, the rubber samples were subjected to study the storage hardening behaviour under accelerated condition using phosphorus pentoxide (P_2O_5).

1. Introduction

Solid natural rubber (NR) is the polymeric substance, which consists of approximately 94% rubber hydrocarbon and 6% non-rubber components i.e., proteins, lipids, carbohydrate, etc [3]. It is normally used as raw material produced in various applications due to their excellent properties such as high tensile strength, tear strength and elasticity. The key parameter affecting the outstanding properties of NR, which make NR different from any synthetic rubber, is the presence of non-rubber components, especially lipids and proteins [4]. A number of reports suggested that rubber chain is associated with proteins at the initiating end and phospholipids at the terminating end [5]. The existence of associated proteins and phospholipids was reported to be the cause of naturally occurring network in solid NR leading to the excellent mechanical properties [1]. However, these associated non-rubbers were also bring about gel formation affecting the change of properties of NR during prolonged storage, which was called storage hardening. It was reported that the important factor affecting gel formation in NR during storage is phospholipid [3]. Nevertheless, lipids found in NR consists of neutral lipids, polar lipids i.e., glycolipids and phospholipids. These lipids in which found in NR are probably to affect the gel

IOP Conf. Series: Materials Science and Engineering **773** (2020) 012040 doi:10.1088/1757-899X/773/1/012040

formation resulted in the storage hardening in NR, which have not been investigated yet. Therefore, in the present work, the effect of lipid types on the gel formation in NR was investigated.

2. Experimental

Fresh NR (FNR) latex used in this work was supported from Thai Rubber Latex Co. Ltd, which has approximately 30- 40 % dry rubber content (DRC). The FNR latex was cast on a plate and dried at room temperature (RT). Deproteinized NR (DPNR) was prepared by enzymatic deproteinization. NR latex was incubated with 0.04% w/v proteolytic enzyme in the presence of 1% w/v sodium dodecyl sulphate (SDS) solution at 37°C for 12 h. Then, the latex was centrifuged twice at 13,000 rpm for 30 min, which cream fraction was collected and re-dispersed in DI water. After that, the DPNR latex was cast and dried at RT. Acetone extracted DPNR (AEDPNR) represents the rubber sample, which neutral lipids and fatty acids were removed by acetone extraction [5]. The dried DPNR sample was cut into small pieces and subjected to soxhlet extraction with acetone for 24 h. The remaining rubber was dried at RT. Chloroform/methanol DPNR (CMDPNR) was prepared by immersed the AEDPNR in 2:1 chloroform:methanol solution at RT for 24 h to remove polar lipids in the rubber surface [5]. The obtained rubber sample was dried at RT. Transesterified DPNR (TEDPNR) was carried out in 2.5% toluene solution with freshly prepared sodium methoxide at RT for 3 h. The obtained rubber was purified by precipitation with excess amount of methanol then dried in the oven at 60°C. The prepared rubber samples were subjected to storage under accelerated condition using P_2O_5 at 60°C for 72 h.

For the characterization. FTIR (JASCO FT/IR 4100) in transmittance mode with a KBr disk was used to characterize functional group in the rubber samples and determined content of long-chain fatty acid ester groups based on calibration through a mixture of synthetic *cis*-1,4-polyisoprene and methyl stearate. Weight-average molecular weight ($\bar{M}_w$), number-average molecular weight ($\bar{M}_n$) of the rubber samples were analyzed by gel permeation chromatography (GPC, JASCO/CO 1560) using refractive Index detector. The nitrogen content was determined by a LECO Nitrogen Analyzer (model FP 528) with a combustion mode. The ^{1}H-NMR measurement using BRUKER UltraShield^{TM} spectrometer was executed at frequency 400 MHz for investigating the molecular structure of the prepared samples. Furthermore, gel content was determined by toluene solubility measurement.

3. Results and discussion

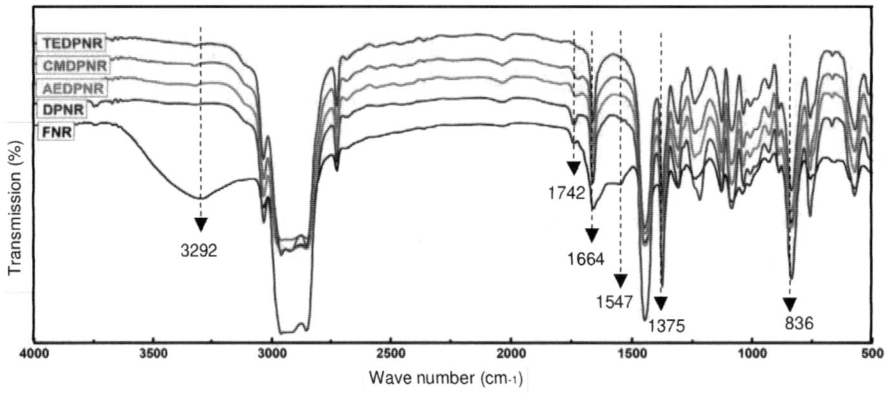

Figure 1. FTIR spectra of rubber samples.

FTIR spectra of rubber samples with different types of non-rubber component, were compared as shown in figure 1. The rubber chain peak was appeared at 1664, 1375 and 836 cm^{-1}, which assigned to C=C stretching, $-CH_3$ symmetric deformation and C–H out-of-plane bending of *cis*-1,4-addition, respectively [6]. The characteristic peaks of protein at 3292 (N–H stretching) and 1547 cm^{-1} (N–H bending and C–N stretching) [6] were disappeared after deproteinization, which were corresponding to the decrease in nitrogen content of DPNR comparing with FNR as shown in table 1. This result confirmed that proteins in FNR were removed by using enzymatic deproteinization.

Table 1. Basic characteristics of rubber samples

Sample	Nitrogen content (%w/w)	Ester content (mmol/kg rubber)	$\bar{M}_w$ (10^6)	$\bar{M}_n$ (10^5)	$\bar{M}_w/\bar{M}_n$
FNR	0.60	48.72	1.67	3.50	4.77
DPNR	0.04	20.47	1.72	3.17	5.42
AEDPNR	0.02	11.44	0.91	2.21	4.09
CMDPNR	0.01	8.83	0.39	1.15	3.39
TEDPNR	0.01	~0	0.99	2.28	4.35

The ester content, estimating from the intensity ratio of C=O stretching at 1742 cm^{-1} to C=C stretching at 1664 cm^{-1}, was shown in table 1. It can be seen that ester content decrease from 20.47 to 11.44 mmol/kg rubber in the AEDPNR comparing with DPNR. From this result, it might be said that neutral lipids and free fatty acids in DPNR were removed after acetone extraction. In addition, the remaining polar lipids were eliminated after immersed in the mixture of chloroform:methanol resulting in the decrease of ester content in the CMDPNR comparing with AEDPNR. Meanwhile, TEDPNR showed no ester content, which suggested that lipids were decomposed after transesterification.

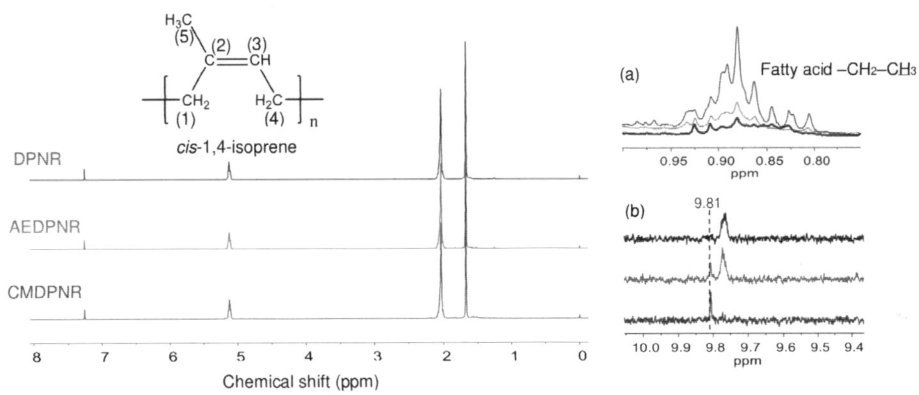

Figure 2. ^{1}H NMR spectra of DPNR, AEDPNR and CMDPNR in the ranges of 0-8.00 ppm, extended ^{1}H-NMR spectra in the ranges of (a) 0.75-1.00 ppm, (b) 9.35-10.05 ppm

Molecular structures of DPNR, AEDPNR and CMDPNR were further characterized using ^{1}H-NMR as shown in Figure 2. Three signals of *cis*-isoprene units were clearly observed at 1.68, 2.04 and 5.12 ppm which related to C-5–CH_3–, C-4–CH_2– and C-3 methine proton (H–C=C), respectively. The multiple signals resonated at 0.88 ppm was assigned to the methyl proton (–CH_3) in long chain fatty acid [7]. It was clearly seen that their intensities tend to decrease after solvent extraction, which corresponding to the decrease of ester content in rubber. However, the small signals appeared at 9.81 ppm was correlated to –CH=O [8]. It indicates that there was the degraded product as aldehyde in AEDPNR and CMDPNR samples, implying the reduction of molecular weight in AEDPNR and especially CMDPNR. It might be because the proteins, which acted as the natural antioxidant [9], was eliminated, so an oxidation reaction might easily occur during solvent extraction.

Table 2. Gel content of FNR, AEDPNR, CMDPNR and TEDPNR under accelerated storage test

	Gel content (%w/w)	
Sample	**storage at 0 h**	**storage at 72 h**
FNR	14.41	20.00
AEDPNR	0	2.70
CMDPNR	0	1.01
TEDPNR	1.78	0

Furthermore, the rubber samples with different type of remaining lipids including FNR, AEDPNR, CMDPNR and TEDPNR were subjected to study the gel formation under accelerated storage over P_2O_5 at 60°C. The DPNR sample was not interested to characterize gel formation in this work because our previous work was reported that proteins are not the important factor affecting the gel formation in NR [3]. The gel content of samples after storage at 0 h and 72 h was reported as shown in table 2. It can be seen that gel content of FNR increased from about 14 %w/w to 20 %w/w. On the other hand, TEDPNR showed a decrease in gel content after storage, which was the lowest value comparing with the other samples. This confirmed that lipid is the important parameter for gel formation in NR. To further consider types of lipid affecting gel formation, AEDPNR and CMDPNR showed low value of gel content, which might be the effect of the occurring of degradation in these samples as described in the $_1$H-NMR result. However, it can be seen that AEDPNR has slightly higher gel than CMDPNR. It might be said that types of lipid has the effect on the gel formation in NR. However, it will have to request other experiments to support this conclusion, which is carried on.

4. Conclusions

In conclusion, the treatment of NR *via* deproteinization, solvent extraction and transesterification resulted in different types of non-rubber components remaining in the rubber sample. Proteins were successfully removed using enzymatic deproteinization, which was confirmed by FTIR result and N-content. Furthermore, the $_1$H-NMR result and ester content suggested that lipids can be removed after solvent extraction. However, some degraded products were observed because proteins, the natural antioxidant, were eliminated. In addition, the gel formation, which correlated to storage hardening behaviour in NR, was not discovered in TEDPNR sample after storage for 72 h, which confirmed that lipids are the major factor affecting branching formation in NR. Furthermore, the gel content of AEDPNR slightly higher than CMDPNR, which suggested that types of lipid might has the effect to the gel formation in NR.

5. References

[1] Amnuaypornsri S, Kawahara S, Toki S, Hsiao BS, Hikosaka M and Sakdapipanich J 2012 *KGK Kautschuk Gummi Kunststoffe*. **65** 46.
[2] Liengprayoon S, Chaiyut J, Sriroth K, Bonfils F, Sainte-Beuve J, Dubreucq E and Vaysse L 2013 *Eur J Lipid Sci Technol.* **115** 1021
[3] Amnuaypornsri S, Nimpaiboon A, Sakdapipanich J 2009 *KGK Kautschuk Gummi Kunststoffe*.
[4] Xu L, Huang C, Luo M, Qu W, Liu H, Gu Z, Jing L, Huang G and Zheng, J 2015 *RSC Adv.* **5** 91742
[5] Kumarn S, Churinthorn N, Nimpaiboon A, Sriring M, Ho C.-C., Takahara A and Sakdapipanich J 2018 *Langmuir*.
[6] Rolere S, Liengprayoon S, Vaysse L, Sainte-Beuve J and Bonfils F 2015 *Polym Test.* **43** 83
[7] Kawahara S, Kakubo T, Sakdapipanich JT, Isono Y and Tanaka Y 2000 *Polymer.* **41** 7483
[8] Ibrahim S, Daik R and Abdullah I 2014 *Polymers.* **6** 2928
[9] George KM, Alex R, Joseph S and Thomas KT 2009 *J Appl Polym Sci.* **114** 3319

Acknowledgement

The authors would like to thank The Center of Excellence for Innovation in Chemistry (PERCH-CIC), Commission on Higher Education, Ministry of Education for the financial support. Sincere appreciation is extended to Thai Rubber Latex Co., Ltd. for generous supplies of NR latex.

IOP Conf. Series: Materials Science and Engineering **773** (2020) 012041 doi:10.1088/1757-899X/773/1/012041

Studies on the effect of purified natural rubber latex and accelerators on rubber allergens in natural rubber dipping product

J Wiriyanantawong[1], P Havanapan[2], C Krittanai[2] and J Sakdapipanich[1,2,*]

[1]Department of Chemistry and Center of Excellence for Innovation in Chemistry, Faculty of Science, Mahidol University, Bangkok 10400, Thailand.
[2]Institute of Molecular Biosciences, Mahidol University, Nakhon Pathom 73170, Thailand.

*Corresponding author: jitladda.sak@mahidol.ac.th

Abstract. Natural rubber (NR) glove is a worldwide use product of NR latex, which is mainly used for protective purposes. However, allergic reactions to NR gloves causing by the residual proteins in NR latex and chemical used in the manufacturing process, which are called as rubber allergens, are still a significant concern. Thus, the present work is an attempt to minimize such the rubber allergens from NR dipping products, herewith, finger cot is used as a model for this study. Purified NR latices were prepared by an urea treatment and saponification method, called as deproteinized NR (DPNR) and saponified NR (SPNR), respectively. Both of DPNR and SPNR were found to have low nitrogen content. Fourier-transform infrared spectroscopy (FTIR) was used to identify functional groups as a result of the absence of the amine functional group, which can be referred to proteins. For a compounded NR, the various purified NR latices and rubber accelerators were used in the compounded formula. The existing and residual proteins of each sample were analysed by Bradford's assay and sodium dodecyl sulphate polyacrylamide gel electrophoresis (SDS-PAGE) method, respectively. The DPNR and SPNR were found to contain lower protein profiles. The residual rubber accelerators released into artificial sweat were extracted and analysed by high-performance liquid chromatography (HPLC) technique.

1. Introduction

Natural rubber (NR) from *Hevea brasiliensis* tree is an important naturally occurring rubber source in industry due to its outstanding properties, i.e. excellent green strength, excellent tear resistance and low heat build-up. There are several dipping products made from NR latex such as gloves, condoms, pacifier and catheter, etc. However, the usage of these products sensitize the users to NR allergy, which reaction can vary from mild to severe symptoms. The most common and least-threatening reaction type is a non-immunological irritant contact dermatitis. For more-severe type, cell-mediated contact dermatitis or type IV allergy is an allergic reaction to chemical used in manufacturing of NR products. Rubber accelerator, especially thiurams and dithiocarbamates groups, is one of a chemical used that have been reported as a rubber allergen [1-2]. Another severe allergic reaction type that can develop to a life-threatening symptom cause by latex hypersensitivity is called immunoglobulin E (IgE)-mediated latex allergy or type I allergy. Due to the ability of some proteins, which is one of a non-rubber composition, in NR latex can trigger type I immune responses individually sensitive people. There are many allergic proteins in NR latex and the most common ones were rubber elongation factor (REF) (Hev b 1), beta-1,3-glucanse (Hev b 2), Small rubber particle protein (SRPP) (Hev b 3), Acid latex protein (Hev b 5), Hevein (Hev b 6.02) and class I chitinase (Hev b 11) [3-4].

Therefore, this research aims to study the two important rubber allergens that can cause NR allergy, which are residual proteins and accelerators used in a sulphur vulcanization system of NR dipping product. Purified NR latex was prepared and subjected to a study of the effect of purified method on the minimization of allergic proteins in NR, while NR dipping product vulcanized with dithiocarbamate accelerators, zinc diethyldithiocarbamate (ZDEC) and zinc dibutyldithiocarbamate (ZDBC), were prepared to investigate a releasing of residual accelerators to artificial sweat.

2. Experimental

NR latices in this experiment were prepared from concentrated NR (CNR) latex with 61% dry solid content (DRC). Deproteinized NR (DPNR) were prepared from deproteinization process by urea and the presence of sodium dodecyl sulphate (SDS) as a surfactant under ambient condition [5]. Saponified NR (SPNR) were prepared from saponification process using NaOH and the presence of Triton-X® 100 as a surfactant under 70°C [6]. For the compounded NR, the ingredients are presented in the Table 1.

Table 1. Formula ingredients of compounded NR latices in phr (part per hundred of rubber)

Formula	CNR	KOH	K-laurate	Sulphur	ZDEC	ZDBC	Winstay L	ZnO
VZDEC	100	0.3	0.2	0.5	0.75	-	0.5	0.25
VZDBC	100	0.3	0.2	0.5	-	0.75	0.5	0.25

FTIR (JASCO FT/IR-4100) in transmission mode with film method, which samples were cast on KBr plate of liquid cell was used to determine the functional groups, which can be referred to a protein, in purified samples. In the same way, nitrogen content of purified samples was determined using the combustion method *via* nitrogen analyser (LECO FP-528). To study the existing rubber allergens, the proteins in both purified and vulcanized latices were extracted and observe protein pattern by a SDS-PAGE method [7]. After the latices was obtained, proteins were extracted by using acetone/methanol as precipitated co-solvent, while the extractable proteins from the dried NR samples were stirred into SDS aqueous solution for 12 h then the proteins were precipitated by co-solvent system. The residual dithiocarbamate accelerators were extracted in an artificial sweat for 24 h under 37°C then converted into copper complexes before analysis by HPLC technique [8]. Chromatographic analyses were performed using HPLC (Dionex, Ulitimate 3000) on a reverse phase column eluted with a solution of methanol and water in the ratio 90:10 (v/v) as a mobile phase.

3. Results and discussion

FTIR spectra of CNR and purified NR, DPNR and SPNR, were shown in Figure 1. The broad band centred at 3200 cm^{-1} corresponding to –N-H stretching, which is the functional group of proteins, can be clearly observed in the CNR sample. However, the absence of this broad band was observed in DPNR and SPNR sample. It was also clear that nitrogen content of CNR, DPNR and SPNR samples were 0.399, 0.063 and 0.032% respectively. Indicating that the obtained DPNR and SPNR were purified latices that have lesser protein containing. Moreover, SPNR prepared from saponification method, which is a chemical reaction of base catalysed hydrolysis of ester and amide groups, was expected to show lower protein than urea-treatment, which is a physical interaction of urea-induced denaturation of protein, of DPNR.

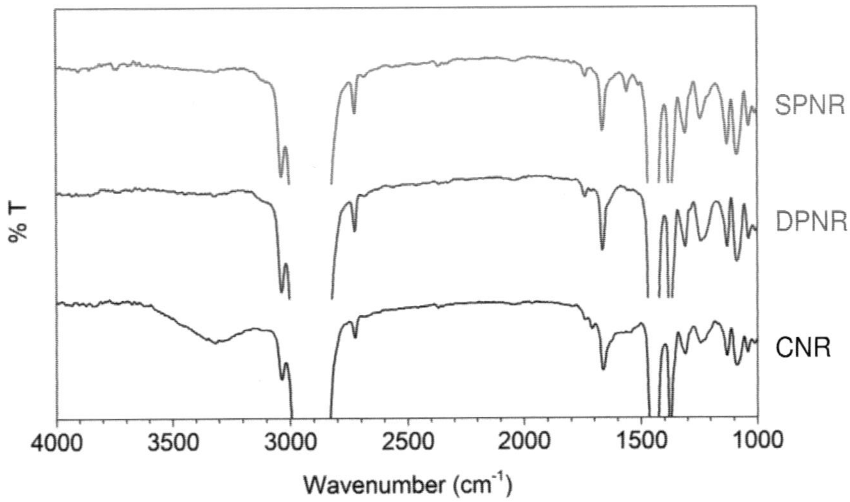

Figure 1. FTIR spectra of rubbers from CNR, DPNR and SPNR lattices.

The extractable allergic protein, which is one of rubber allergens, from NR latices and dried NR samples are shown in Figure 2. It was found that protein from CNR sample showed lower protein band than fresh NR (FNR) sample that showing the beta-1,3-glucanase, class I chitinase, SRPP and REF proteins. In SPNR samples, there are no detectable proteins. It can be indicated that SPNR prepared from saponification method is the highly purified NR that free from allergic proteins. Moreover, the latex and dried sample of the obtained vulcanized NR from CNR showed the same protein band around 14 kDa, which is REF protein.

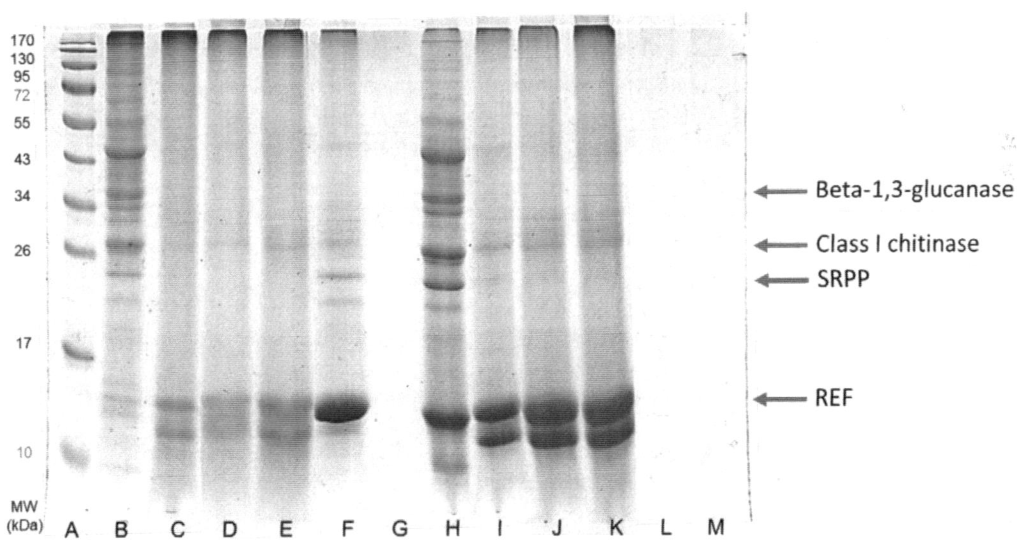

Figure 2. SDS-PAGE profiles of extracted proteins from various NR samples, Lane b-g are originated from latex sample, while lane h-m are from dried rubber. Lane a) protein marker, b) fresh NR latex, c) CNR latex, d) VZDBC latex, e) VZDEC latex, f) DPNR latex, g) SPNR latex, h) dried fresh NR, i) dried CNR, j) dried VZDBC, k) dried VZDEC, l) dried DPNR and m) dried SPNR.

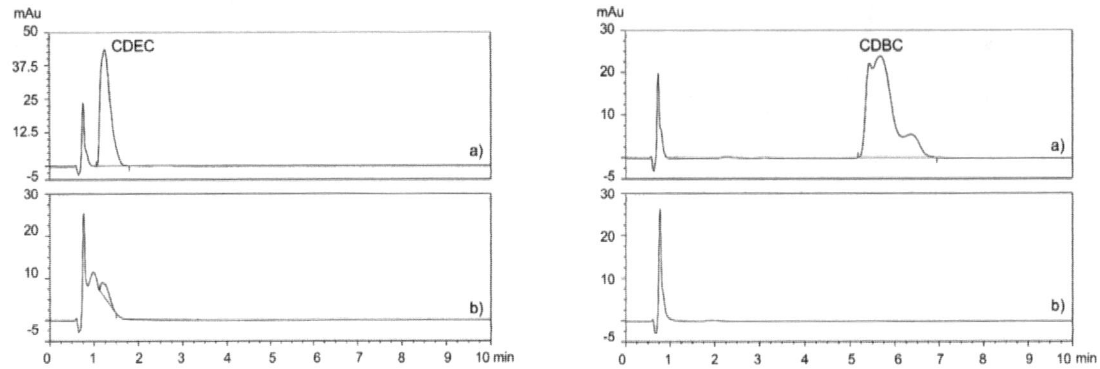

Figure 3. HPLC chromatograms of copper complex of ZDEC (left) and ZDBC (right) at λ 435 nm with the comparison of a) standard accelerators and b) accelerator-released into sweat of samples.

The residual accelerator-release into the artificial sweat was investigated by HPLC analysis. From Figure 3, copper complex ZDEC and ZDBC were observed at 1.3 and 5.7 mins respectively. The extent of ZDEC-release into artificial sweat, at pH 6.5 for 24 h under 37°C, from prepared dipping sample was 4.63 μg/g. Meanwhile, the releasing of ZDBC could not be found from the ZDBC-dipped sample. This implies that there is more occasion of the accelerator-release from a NR dipped product using ZDEC as an accelerator than ZDBC. Therefore, NR dipped products using ZDBC as an accelerator were recommended to use due to the non-released allergic accelerator that can expose to human skin.

4. Conclusions

In this research, DPNR and SPNR showed lower protein functional groups as a result of FTIR, and nitrogen content comparing to that of the CNR. SDS-PAGE analysis showed the lower allergic protein bands of CNR, DPNR and SPNR than FNR. Moreover, it can be concluded that the obtained SPNR is the highly purified NR that free from allergic proteins. Further studied on the released-accelerator of NR dipped product from HPLC technique, ZDBC could not be found from the artificial sweat extraction. Therefore, the prepared NR dipped product from purified latex and ZDBC accelerator, instead of ZDEC, was suggested to be a product of choice that can minimize the rubber allergens.

5. References

[1] Nettis E, Assennato G, Ferrannini A and Tursi A 2002 *Clin. Exp. Allergy* **32** 441.
[2] Uter W, Werfel T, White IR and Johansen JD 2018 *Int J Environ Res Public Health* **15** 1108.
[3] Rietschel LR, Fowler FJ and Fisher AA 2008 (6th Eds.) Fisher's Contact Dermatitis, *Hamilton*, 862 pages.
[4] Sussman GL, Beezhold HD and Kurup PV 2002 *J. Allergy Clin. Immunol.* **110** 33.
[5] Klinklai W, Saito T, Kawahara S, Tashiro K, Suzuki Y, Sakdapipanich JT and Isono Y 2004 *J Appl Polym Sci* **93** 555.
[6] Yunyongwattanakorn J, Tanaka Y, Sakdapipanich J and Wongsasutthikul V 2008 *Rubber Chem. Technol.* **81** 121.
[7] Havanapan P, Bourchookarn A, Ketterman JA and Krittanai C 2016 *J Proteomics* **131** 82.
[8] Abraham E, Parameswaran R, Joseph R, Mohanan PV, Remakumary VM 2005 *Rubber Chem. Technol.***78** 674.

Acknowledgement

The authors would like to acknowledge the financial support from the Centre of Excellence for Innovation in Chemistry (PERCH-CIC), Ministry of Higher Education, Science, Research and Innovation, Mahidol University. The SDS-PAGE method were support with the helping of Mrs. Nuanwan Phungthanom at Institute of Molecular Biosciences, Mahidol University. Sincere appreciation is extended to the Thai Rubber Latex Group Public Company Limited for their kind support of NR latex.

IOP Conf. Series: Materials Science and Engineering **773** (2020) 012042 doi:10.1088/1757-899X/773/1/012042

Thermal and barrier properties of poly(butylene adipate-co-terephthalate) incorporated with zeolite doped potassium ion for packaging film

P Threepopnatkul[1],*, K Wongsuton[1], C Jaiaue[1], N Rakkietwinai[1], and C Kulsetthanchalee[2]

[1]Department of Materials Science and Engineering, Faculty of Engineering and Industrial Technology, Silpakorn University, Nakhon pathom 73000, Thailand.
[2]Event and Exhibition Design Department, Suan Dusit University, Bangkok 10300, Thailand.

*Corresponding author: poonsubt@yahoo.com

Abstract. In this research, the works were focused on development of film packaging of poly(butylene adipate-co-terephthalate) (PBAT) incorporated with modified zeolite using (3-aminopropyl) trimethoxysilane. PBAT composites were compounded by melt mixing using twin screw extrusion then molded into film by cast film extrusion process. Effects of zeolite 13X at 1 wt% and zeolite doped potassium ion (K+) on thermal properties, permeability of gases (oxygen, carbon dioxide and water vapor) and also absorption of ethylene gas of PBAT composites films have been studied. Typically, zeolite would act as a nucleating agent. Therefore, adding zeolite would increase the degree of crystallinity of PBAT. Comparing zeolite with and without cation exchange, the degree of crystallinity in film tended to increase for the one doped with potassium ion. Either zeolite with or without potassium exchange had no impact on thermal stability of composite films. Gas permeation of PBAT/zeolite film possessed lower permeation of oxygen and carbon dioxide whereas the water vapor transmission showed relatively higher rate than the neat PBAT. After cation exchange with zeolite, the permeation of oxygen and carbon dioxide has reduced 21.04% and 21.90% respectively but the effect of K+ exchange had no effect on water vapor transmission compared with PBAT. Film of PBAT and zeolite doped with cation could absorb ethylene gas more than the one with zeolite.

1. Introduction

Low density polyethylene (LDPE) is one of the most used polymers to fabricate the fruits packaging with the drawbacks of long period of time to decompose. Researchers are interested in poly(butylene adipate-co-terephthalate) (PBAT) which is a biodegradable polymer with mechanical properties similar to LDPE. However, properties only are not enough to lengthen the fruits shelf life. There are a number of choices to extend the shelf life, for example, by improve barrier properties of food packaging using fillers like activated carbon, titanium dixode and blending polymer [1]. In this research, adding zeolite into PBAT is one option to control oxygen (O_2), carbon dioxide (CO_2), water vapor and ethylene gas (C_2H_4) permeated through the film packaging. Normally, zeolite is used as catalyst in petroleum for selective absorbent of gaseous molecules. Zeolite13X itself has effective pore opening of 8 angstroms. Therefore, it can absorb O_2, CO_2, water vapor and C_2H_4 which have critical diameters of 2.8, 2.8, 3.2 and 4.2 angstroms. Zeolites not only improve gas permeability of polymer but also thermal properties. However, the polarity difference of PBAT and zeolite could affect the dispersion of zeolite in polymer matrix. (3-aminopropyl)trimethoxysilane was used as the coupling agent to improve the compatibility between zeolite and PBAT. Potassium ion (K_+) is one of effective gas ethylene adsorbent [2]. Potassium nitrate (KNO_3) was selected as the ethylene adsorbent to improve ethylene adsorption of zeolite. [3]

As mention earlier, researchers are focusing on development of film packaging of poly(butylene adipate-co-terephthalate) incorporated with modified zeolite using (3-aminopropyle)trimethosysilane. Effects of zeolite 13X at 1 wt% and cation exchanged zeolite with potassium ion on thermal properties and permeability of gases (oxygen, carbon dioxide and water vapor) and also adsorption of ethylene gas of PBAT composites films have been thoroughly studied.

2. Experimental

Poly(Butylene adipate-co-terephthalate), PBAT Ecoflex F blend C1200 was used as polymer matrix. It was purchased from Polymats Co., LTD Thailand. Zeolite 13X with size of 38-45 micron and (3-aminopropyl) trimethoxy silane was purchased from Sigma-Aldrich. Potassium nitrate was purchased from Ajax Finechem. For surface functionalization of zeolite, zeolite was added to mixed solvent of methanol and deionized (DI) water 2:1 v/v with (3-aminopropyl) trimethoxy silane 1 wt% of zeolite and stirred at room temperature for 5 hours. After that reaction mixture was filtered and washed by methanol and DI water to remove non react silane. Zeolite powder was dried in over at 80°C overnight. For preparing cation exchange of zeolite with potassium ion, potassium nitrate 1.25 wt% was dissolved in DI water. Zeolite powder was mixed with potassium nitrate solution and stirred for 24 hours at room temperature. Zeolite with potassium ion exchanged was filtered, washed and dried overnight at 80°C. For preparation of PBAT/Zeolite film, PBAT and 1 wt% zeolite were compounded by twin screw extruder, SHJ-25 Enmech model, with temperature profile 120-175°C. Die temperature is 180 °C at 30-45 rpm screw speed. Extrudate was cooled down to room temperature and was pelletized to obtain PBAT/Zeolite pellets and then were processed to film using cast film extruder. Temperature profile was set 135-160°C from feeder to die with 60 rpm screw speed. PBAT/zeolite films thickness were gauge-controlled into the range of 0.07 mm ± 0.02 mm.

For characterization of PBAT/zeolite film, Differential Scanning Calorimetry (DSC) DSC Perkin Elmer, were carried under nitrogen atmosphere. The sample was heated from 25°C up to 200°C at rate of 10 °C/min in Heat-Cool-Heat mode. Crystallinity degree, X_c, was calculated using Equation (1) where ΔH_m is the heat of melting and ΔH_m^0 is the heat of fusion of 100% crystalline PBAT, considered as 114 J/g.

$$X_c = \frac{\Delta H_m}{\Delta H_m^0} \times 100 \qquad (1)$$

For oxygen and carbon dioxide permeability test, films were cut into square shape 2 inch x 2 inch then attached to gas permeation tester foil for Gas Permeation Tester (GDP-C, Brugger). Water vapor transmission rates (WPVR) of films were determined according to the standard method ASTM E96. Film samples were placed on cups containing 15 g of silica gel and place in desiccators contain NaCl solution to control humidity 75% RH. Cups were weighed every hour for nine hour and then every 24 hours until cups had constant weight. For ethylene adsorption testing, films were prepared into square shape 5 inch x 5 inch then were placed into 10 ml of container which saturated with 5 ppm of ethylene gas. Sampling 2 ml of ethylene gas from container using syringe then testing using Gas chromatography - Mass spectroscopy using Flame Ionization Detector (FID) mode, carrier gas is helium with flow gas rate and temperature is 1.20 ml/min and 40 °C respectively.

3. Results and discussion

3.1. Thermal properties of film

Table 1 shows the melting temperature (T_m) and % crystallization (X_c) of PBAT film. PBAT has % crystallinity 8.16% while PBAT with zeolite 13X has higher percentage of crystallinity than PBAT. This is due to zeolite acting as nucleating agent. Regarding the effect of zeolite treatment by doping K_+, crystallinity of PBATZ13X_1%_K+ was lower than the one of PBAT composite but higher than the one of neat PBAT. The tabulated data implied the treated K+ zeolite could act as nucleating agent when compared crystallinity with neat PBAT. Naturally, zeolite 13X composed of sodium ion (Na_+) in the structure. Treated zeolite 13X by doping potassium ion (K_+) lessened the capability as nucleating agent. This could be the fact of potassium ion 1.33 angstrom as compared to sodium ion size of 0.97 angstrom. And thus treated zeolite with doping potassium ion possesses less surface area for crystallinity induction of PBAT. In addition, it is found that melting temperature of PBAT composite was lower than the one of neat PBAT. This might be more crystals nucleated by zeolite in the polymer composite. Thus the crystal size of PBAT composite would be smaller than the neat PBAT which then affected the lower T_m. However, modified zeolite with potassium ion had no effect on melt temperature of PBAT.

Table 1 DSC results of film that was added zeolite 5A and zeolite 13X at 1, 3 and 5 wt%.

Sample	T_m (°C)	Delta H (°C)	X_c (%)
PBAT	121.5	9.31	8.16
PBATZ13X	119.8	18.01	15.80
PBATZ13X_K+	119.7	11.11	9.74

3.2. Barrier properties of film

Table 2 shows permeation of oxygen and carbon dioxide through PBAT composite film compared with neat PBAT. Adding zeolite into PBAT will reduce the permeability of oxygen and carbon dioxide due to zeolite particles created the longer pathway for gases to diffuse through the film. Moreover, zeolite pores could absorb gases with smaller molecules than the pore sizes. Therefore, permeation rate of both oxygen and carbon dioxide through film were decreased. In addition, crystalline of polymer could affect the permeation rate. More crystalline phase could reduce the rate of gas permeatility through tested film [4]. Regarding the effect of cation exchange in zeolite, PBAT films incorporated with modified zeolite had both of oxygen and carbon dioxide permeation rates lower than unmodified zeolite. This could be explained by the reduction of pore size from the introduction of potassium ion of which diameter was larger than sodium (Na_+) ion. Potassium ion could form donor-acceptor complexes or pi-complexes with oxygen and carbon dioxide. Potassium ion doped in modified zeolite 13X had likely to have natural affinity with oxygen and carbon dioxide rather than sodium ion in such modified zeolite and thus it illustrated more effectiveness on permeability of both oxygen and carbon dioxide gases through PBAT film.

Table 2 Oxygen and carbon dioxide permeation rate of PBAT film and PBAT composites film.

Sample	O_2 permeability ($cm_3mm/m_2d.atm$)	O_2 permeability (barrer)	CO_2 permeability ($cm_3mm/m_2d.atm$)	CO_2 permeability (barrer)
PBAT	182.17	2.77	917.15	13.97
PBATZ13X	151.78	2.31	826.77	12.59
PBATZ13X_K+	143.83	2.19	744.09	11.33

Water vapor transmission rate (WVTR) of PBAT composites films were shown in Table 3. Neat PBAT film possessed lower wather vapor transmission rate than PBAT composite film. This could be the lower hydrophilicity of PBAT compared to zeolite which composed of silica and alumina in which the pore structures were made. These pores will allow those smaller molecules to pass through. In this case, water molecular size of 3.2 angstrom with bending shape molecule can diffuse through zeolite in PBAT better than neat PBAT. Diffusion and adsorption abilities depend on many factors. One of important factors is so called active surface area. [5] Zeolite 13X is approximately 8 angstrom of pore size. Zeolite 13X possesses higher surface area which can lead to high ability of water adsorption by hydrogen bonding between water molecule and zeolite. However, PBAT film incorporated with modified zeolite 13X with doping K+ had not much difference in WVTR. Therefore, PBAT film incorporated with zeolite 13X spent less time in order to diffuse through film compare to neat PBAT.

Table 3 WVTR and percentage of ethylene gas absorption by PBAT films.

Sample	WVTR (g /m2 day)	% Ethylene gas absorption			
		Day 2	Day 3	Day 7	Day 14
PBAT	33.83	4.00	21.86	34.88	49.35
PBATZ13X	44.92	11.99	26.19	36.46	52.52
PBATZ13X_K+	45.95	25.40	35.00	43.94	53.34

Table 3 shows the barrier property of ethylene gas absorption of neat PBAT compared with PBAT composite film of 2-14 days. Fruits and vegetables would normally release ethylene gas resulting in premature ripening. If packaging film could absorb the ethylene gas, it would highly likely prolong the shelf life of packaged products. PBAT composite film in cooperative with modified zeolite by potassium ion could certainly absorb more ethylene gas than PBAT composite and neat PBAT, respectively. This could be explained by the formation of pi complex between ethylene gas and K+ ion. Alkene from ethylene which acts as electron donor can donate its pi bonding electron to metal or so called electron acceptor.

4. Conclusion

This research focused on the effect of modified zeolite on thermal and barrier properties of biopolymer film made of PBAT. From DSC results, zeolite itself would initiate a higher percentage of crystallinity in PBAT film because zeolite could act as nucleating agent. The results of permeability test of O_2 and CO_2 indicated that zeolite 13X caused the PBAT film to possess a relatively lower O_2 and CO_2 permeability than the neat PBAT. In addition, PBAT incorporated with zeolite 13X led to higher ability of ethylene gas absorption thorough PBAT film. Regarding the treated zeolite 13X by doping potassium ion in PBAT film, PBAT film with treated potassium ion zeolite had no effect in melting temperature but it would have the effect on decrement of crystallinity. Percent crystalline of PBATZ13X_K_+ would be decreased once compared with PBAT film with untreated K_+. The results of permeability test of O_2 and CO_2 indicated that treat K_+ zeolite 13X had the effect on PBAT film of lowering both of O_2 and CO_2 permeability once compared with film of untreated K_+ zeolite. In regards to the water vapor transmission, zeolite with K_+ exchange had no significant effect on WVTR. This might be because the pore size play more significant role than the cation exchange effect.

5. References

[1] Ketkul K, Threepopnatkul P, Aussawasathien D, and Hrimchum K 2017 *Key Eng Mater* **751** 344
[2] Erdogan B, Sakızc M, Yorukogullar E. 2008 *Appl Surf Sci* **254** 2450
[3] Sue-aok N, Srithanratana T, Rangsriwatananon K, Hengrasmee S 2010 *Appl Surf Sci* **256** 3997
[4] Zarshenas K, Raisi A, Aroujalian A 2016 *J Membr Sci* **510** 270
[5] Kim KM, Oh HT, Lim SJ, Ho K, Park Y, and Lee CH 2016 *J Chem Eng Data* **61** 1547

Acknowledgement

The authors gratefully acknowledge the financial support and materials of the following: SURIC (Silpakorn University Research, Innovation and Creativity Administration Office, Thailand) and Department of Materials Science and Engineering, Faculty of Engineering and Industrials Technology, Sipakorn University.

IOP Conf. Series: Materials Science and Engineering **773** (2020) 012043 doi:10.1088/1757-899X/773/1/012043

Effect of composition ratios and mixing steps on properties of BR/NR/NBR blends and blends foam

C Thongpin[1,*], K Jumpathong[1], R Jaroenjai[1], W Bunwanna[1], and J Jareanta[1]

[1]Depertment of Materials Science and Engineering, faculty of Engineering and Industrial Technology, Silpakorn University, Sanamchandra Palace Campus, Nakornpathom, 73000, Thailand.

*Corresponding author: THONGPIN_C@su.ac.th, cmaterials@hotmail.com

Abstract. This research is aimed to prepare foam from BR/NR/NBR blends which should integrate key property of these three rubbers. BR possesses high abrasion resistance but low tensile strength while NR possesses high tensile strength [1]. The addition of NBR was advantage for non-polar solvent resistance [2]. The effect of the mixing steps of rubber with vulcanizing additives on vulcanization of rubber and on mechanical properties and foaming of BR/NR/NBR were reported. The ratios of BR/NR/NBR were various as 60/20/20, 40/40/20 and 20/60/20. The vulcanizing additives used for rubber compounding were 5 phr ZnO, 2 phr stearic acid, 1.5 phr MBTS and 2 phr sulfur. The first part of the work was to study the effect of mixing step on rubber blends. Compounding system I was carried out by masticated BR, NR and NBR on a two-roll mill after that vulcanizing agent was added consecutively obtaining BR/NR/NBR rubber compounds with various rubber compositions. Rubber compounds were then characterized and vulcanized in order to investigate vulcanizate properties. In System II, three rubbers were mixed separately with required vulcanizing additives on the two-roll mill, obtaining BR, NR and NBR compound containing the same content of vulcanizing agent. Then the compounds were blended to obtain compound blended for the same blend ratios in System I. The blended rubber compounds were then characterized and vulcanized for their properties. In term of rubber blend foaming these two mixing systems were also used to compound the rubbers by addition foaming agent into the compound during mastication or blending of rubber on the two-roll mill. The results showed that mixing step affected the properties of rubber compound and vulcanizate in rubber blend. As various rubbers were mixed in different step resulting in different viscosity and hence affect foaming of rubber and subsequently properties of rubber foam.

1. Introduction

Rubber foam, also known as cellular, sponge or expanded rubber, is mainly manufactured with foaming or blowing agent to produce air gap in rubber matrix. Rubber foams are well known for their endurance and shock absorber. In fact, rubber foam and rubber sponge can be used interchangeably. There are various applications of rubber foam including cushioning in automobile seat, shock absorber, furniture, wall insulation, shoes and heel for footwear and so on. Natural rubber not only shows distinctive features in high tensile and tear strength, high resilience [3] and good tacking properties, but also shows excellent dynamic properties with low hysteresis with high resistance to abrasion. Nonetheless, NR also has disadvantages which it could be easily deteriorated under oxygen, ozone and heat due to the unsaturation in the chains. In addition, NR is not resistant to petrol and organic solvent. Blending NR with other rubber or polymers can survive both processing and its disadvantage such as blending with

EVA and crosslinking by irradiation [4] could improve foam processing and foam properties such as foam density and rebound resilience. High polar rubber such as NBR can be foam and applied mainly for gaskets, heat insulation, soundproof and vibration dampening applications. The advantages of nitrile rubber foam are smooth, even and soft outer skin. The surface of NBR foam offers flexible and resilience touch [5]. Butadiene rubber (BR) is the most elastic rubber and offers good abrasion and tear resistance as well as low heat buildup including low rolling resistance. In order to integrate properties of NR and NBR sponge and advantaged properties of BR, blend of BR/NR/NBR sponge will be studied in this part of the research. Sulfur curing system will be applied to the rubber blend. The solubility of the curatives into each rubber phase should affect curing and blowing of rubbers hence two compounding system will be compared.

2. Experimental

Materials: NR grade STR5L (standard Thai rubber) was purchased from PJ RUBBER Co., Ltd. Butadiene rubber, BR, grade 130B was kindly provided from UBE Chemicals, Thailand. Acrylonitrile butadiene rubber, NBR, grade 6240, with 34% acrylonitrile content, was purchased from LG Chem. Ltd., Korea. ZnO, stearic acid, MBT, OBSH and sulfur were kindly provided by Innovation group Thailand.

Preparation of rubber compound: System I, BR/NR/NBR with the ratio indicated in Table 1 were kneaded and masticated on a two-roll mill for 30 minutes. ZnO, stearic acid, MBT and OBHS were added respectively. After well dispersion, sulfur was added and mastication was carried on for 5 minutes obtaining BR/NR/NBR System I compound. In System II, each rubber was kneaded and masticated for 30 minutes before addition of additives written in Table 1. To prepared rubber blend, compound follow the composition stated in Table 1 was weighted to the required rubber ratio and mixed on the two-roll mill obtaining BR/NR/NBR System II compound. Rubber blend compound were then investigated for cure characteristic parameters using Moving Die Rheometer (MDR, Gotech model GT-M2000, Taiwan) at 170°C.

Preparation of rubber foam: Various BR/NR/NBR blends foam was produced in a compression moulding machine at 170°C under the pressure of 1500 psi for 16 min. The rubber sponge obtained was then cut into specimen for various tests. Scanning electron microscopy, Hitashi, TM3030, was used to take rubber cell foam micrographs. The dumbbell-shaped specimens were tensile tested according to ASTM D412 using Instron universal testing machine Model Instron 596, with the extension rate of 500 mm/min. Compression set was performed by applied compression force on to specimens to its 50 % height for 22 h. After determined time, the specimens were left to relax for 30 min then the height of the rubber foam were recorded. Compression set was determined using equation 1:

$$\% \text{ Compression set} = CRI = \frac{100}{t_{s2} - t_{c90}} \quad \text{equation 1}$$

Table 1 Formulation of BR/NR/NBR and sulfur curing additives in phr (part per hundred of rubber).

Ratios of BR/NR/NBR (by weight)	BR	NR	NBR	ZnO	Stearic acid	MBT	OBSH	Sulfur
60/20/20	60	20	20	5	2	1.5	10	2
40/40/20	40	40	20	5	2	1.5	10	2
20/60/20	20	60	20	5	2	1.5	10	2

3. Results and discussion

Mechanical properties of rubber blend prepared via System I and System II are presented in Figure 1. (a)-(b). It was reported by Noordemeer [6] that MBT has preference to be soluble in NBR to BR and EPDM meanwhile sulfur prefer to dissolve in nonpolar rubber to polar rubber. At high NR content, i.e. 60 % by weight, curing rate in NR phase occurred in System I was less than that in System II and resulting in lower tensile strength. Among these 3 three rubbers BR possess lowest curing rate and this causes the lowered curing rate of rubber blend with high BR content in System II. Tensile strength and elongation at break of the rubber blends were increased with NR content as vulcanized NR generally possesses high tensile strength due to its stress induced crystallization.

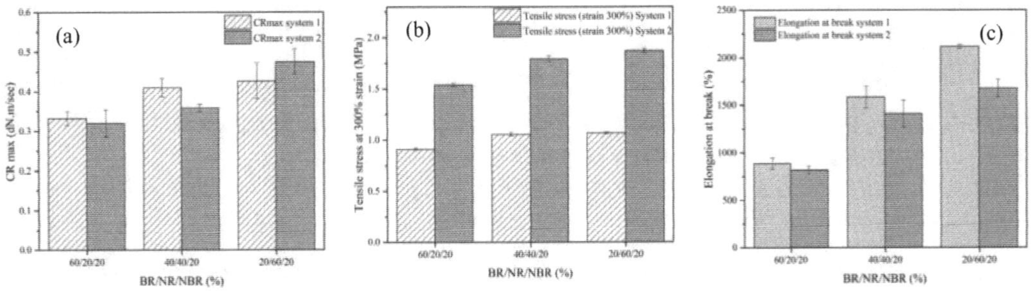

Figure 1. Maximum cure rate (a), tensile strength at 300 % strain (b) and elongation at break (c) vulcanized rubber blends.

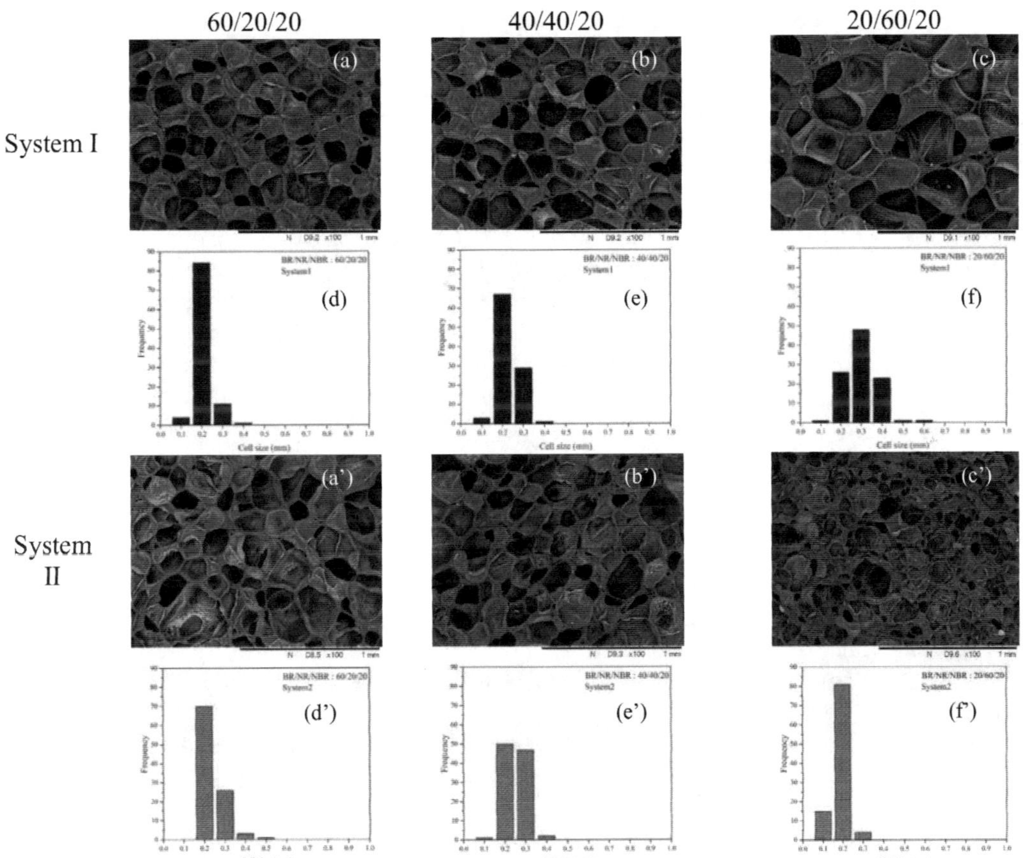

Figure 2 Morphology and cell size distribution of BR/NR/NBR foam prepared via System 1, (a)-(f) compared via System II (a')-(f').

The morphology of BR/NR/NBR blends foams and cell size distributions are shown in Figure 2 (a)-(f) and cells size distribution are shown in Figure (a')-(f'). For rubber blend foam prepared via System I with various BR and NR as shown in Figure 2 (a)-(c). Cell size of rubber foam was found increased with the content of NR. This was due to the Mooney viscosity of NR compound was 13 Mooney units whereas BR and NBR compounds were found 25 and 40 Mooney unit respectively. High NR content, 60

% by weight, lead to flow ability of rubber compound during compression moulding. During melt, blowing agent was thermally decomposed and produced N_2 gas [7] then crosslinking rubber was expended. BR/NR/NBR with low Mooney viscosity would be easily expanded resulting in larger cell foam as seen in Figure 2 (c). For rubber blend foam compounded via System II, rubber cell size at high NR compound content, i.e. 60%, shows the smallest cell size. As mentioned earlier that rubber blend with the ratio of 20/60/20 in System II exhibited maximum cure rate compared with the other ratios due to the high NR content and curatives has already dispersed in NR compound before blending of the rubber compounds and this cause highest delta toque (i.e. crosslink density) consequently less expansion of the cell foam was performed. Tensile stress at 300 % strain and compression set of the foams were elucidated in Figure 3 (a)-(b). It has been known that BR and NBR vulcanizates performed less elastomeric behaviour than NR [8] and this was responsible for compression set of BR/NR/ENR foam.

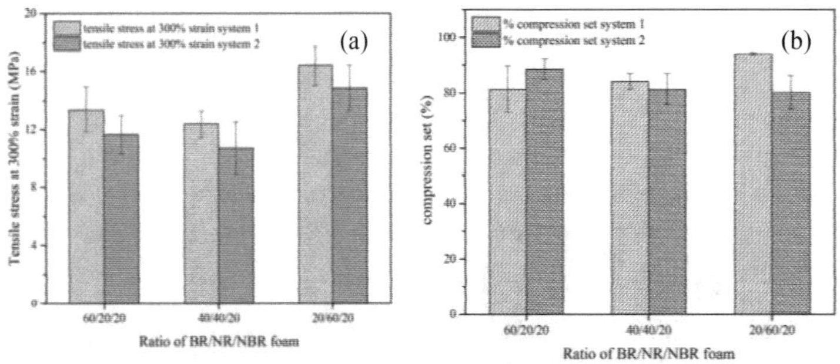

Figure 3 (a) Tensile stress at 300 % strain and (b) compression set of BR/NR/NBR foam prepared from System I compared with System II.

4. Conclusions

Ternary rubber blend of BR/NR/NBR were prepared via 2 compounding method named as System I and System II and foaming of the blends were also prepared. The morphology of the foam showed that blends ratio and step of mixing affected cure characteristic, mechnical properties of rubber blends. Mechnical properties of rubber blend prepared via system I are higher than that prepared via system II. The highest delta torque of comound prepared via System II was the highest and this resulted in the smallest cell size of the foam consequently low compression set. It was clear that the mixing step affected the properties of rubber compound and vulcanizate in rubber blend. As various rubbers were mixed in different step resulting in different viscosity and hence affect foaming of rubber and subsequently properties of rubber foam.

5. References

[1] Ismail M-N, El-Sabbagh S. H., and Yehia A.A. 1999 *Elastomer and plastics*. **31** 255-270.
[2] Matador Rubber 2007 Education and Culture Lifelong Learning Program Leonardo Da Vinci.
[3] Salleh S, Ahmad Z and Ismail H 2016 *Procedia chemistry* **19** 346-350.
[4] Salmazo LO, Lopez-Gil A, Ariff Z.M, Job A.E and Rodriguez-Perez M.A 2016 *Industrial Crops and Products* **89** 339-349.
[5] Zhao X, Niu K, Xu Y, Peng Z, Jia L, Hui D and Zhang L 2016 *Composites Part B* **107** 106-112.
[6] Guo R, Talma A.G, Datta R.N, Dierkes W.K and Noodermeer J.W.M 2008 *European Polymer Jornal* **44** 3890-3893.
[7] Lopattananon N, Wangpradit N, Nakason C and Kaesaman A 2014 *Journal of Rubber Research* **17** 80-95.
[8] V Raji V, Ramakrishnan S, Sukumar R, Brahmakumar M and Ravindranatha Menon AR 2015 *Society of Chemical Industry* 10 4956.

Acknowledgement

The authors would like to thank Department of Materials Science and Engineering, Faculty of Engineering and Industrial Technology, Silpakorn University for their support throughout this research.

Study of rubber composites between natural rubber and Mahogany Shell Powder (MHSP) and potential for pavement block

J Jaroenta, R Siangsawad, R Onkate, V Jearjiratikul, and C Thongpin*

1 The department of Materials Science and Engineering, Faculty of Engineering and Industrial Technology, Silpakorn University, Sanamchandra Palace Campus, Nakornpathom, 73000, Thailand.

*Corresponding author: THONGPIN_C@su.ac.th, cmaterials@hotmail.com

Abstract. This research was aimed to study the feasibility of using mahogany fruit shell powder as filler in rubber and investigated in properties of obtained vulcanized rubber. As mahogany fruit shell contains 75.61% holocellulose, 13.54% lignin and 11.02% extractives [1]. Hence the fruit shell should be able to act as reinforcement in rubber materials [2]. In this study, mahogany fruit shell was milled into small particles with the particle size of approximately 60 micrometres follow by ASTM D 422 - Standard Test Method for Particle-Size Analysis of Soils. The particle was characterized for their thermal stability. The morphology of mahogany fruit shell powder, MHSP, was flake like particle and can be used as reinforcement fillers in natural rubber. As NR is generally sensitive to O_2 thus ethylene propylene diene monomer, EPDM, would be laminated on the top of NR/MHSP composites. The rubbers were formulated with efficient vulcanization system and contain 10, 20, 30, 40 and 50 phr of MHSP. The rubber compounds were characterized for cure characteristic parameter such as scorch time, cure time, minimum and maximum torque. Rubber vulcanizates were prepared using compression molding and characterized for their morphology, mechanical properties, thermal degradation, hardness and weathering resistance. In order to improve environment stability and abrasion resistance of the NR/MHSP composites, EPDM/MHSP were prepared and laminated on NR/MHSP and vulcanized. The laminated vulcanizated composites possessed the properties follow Thai Industrial Standard for rubber paving blocks could be obtained. The results also showed that NR filled with 50 phr of MHSP in NR and in EPDM gave the properties follow the standard and has high potential for outdoor rubber paving block.

1. Introduction

Vulcanized NR has been known for its excellent properties and extensively used in tire applications. In the rubber industry fillers are added to improve rubber compounds. The fillers are aimed either to improve electrical properties, mechanical properties, thermal resistance or cost reduction. Carbon black is the most important reinforcing filler used in the rubber industry. However, it causes pollution and also gives black colour to the rubber. With the awareness of environmentally global protection, biocomposites between rubber and natural based filler have been interested [3]. Natural fillers can be varied from natural fibres to particulate producing natural fibre/NR and particulate/NR composites. The application of rubber biocomposites have been ranged from aerospace, automotive to building materials. Rubber tile for paver block or pavement block could be useful for ageing society due to its great properties such as shock absorption preventing elder age injuries. Floor and walkway tile, sport and play surface and sheet rubber are manufactured using crumb rubber particle using polyurethane as binder [4]. In this research we compound NR with mahogany fruit shell powder (MHSP), as natural filler, in order to obtained required hardness of vulcanized NR follow rubber tile standard and applied for rubber pavement block. Mahogany fruit shell contains 75.61% holocellulose and 13.54% lignin. The rubber block will be laminated with MHSP/EPDM vulcanizate in order to protect MHSP/NR composite from environmentally degradation.

2. Experimental

Materials: NR STR5L and EPDM 4725P containing ethylene and ENB of 70 and 4.9 % by mass respectively, were obtained from CHAREON TUT Co. Ltd. Thailand. Curing additives which are ZnO, stearic acid, MBT, DPG, CTP, 6PPD and sulphur were kindly supplied by Chemical Innovation group, Thailand. MHSP/rubber composites compound was prepared using efficient vulcanization system. Ground samples of Mahogany were separated according to particle size using a sieve shaker. Mesh size of sieves was 0.3, 0.15, 0.106 and 0.075 mm. Each sample was placed in the top sieve with the largest mesh and shaken for 5 min at an amplitude setting of 2 mm, disassembled and stirred lightly, then shaken for additional 5 min. The content of MHSP with particle size was varied from 0, 10, 20, 30, 40 and 50 phr. For EPDM was compound using efficient vulcanization and filled with MHSP at the same content as former rubber composites. Rubber compounds were tested using MDR, Gotech2000, Taiwan, at 160 °C for cure characteristic parameters. Vulcanization was carried out in compression moulding at 160 °C under pressure of 1500 psi. The vulcanized rubber composites were evaluated for their tensile properties and hardness using UTM, Instron model 5965, USA. Weathering resistance of the composites were evaluated according to ASTM G154 and the degree of colour fastness was measured using colour meter. The highest hardness MHSP/NR was selected and laminated with EPDM composite with the laminated thickness of 1-2 mm to obtain rubber tile 4 mm thick according to Thai Industrial Standard [5]. Hardness, Weathering resistance was also evaluated.

3. Results and discussion

Cure characteristic of MHSP/NR and MHSP/EPDM are presents in Figure 1 (a)-(f).

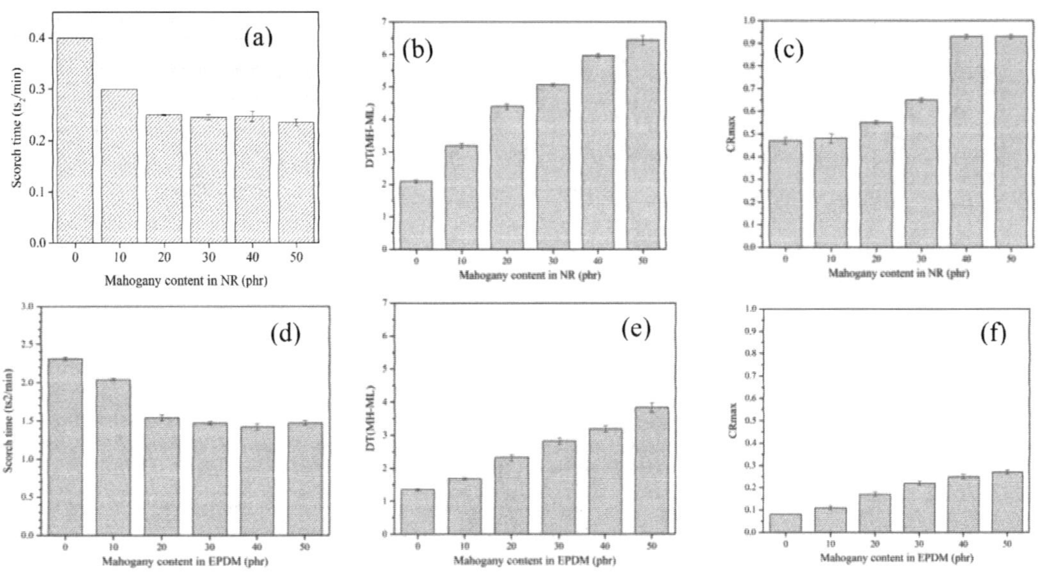

Figure 1 Scorch time, delta torque and maximum cure rate of MHSP/NR (a)-(c) and MHSP/EPDM (d)-(f).

The results show that the addition of MHSP into both rubbers can accelerate vulcanization reaction, as can be seen by the shortening of scorch time and maximum cure rate index. This should be due to the organic substance contained in mahogany fruit shell such as fatty acid [6] which can act as co activator in sulfur vulcanization. In term of mechanical properties, MHSP could act as reinforcement in NR and EPDM and resulted in the increased in secant modulus as shown in Figure 2(a) and (d). In order to obtain hardness of rubber tile higher than 50 Shore A, according to Thai industrial standard, the content of MHSP added into rubbers should be higher than 40 phr, as shown in Figure 2 (c) and (d).

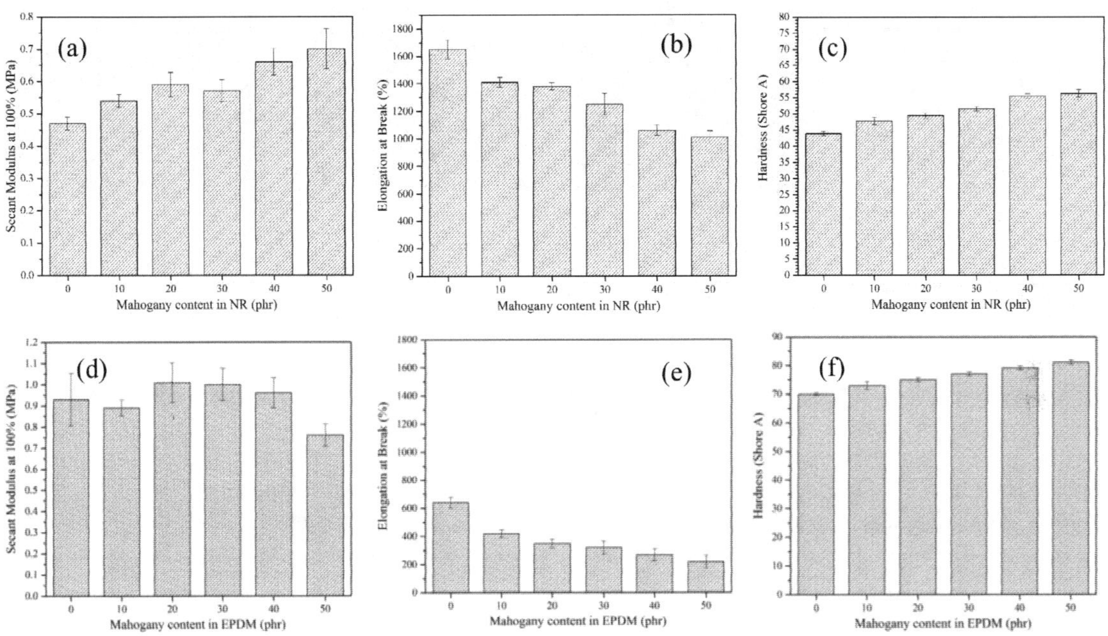

Figure 2 Secant modulus at 100 % strain, elongation at break and hardness of MHSP/NR (a)-(c) and MHSP/EPDM (d)-(f).

Environmental resistance of MHSP/rubber composites were also examined and reported as degree of colour fastness depicted in Table 1. The results show that MHSP/NR composite with high MHSP content has lower degree of fastness than low MHSP content this should be due to the lignin content in MHSP. In fact, the degree of fastness should be higher than 4 and all composites have passed the standard of rubber tile.

Table 1 Degree of colour fastness of rubber/Mahogany fruit shell powder composite after weathering test (environmental resistance).

Formulation	Degree of colour fastness	Formulation	Degree of colour fastness
NR	3	EPDM	1
NR/MHSP 10 phr	4	EPDM/MHSP 10 phr	2
NR/MHSP 20 phr	4	EPDM/MHSP 20 phr	3
NR/MHSP 30 phr	4	EPDM/MHSP 30 phr	(2-3)
NR/MHSP 40 phr	(3-4)	EPDM/MHSP 40 phr	3
NR/MHSP 50 phr	(3-4)	EPDM/MHSP 50 phr	3

MHSP/NR composite and MHSP/EPDM composites contained MHSP 50 phr were selected to prepare laminated composite and evaluated for their hardness and degree of colour fastness. The thickness of EPDM composite laminate was varied from 1 to 2 mm. The hardness was found at 75 and

77 shore A respectively. Also, the degree of colour fastness was found at 3-4 in both 1 and 2 mm EPDM composite laminate thickness showing that both types of EPDM laminated MHSP/NR composite are qualified according to Thai Industrial standard for rubber tile. It is also useful to confirm that after weathering test hardness of both laminated rubber composite was not changed due to the environmental resistance of MHSP/EPDM composite. However, physical appearance was slightly changed due to the MHSP itself as shown in Figure 3. This colour change followed Thai industry standard for rubber tile as mentioned earlier.

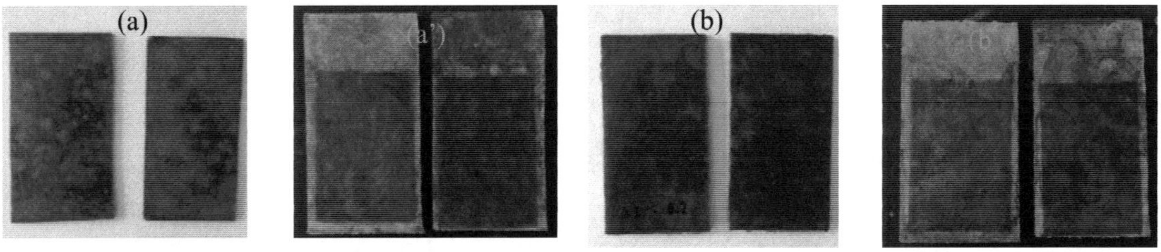

Figure 3 Digital picture of MHSP/NR composite laminated with MHSP/EPDM composite the laminated thickness of 1 mm (a, a') and 2 mm (b, b') before (a, b) and after (a', b') weathering test.

4. Conclusions

In this research, rubber composites using mahogany fruit shell powder as reinforcing filler were successfully prepared. The composite with MHSP 50 phr possess required hardness for rubber tile, according to Thai Industry Standard. The composite was applied for rubber tile by laminate MHSP/EPDM on to MHSP/NR in order to protect MHSP/NR from environmental deterioration. The lamination thickness that was examined was 1 and 2 mm and found that hardness of both types of tile still follow the standard. In addition, the degree of colour fastness also agrees well with the standard. The results confirmed that the laminated composite are able to be used for pavement block particularly advantage for ageing society.

5. References

[1] Colares C.J.G, Pastore T.C.M, Coradin V.T.R, Camargos J.A.A, Moreira A.C.O, Rubim J.C and Braga J.W.B 2015 *Journal of the Brazilian Chemical Society* **26** 1-9.
[2] Sarki J, Hassan S.B, Aigbodion V.S and Oghenevweta J.E 2011 *Journal of Alloys and Compounds* **509** 2381-2385.
[3] Mahir N, Ismail H and Othman N 2016 *Journal of Polymer Materials* **33** 233-243.
[4] Kiatkamjornwong S and Pairpisit K 2004 *Journal of Applied Polymer Science* **92** 3401-3416.
[5] https://is.gd/vd6e9h, December 2019.
[6] Ibrahim H, Fasanya, Opeoluwa O, Hayatudeen A and Evidence O 2018 *Nigerian Journal of Technological Research* **13** 45-49.

Acknowledgement

The authors would like to thank the Department of Materials Science and Engineering, Faculty of Engineering and Industrial Technology, Silpakorn University for their support throughout their work.

Bio-based thermoplastic vulcanizates from natural rubber (bioplastic/NR)

Chanchai Thongpin[1,2,*], Kwanchai Buaksuntear[1], Theraphat Tanprasert[1] and Jedtarin Charoenta[1,2]

[1]Department of Materials Science and Engineering, Faculty of Engineering and Industrial Technology, Silpakorn University, Sanamchandra Palace Campus, Nakornpathom, 73000, Thailand.
[2]Centre of Excellence on Petrochemical and Materials Technology (PETROMAT), Chulalongkorn University, 7th floor, Chulalongkorn University Research Building, Phayathai Rd., Bangkok 10330, Thailand

*Corresponding author: THONGPIN_C@su.ac.th, cmaterials@hotmail.com

Abstract. Thermoplastic vulcanizates (TPV) is a class of thermoplastic elastomer of which generally consists of polyolefin phase, mainly PP or PE, with dynamically cured elastomer phase such as EPDM. The elastomer should be well dispersed in the polyolefin matrix which facilitates thermoplastic like processing of elastomer. The properties of TPV are mainly governed by properties of thermoplastic matrix and vulcanized rubber phases including curing system used for dynamic vulcanization. As environmental problem has been an important issue, application of bioplastic in thermoplastic elastomer is an interesting issue. The blending of NR with various has been studied in our work and the system should classify as bioplastics/NR. Dynamic vulcanization of NR during melt blending of NR with various bioplastics was also studied. In this report, the melt blending system of PBAT/NR; PCL/NR and PBS/NR were carried out. The batch melt blending in an internal mixer were performed with various NR and bioplastics compositions. Dynamic vulcanization of NR phase, using Luperox® 101 as curing agent, was also occurred in the mixer during melt blending. Except for PBS/NR TPV system where dicumyl peroxide was used as curing agent. This will advantage for foam preparation in further foaming step. The results showed that PBAT/NR TPV with NR content as high as 70% possessed tension set less than 20% and elongation at break under tension was higher than 600%. For PCL/NR system at the same composition, the tension set was also found at 20%, elongation at break under tension was as high as 1000%. The TPV system of PBAT/NR and PCL/NR were also extrudable. Tension set for NR/PBS TPV system was found at about 30% and the elongation at break under tension was about 600%. From the mentioned basic properties of biobased TPV in our researches, it presents that they could propose the promising qualified TPV, in term of reprocess ability and advantage for the sake of biodegradability.
Note: PBAT, PCL and PBS stand for poly(butylene adipate-co-terepthalate, polycarprolactone, and poly(butylene succinate) respectively.

1. Introduction
Elastomer is a class of polymeric materials that can be deformed at room temperature and can be reverted quickly to nearly original size and form after unloading. In fact, elastic recovery under

compression and tension is its predominant properties. Generally, elastomer can be stretched as twice as its original length. Resilient property is also an important property of elastomer. Elastomer could be thermoplastic and thermoset polymers. Thermoplastic elastomers can be processed using conventional processing techniques for thermoplastics such as extrusion and injection molding. The scrap from the processes can be recycled. There are various classes of thermoplastic elastomers which are styrenic block copolymers (TPS), thermoplastic polyolefin elastomers (TPO), thermoplastic polyurethans (TPU), thermoplastic copolyester (TPC), thermoplastic polyamides (TPA) and thermoplastic vulcanizates (TPV). As it is generally known that rubber, both synthetic and natural rubber, show viscoelastic properties which can be applied for specific applications especially for vibration and shock isolation. Coran and Patel [1] reported that blending EPDM with thermoplastic polyolefin with fully cured EPDM referred as thermoplastic vulcanizate having superior strength, high temperature mechanical properties and good compression set. There has been number of researches studied on TPV prepared from thermoplastic and natural rubber (NR) including polyolefin/NR and polyester/NR [2, 3]. Even though NR is not 100% biodegraded (only protein attached to polymer chain could be biodegraded) TPV prepared from bioplastic/NR blend with bio-based NR should be able to encourage circular economy in Thailand.

2. Experimental

2.1. Materials

Natural rubber (NR) STR5L used in this research is high in its specification with low contamination and color and was purchased from Chareon Tut Co., Ltd., Thailand. It was purchased from PJ Rubber Co., Ltd (Thailand). Poly (butylene adipate-co-terephthalate), (PBAT) or Ecoflex® F Mulch C1201 with MFR 2.7-4.9 g/10 min (at 190°C; 2.16 kg), density 1.25-1.27 g/cm$_3$, melting point 110-120°C was purchased from POLY-MATS Co., Ltd. (Thailand). Granule form of PCL (CAPA 6800) was purchased from Perstorp, UK. PCL has a molecular weight and melt flow index (MFI) is 80,000 g/mol and 3 g/10 min at 160°C/2.16 kg, respectively. The PBS (BioPBS™ FZ71PM) was purchased from PTT MCC Biochem Company by Thailand of which the melting point is 115°C, density is 1.26 g/cm$_3$. Its melt flow index (MFI) is 22 g/10 min.

2.2. Preparation of TPV

Bioplastic which are PBAT, PCL and PBS were dried at 50°C under vacuum for 6 h before use. NR was masticated on a two-roll mill for 30 min before mixing with PBAT. PBAT/NR and PCL/NR were melted mixing in an internal mixer (PLASTI-CORDER, Brabender, Germany) at 140°C with the various rotor speeds. The mixing was performed at 140°C until plasticisation of thermoplastic had occurred for 2 min and then 1, Germany) with the blend ratios of PBAT/NR as follows 60/40, 55/45, 50/50, 45/55, 40/60, 35/65 and 30/70% by weight. The blends were melted at and Luperox® 101 (Lup) as a curing agent was added. The rotor speed of mixing was also varied. PBAT/NR vulcanizates were then being compressed to sheets by compression molding in a hot-plate hydraulic press (Scientific, Labtech Engineering, Thailand) at 140°C. After hot pressing, the sheet was allowed to cool to room temperature under pressure of 1,000 psi.

NR/PCL: Dried NR and dried PCL were also melted blended in at 140°C with the rotor speed of 80 rpm using also Luperox101 as curing agent. The mixing time was carried on for 15 min. The compositions of NR/PCL were as follow: 70/30, 65/35, 60/40, 55/45, 50/50, 45/55, and 40/60% by weight. The obtained NR/PCL vulcanizates was then being compressed in a compression-molded with the temperature of 140°C and the pressure of 1,000 psi.

NR/PBS: In this system, NR was masticated and being compounded with 1.5 phr. of dicumyl peroxide (DCP) before melted blending with PBS in an internal mixer at 140°C for 15 min. before dispatching the TPV from the internal mixer chamber. The ratios of NR/PBS are as follows: 50/50, 60/40, 70/30 and 80/20% by weight. The obtained TPV were then compress in a compression molding at 150°C under the pressure of 1,000 psi.

IOP Conf. Series: Materials Science and Engineering **773** (2020) 012045 doi:10.1088/1757-899X/773/1/012045

2.3. Characterization

The tensile properties, i.e. initial modulus, tensile strength and elongation at break of the TPV were performed under universal testing machine (UTM; Instron 5969, USA). Dumbbell specimens were tested at room temperature according to ASTM D638. Furthermore, tension set were also determined according to ASTM D412 and performed on dumbbell specimen. The dumbbell specimens were tensioned to 200% and kept under tension for 10 min. The length of the specimen was measured and compared between before and after elongation. The tension was then calculated followed equation (1):

$$\%\text{Tension set} = \frac{l_2 - l_0}{l_1 - l_0} \times 100 \qquad \text{equation (1)}$$

3. Results and discussion

Mechanical properties of TPV are generally governed by many factors such as types of thermoplastic, vulcanizing system in rubber phase and importantly the phase morphology of thermoplastic and vulcanized rubber in TPV [4]. The morphology of the TPV was not only governed by the melt viscosity of thermoplastic but also the speed of rotor in mixing chamber. PBAT/NR TPV system was selected to report the effect of rotor speed of mixing as shown in Figure 1 which presents tensile strength, Figure 1 (a) and tension set, Figure 1 (b), of TPV obtained from tensile testing of TPV containing various ratios of PBAT and NR. The results show that mixing with low rotor speed of 60 and 80 rpm, PBAT/NR vulcanizates could not be elongated due to the large particle size of crosslinked NR particles. In fact, stress-strain curves of PBAT/NR TPV mixing with rotor speed of 100 rpm showed strain hardening behaviour under tensioning. The tensile strength behaviour of PBAT/NR also implied that at low rotor speed of mixing, i.e. 60 and 80 rpm, poor dispersion of NR phase was obtained and resulted in the low tensile properties [5]. Mixing PBAT and NR at high rotor speed, i.e. 100 rpm, the size of NR phase was reduced and completely dispersed led to the improvement in the tensile properties. It should be point out here that, at high rotor speed of mixing, shear heating and heat produced during dynamic vulcanization were also affected properties of TPV [5]. It was noticeable (the results could not be shown) that at 100 rpm, crosslinking (torque increment) occurred earlier than that at 60 and 80 rpm. In term of elastomeric property, tension set of the TPV should be lower than 50% meaning that the produced materials should not show permanent deformation higher than 50 % [6]. It was found that PBAT/NR with higher NR contents shows lower tension set. Also, the higher rotor speed resulted in lower tension set. This advantaged for elastomer applications.

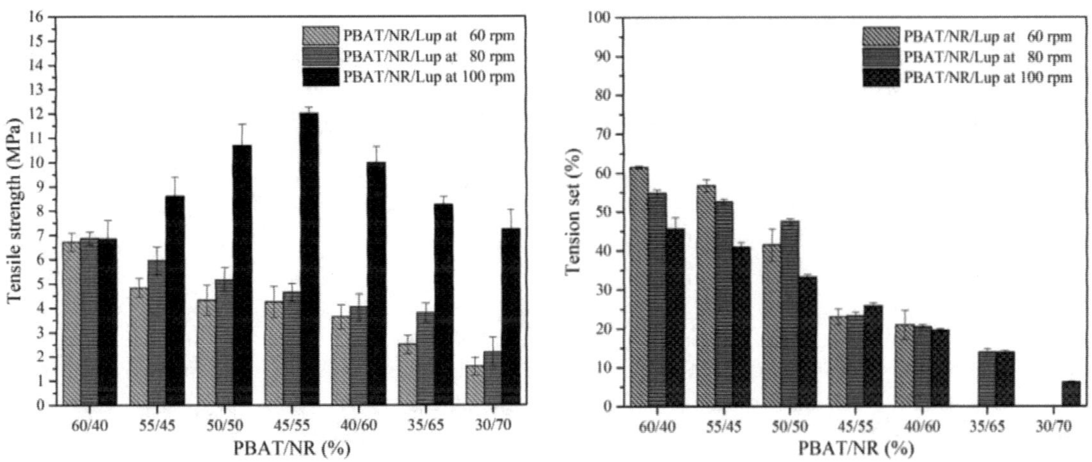

Figure 1. Tensile strength and tension set of PBAT/NR TPV dynamic vulcanized in internal mixer with Luperox101® 1 phr, various PBAT/NR ration and various rotor speed.

The degree of crosslinking in rubber phase is also important for TPV properties [7]. We investigated the effect peroxide content on tensile properties and tension set of TPV prepared from dynamic vulcanization of PCL/NR. As melting temperature of PCL is as low as 61 °C [8], we investigate the effect of peroxide content from 1, 1.5 to 2 phr by melt mixing NR and PCL at 140 °C with rotor speed of 80 rpm for 10 minutes in an internal mixer. At this condition PCL was not yet crosslinked by peroxide. The tensile properties and tension set are presented in Figure 2. In order to avoid the effect of stress induced crystallization and yielding of PCL phase, tensile stress at 800 % strain are reported. Tensile strength of PCL/NR TPV was decreased with NR content this is due to the high crystallinity of PCL (50% crystallinity) [8]. The increased in Luperox101 led to the increased in crosslink density and hence stress hardening of crosslink NR phase. Therefore, tensile strength of NR/PCL TPV was increased with the content of Luperox101. It must be noted that the properties of NR/PCL would be different if the temperature and time of mixing were increased due to the crosslinking of PCL phase. In term of tension set, crosslink density affected the properties of elastomer, i.e. elasticity of rubber phase as shown by the decreased of tension set. In this particular TPV system, PCL present as continuous phase in all composition ratios.

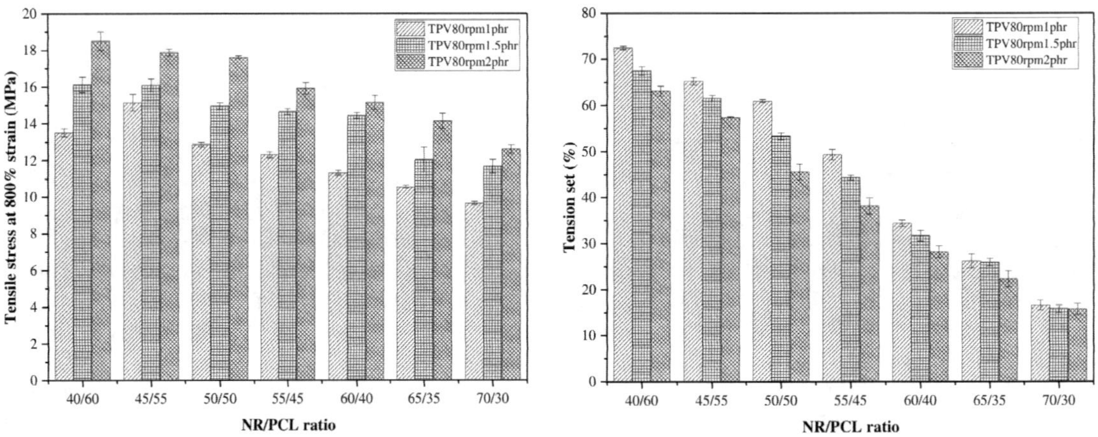

Figure 2. Tensile strength at 800% strain (a), tension set (b) of NR/PCL TPV dynamically vulcanized with various Luperox101 contents.

NR/PBS dynamically vulcanized using DCP as a curing agent was also investigated. This system was carried out in order to prepare TPV for NR/PBS foam sponge. Tensile strength of NR/PBS was increased with PBS content due to its high crystallinity (more than 60%) [9]. It was found that high rotor speed of mixing, high tensile strength NR/PBS TPV was obtained. It was noticed from the torque-time mixing curve that at higher rotor speed curing has occurred earlier than lower rotor speed. This was due to the shear heating during shearing in an internal mixer and leading to high crosslinking and hence high elasticity. This is responsible for the low tension of this system.

4. Conclusion

Thermoplastic vulcanizates derived from various Bioplastic/NR were successfully prepared and investigated for their tensile properties and tension set. It was clear that phase morphology between bioplastic and NR are very important for their properties especially elastic recovery reported by their tension set. Rotor speed of 100 rpm was found to be suitable for mixing bioplastic/NR during which dynamic vulcanization had occurred. Although higher rotor speed of mixing may be performed but shear heating occurred during mixing had to be attentive by the excess crosslinking and degradation of bioplastic. Indeed, degree of crosslink in NR phase can be increased by the peroxide content nonetheless shear heating occurred during melt mixing could also accelerate crosslinking reaction during mixing.

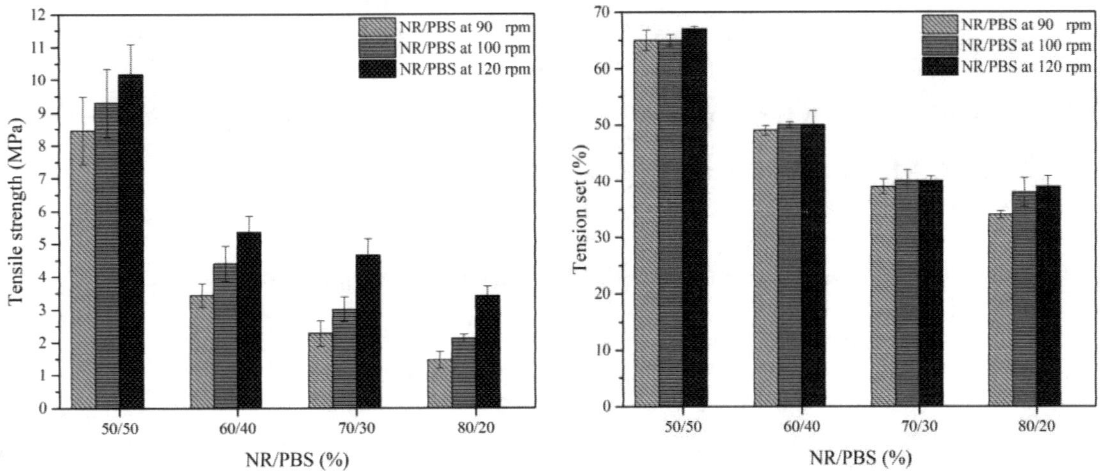

Figure 3. Tensile strength and tension set of PBS/NR TPV dynamic vulcanized in internal mixer with DCP 1.5 phr, various PBS/NR ration and various rotor speed.

5. References

[1] Coran A-Y and Patel R 1980 *Rubber Chemistry and Technology* **53** 141-150.
[2] Mondal M, Gohs U, Wagenknecht U and Heinrich G 2013 *Radiation Physics and Chemistry* **88** 74-81.
[3] Liu F, Liao K and Wang Z 2019 *Soft Materials* **17** 1-10.
[4] Naskar K and Noordermeer J.W.M. 2005 *Progress in Rubber Plastics and Recycling Technology* 1 21.
[5] Azevedo De Carvalh A-P, G. Soares B, Pontes K and Sirqueira A 2018 *Materials Research* **5** 21.
[6] Thongpin C and Tanprasert T 2019 *Key Engineering Material* **798** 337-342.
[7] Chatterjee T, Wiessner S, Naskar K and Heinrich G 2014 *eXPRESS Polymer Letters* **8** 220-231.
[8] Guang-Mei C, Tie-Mei Z, Lei C and Yi-Ping H 2010 *Asian Journal of Chemistry* **22** 5902-5912.
[9] Phua Y-J, Chow W-S and Mohd Ishak Z-A 2011 *eXPRESS Polymer Letters* **5** 93-103.

Acknowledgement

The authors would like to thank department of materials science and engineering for the supports throughout the works. The authors are also in debt to PETROMAT for the financial support.

IOP Conf. Series: Materials Science and Engineering **773** (2020) 012046 doi:10.1088/1757-899X/773/1/012046

Bio-based composite from poly(butylene succinate) and peanut shell waste adding maleinized linseed oil

N Hongsriphan, P Kamsantia, P Sillapasangloed, and S Loychuen

Department of Materials Science and Engineering, Faculty of Engineering and Industrial Technology, Silpakorn University, Sanam Chan Palace Campus, Nakhon Pathom 73000, Thailand

E-mail: Hongsriphan_N@su.ac.th

Abstract. Nowadays, the biobased plastic products have become one of the worldwide topics that people give the attention. Applications of bio-based poly (butylene succinate) (PBS) is interesting since it is fully biodegradable. However, the resin cost is expensive compared to olefins so that it is not widely used. This research attempted to produce cost-effective composite sheets from PBS and peanut shell powder (PSP) with particle size of 100 mesh in the weight ratio of 70/30, 60/40, and 50/50 wt% using a twin-screw extruder and then a compression molding. In addition, maleinized linseed oil (MLO) of 3 phr was used as a compatibilizer for the composites. It was found that the obtained composites had higher Young's modulus and Shore D hardness with respect to the PSP content, but elongation at break was reduced. The impact resistance by means of the falling dart impact test also reduced with the higher PSP content. Adding MLO into the PBS/PSP composites increased elongation at break and impact resistance, but reduced the rigidity due to plasticizing effect. Due to lignocellulosic nature of PSP, the thermal stability of the composites was decreased and MLO did not have significant influence on it. After the weathering testing for 60 h, mechanical properties and thermal stability of the composites were reduced significantly implying that these bio-based composites could degrade faster compared to pure PBS sample.

1. Introduction

Bio-based plastic products are interesting since they are sustainable, could be used and disposed with less pollution to the environment. Poly (butylene succinate) (PBS) is one of these biobased plastics that is flexible similarly to olefins and is fully biodegradable. The main disadvantage of this plastics is its high cost so that it is not widely used. In order to achieve the cost-effective products made from PBS, this research mixed high concentration of peanut shell powder (up to 50 wt%) from agricultural waste with PBS pellets to produce the composites using a conventional melt processing. In addition, maleinized linseed oil [1] was used as a compatibilizer to improve brittleness of the composites. Moreover, the composite samples were placed in the accelerated weathering chamber to investigate mechanical properties and thermal stability after the weathering.

2. Experimental

2.1. Materials

Poly(butylene succinate) or PBS (BioPBS™ FZ71PM, MFI of 22 g/10 min (190°C/2.16 kg)) was purchased from PTT MCC Biochem, Thailand. Peanut shell waste was kindly supplied by a local Thai company, which it was grinded and sieved to obtain the particle size of about 100 mesh. Purified linseed oil was purchased from Kenton Intertrade Co. Ltd., Thailand, and used as received. Maleic anhydride was purchased from Fluka™, USA. Dicumyl peroxide was purchased from Sigma-Aldrich, Germany.

2.2. Preparation of maleinized Linseed oil (MLO)
Linseed oil of 300 ml was poured into a 500 ml three-neck round bottle, which a condenser was placed at the middle neck, and nitrogen gas was purged into one neck and flow out the other neck to prevent thermo-oxidation of linseed oil. Maleic anhydride (MA) with the weight about 9% of linseed oil was poured into the bottle, and dicumyl peroxide with the weight of 6% of MA was added. Then, the mixing ingredient was stirred and heated at 150-170°C for 4 h to obtain the MLO.

2.3. Melt blending and hot pressing
PBS pellets and peanut shell powder (PSP) were dried in an air-circulating over at 60°C for 24 h. The composites were compounded with the weight ratio of PBS to PSP 70/30, 60/40, and 50/50 wt% in a co-rotating twin-screw extruder (Hakke, Germany). The barrel/die temperature was 120-130°C, and the screw-speed was 40 rpm. The extrudate was cooled in air and pelletized. MLO of 3 phr was added for the composites with MLO. After thoroughly drying, the samples were prepared by hot pressing into thin sheets (180 x 180 x 1 mm) using a compression molding machine (Labtech Engineering Co, Thailand) at 150°C. Pure PBS samples were prepared at the same procedure.

2.4. Weathering testing
The weathering testing was carried out using an accelerated weathering chamber (Q-LAB Co, Ltd, USA). Five cycles of the weathering were performed which each cycle was set as the following; exposed to UV at 60°C for 8 h, and then being moisture spraying at 50°C for 4 h. Ten samples for each formula were used. After that, the weathered samples were tested their tensile and thermal stability.

2.5. Tensile and Hardness testing
The tensile testing of pure PBS and the composite sheets was carried out in accordance to ASTM-D882 using a universal testing machine (Instron 5969, Instron Engineering Corporation, USA). The specimens were cut by a cutter to obtain the rectangular strip of 2.54 x 15.24 cm. The gauge length was 90 mm. The crosshead speed was 5 mm/min. Five trials were tested which the averages and standard deviations were calculated. In addition, the hardness testing was carried out in accordance to ASTM-D2240 using a Shore D durometer (PTC® Instruments, Pacific Transducer Corp., USA). The loaded weight was 5 kg. Five trials were tested which the averages and standard deviations were calculated.

2.6. Impact resistance characteristics by the falling dart impact tester
Impact resistance of the pure PBS and PBS/PSP composite sheets was evaluated by means of the falling dart impact test (modified method A). The sheet sample was 150 x 150 x 1 mm. The falling dart, installed with either loading weights, 200, 300, or 400 g, was released from the height of 66 cm to hit the sheet samples mounted. The characteristics of impact samples was photographed for analysis.

2.7. Thermal stability analysis by thermogravimetric analysis (TGA)
The pure PLA and the PBS/PSP composite sheets were cut into small pieces (5-7 mg) for thermal stability analysis by a thermogravimetric analyzer (TGA7, Perkin Elmer, USA). The sample was heated under nitrogen atmosphere from 50 to 700°C with a heating rate of 20°C/min. The thermal degradation was recorded and reported.

3. Results and discussion

3.1. Tensile properties and hardness of the PBS/PSP composite sheets

Figure 1 presents tensile properties and hardness of the PBS/PSP composite sheets before and after the weathering. It was found that the composites with higher content of PSP had higher Young's modulus and hardness in the similar trend. This indicated the reinforcing effect of the PSP particles in the PBS matrix. On the other hand, tensile strength and elongation at break of the composites were reduced with higher PSP content. Adding MLO of 3 phr reduced the rigidity of the composites implying the plasticizing effect by MLO [1, 2] in the composites. The PBS/PSP 70/30 wt% with MLO had higher elongation at break about 66% compared to the composite without MLO. The maleic anhydride groups in MLO reacted with the hydroxyl groups in PSP's cellulose to allow better dispersion of PSP particles in the PBS matrix. These polar groups also reacted with the hydroxyl end groups of PBS molecules [2] to allow better molecular mobility. After the weathering, all samples had lower values of mechanical properties. This was attributed to the deterioration of the PBS matrix by the hydrolytic degradation [3].

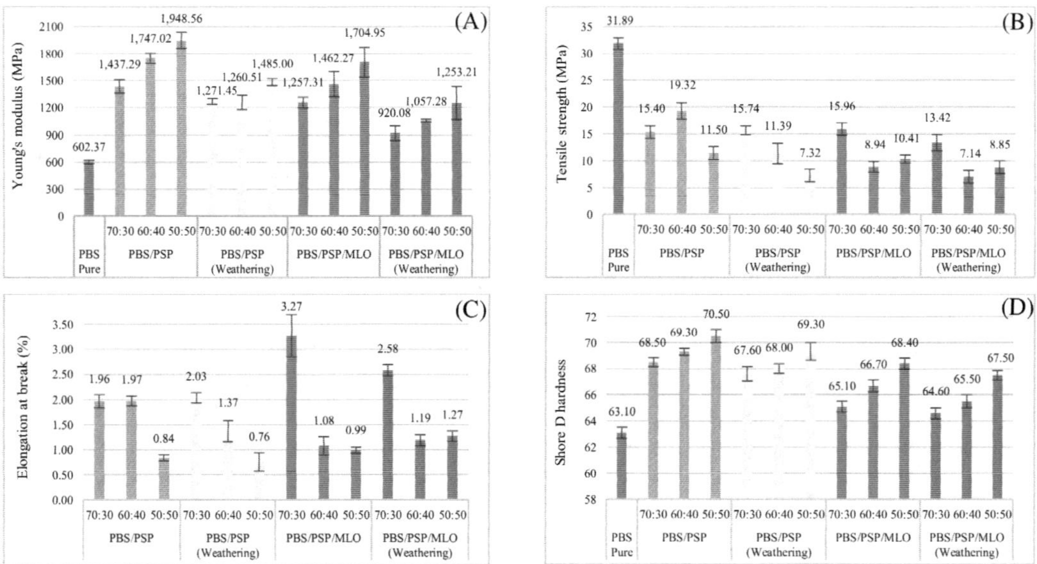

Figure 1. Young's modulus (A), tensile strength (B), elongation at break (C), and Shore D hardness (D) of pure PBS and the PBS/PSP composite sheets before and after the weathering.

3.2. Impact resistance by the falling dart impact testing

Figure 2 shows impact characteristics of pure PBS and the PBS/PSP composites. Using the lowest weight of 200 g, the PBS/PSP 70/30 wt% and 60/40 wt% could withstand the impact showing non-break surface, but the PBS/PSP 50/50 wt% sample was broken indicating the brittleness. With heavier weight; 300 and 400 g, only the PBS/PSP 70/30 wt% with MLO could withstand the higher impact. This correlates to the highest elongation at break of these samples in the tensile testing.

3.3. Thermal stability by thermogravimetric analysis

Table 1 shows thermal degradation temperatures of pure PBS and the PBS/PSP composites before and after the weathering. Since PSP consisted of hemicellulosic, lignin, and cellulose, mixing it into PBS reduced thermal stability of the composites similarly to the report by Liminana *et al* [1]. Adding MLO delayed the thermal degradation of the composites slightly. According to Quiles-Carrillo *et al* [4], this resulted from the nature of vegetable oil that had good thermal stability that could be applied for thermal stabilizer in polymer such as polyvinyl chloride. After the weathering, these degradation temperatures further reduced confirming the polymer degradation under the hydrolytic condition.

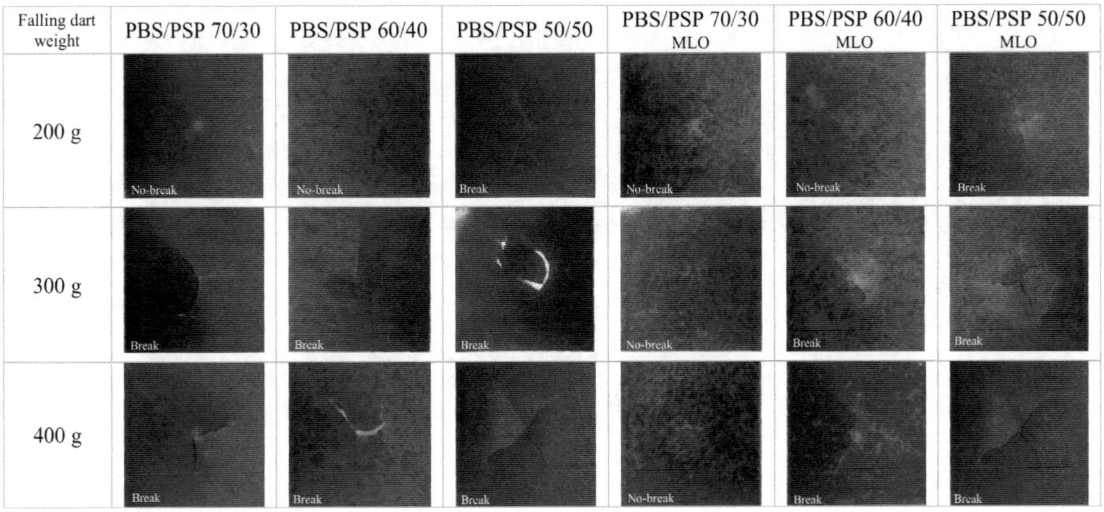

Figure 2. Impact characteristics of the PBS/PSP composite sheets by the falling dart impact testing.

Table 1. Thermal degradation temperatures of pure PBS and the PBS/PSP composites.

Samples		PBS:PSP (wt%)	Onset (°C)	Endset (°C)	Inflection point (°C)	Weight loss (°C)
Pure PBS		100:0	370.15	416.59	400.89	95.54
PSP		0:100	247.18, 403.01	322.69, 443.20	286.09, 416.21	39.67, 38.16
PBS/PSP	Before weathering	70:30	354.39	407.52	395.15	83.70
		60:40	348.63	395.61	376.29	94.60
		50:50	327.47	407.36	389.20	87.94
	After weathering	70:30	340.63	378.30	363.04	90.65
		60:40	343.78	377.65	368.53	91.21
		50:50	338.86	403.16	387.82	75.17
PBS/PSP/ MLO	Before weathering	70:30	358.69	413.35	399.03	84.71
		60:40	353.00	413.90	397.46	81.71
		50:50	332.56	409.93	390.83	73.66
	After weathering	70:30	342.77	410.52	393.53	83.81
		60:40	350.34	408.50	393.25	79.80
		50:50	344.53	408.68	393.00	77.34

4. Conclusion

The bio-based composite sheets from PBS and peanut shell powder were more rigid and harder with respect to higher PSP concentration. Adding MLO improved ductility for the composites. The composite sheets made from PBS/PSP 70/30 wt% with MLO was the toughest that could withstand the highest impact energy in the study. Mechanical properties and thermal stability of the specimens were decreased after placing them in the weathering condition. The composite sheets could be applied to short-life application such as seedling bags which they are biodegradable faster than pure PBS with reduced cost.

5. References

[1] Liminana P, Garcia-Sanoguera D, Quiles-Carrillo L, Balart R and Montanes N *Compos Part B-ENG* 2018 **144** 153

[2] Ferri J M, Garcia-Garcia D, Sánchez-Nacher L, Fenollar O and Balart R *Carbohydr Polym* 2016 **147** 60

[3] Cho K, Lee J and Kwon K *J Appl Polym Sci* 2001 **79** 1025

[4] Quiles-Carrillo L, Blanes-Martínez M M, Montanes N, Fenollar O, Torres-Giner S and Balart R *Eur Polym J* 2018 **98** 402

IOP Conf. Series: Materials Science and Engineering **773** (2020) 012047 doi:10.1088/1757-899X/773/1/012047

The modification of steel belt layer of airless tire for finite element analysis

J Phromjan[1,2], and C Suvanjumrat [1,2,*]

[1] Department of Mechanical Engineering, Faculty of Engineering, Mahidol University, Nakhon Pathom 73170, Thailand.
[2] Laboratory of Computer Mechanics for Design (LCMD), Department of Mechanical Engineering, Faculty of Engineering, Mahidol University, Nakhon Pathom 73170, Thailand.

*Corresponding author: chakrit.suv@mahidol.ac.th

Abstract. The airless tire can be considered as composite structure since it composed of several components with different materials for desired performance. The three main components of airless tire are rubber tread, shear band and polymer spokes. The shear band composed of three complex steel belt layers and should be considered as a sub composite structure. The finite element method can be used in designing and development of airless tire. However the diversity of materials in each component including the complex sub structure resulted in complicate modelling and consuming analysis time. Thus, the large computational resources are required to analyse those components. In this research, the simplification of modelling and analysis of airless tire was attempted by modification of steel belt layers. The Mooney-Rivlin hyperelastic model was used to describe the material properties of tread and spoke. The constitutive model constants were obtained by tensile and compressive test according to ASTM D412 and D575 standards, respectively. The simplified belt, which was developed by homogenization approach, was integrated into shear band component. The airless tire model with modified steel belt layers can be used to reduce the model complexity and analysis time while yields accurate results.

1. Introduction

The non-pneumatic tires (NPTs) which are built by several components have excellent performance according to its construction. To develop tire performance, the finite element method is widely used to study tire behaviour in various condition. Jan and Vladimira [1] presented the different ways of replacement of steel cord belt into pneumatic tire model such as SHELL element [2], reinforcement bar (REBAR) element and homogeneous material. The orthotropic material with homogenization of belt layers has good response to the surrounding environment. Next, Korunovic *et al.* [3] had been performed the finite element method evaluate the pneumatic tire structure. The material modelling of tire reinforcements was concentrated. REBAR element with various constitutive models was compared. The Marlow model was found to be the most suitable. Moreover, the REBAR element was used to replace the steel cord belt into shear band of NPTs [4]. The deformation analysis of finite element model agreed well with the experiment. However, the great time and resource of computation by this model remain the severe problem. The simplified belt by homogenization approach can integrate into shear band component. It can reduce the complex model and analysis time. The NPT which is developed the shear band can use under the rapid time. Particularly, this NPT model will be advanced in the future.

2. Experimental

The characteristic of NPT or airless tire, Tweel 12N16.5 SSL ALL TERRAIN, which is developed by Michelin is presented in Table 1. The Tweel airless tire has three main components. There are rubber tread, shear band and polymer spoke (Fig. 1). The tire stiffness testing was performed to investigate the tire stiffness of Tweel airless tire. The airless tire was mounted at the mounting arm of tire stiffness tester (EKTRON TEK model: PL-2003). The compressive force and tire vertical deformation was recorded while the measurement table was moved and compressed the tire. The compressive load of 1000 kg was applied in this study. The experimental result was indicated that the vertical stiffness of airless tire which was studied is 949.37 N/mm.

Table 1 The characteristic of airless tire model: Tweel 12N16.5 SSL ALL TERRAIN

Tire Size (inch)	Tire Dimension (mm)		Weight (kg)
	Width	OD	
12N16.5	309	860	100

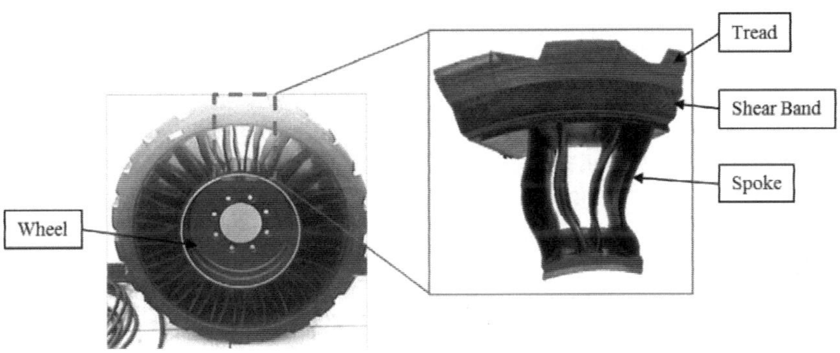

Figure 1. The airless tire model: Tweel 12N16.5 SSL ALL TERRAIN.

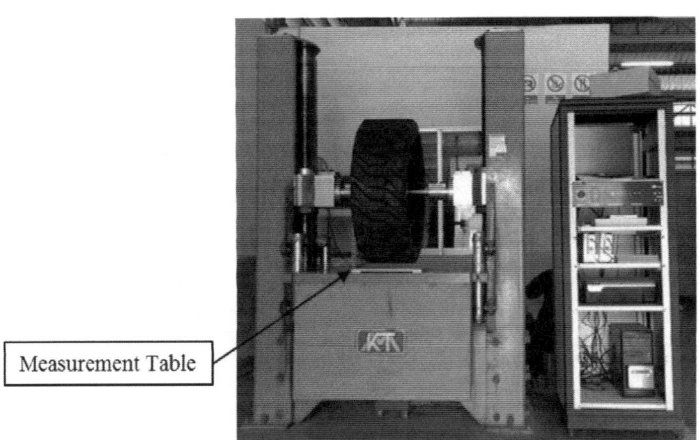

Figure 2. The tire stiffness testing of airless tire.

2.1. Finite element method

The computer aided design (CAD) and computer aided engineering (CAE) was use to create the 3D finite element model of Tweel airless tire. The finite element model of steel belt layer is developed according to the Tweel airless tire which has a complex shear band construction as shown in figure 3. The shear band of finite element model composed wall and belt elements. The tread material property of airless tire model was investigated by compresion test according to ASTM D575 standard, while material property of spoke was investigated by tensile test according to ASTM D412 standard, respectively. The tread, wall and spoke properties were defined by Mooney-Rivlin hyperelastic model. On the other hand, the steel belt is specified to be an elastic isotropic material as described in table 2. The airless tire model was compressed with the compressive load of 1,000 kg. This simulation performs by moving the rigid flate plate to press airless tire model according to the testing of tire stiffness tester (Figure 4).

Table 2 The material properties of airless tire model

Components	Mooney-Rivlin constant		Young's Modulus (GPa)	Poisson's Ratio
	C_{10}	C_{01}		
Tread	1.19085	0	-	-
Wall	1.19085	0	-	-
Steel Belt	-	-	200	0.3
Spoke	0.5063	4.2552	-	-

3. Results and discussion

The compressive load and vertical deformation of airless tire model was recorded while the airless tire model is compressed by rigid flat plate. The simulation result has a good agreement with the experimental result (EXP) from tire stiffness testing. The comparison of simulation model and experiment is presented in Figure 5. The result of using homogenization approach with isometric material different as the result of using REBAR element to replace the steel belt as shown in Figure 5. The vertical stiffness of airless tire finite element model by using homogenization approach and REBAR element were 905.77 N/mm and 814.93 N/mm, respectively. The summary of simulation result is described in Table 3.

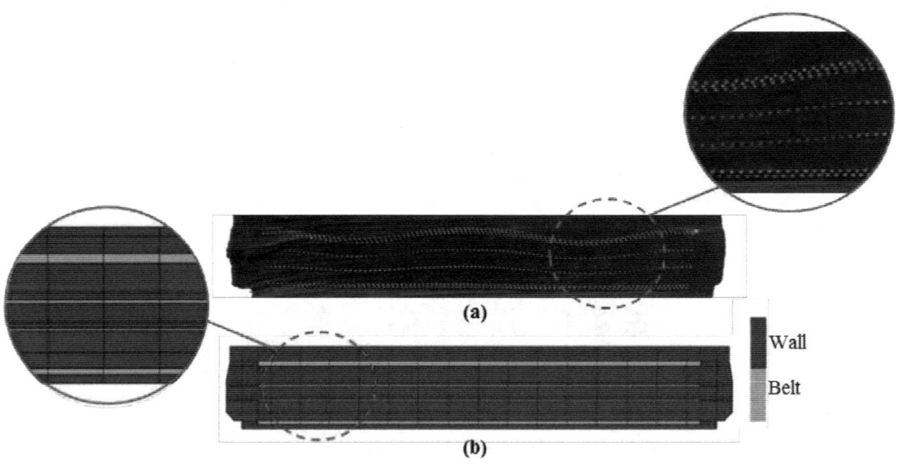

Figure 3. The cross section shear band of (a) Tweel airless tire and (b) airless tire model.

Figure 4. The finite element model of compressed airless tire.

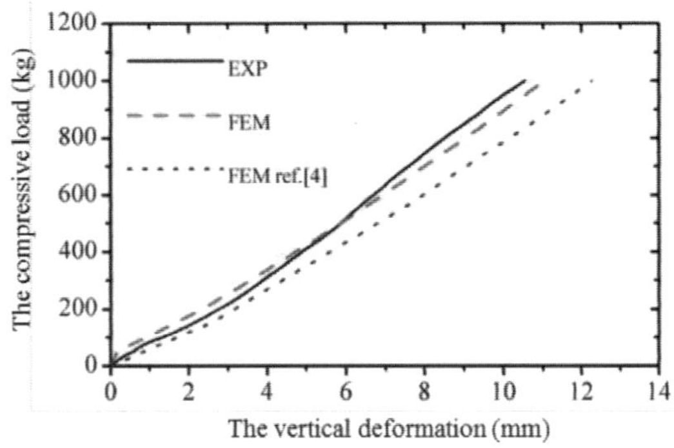

Figure 5. The vertical deformation and compressive load of compressed airless tire.

Table 3 Summary of simulation result.

Model	Simulation Time (min)	Error (%)
REBAR element [4]	40	14.16
Homogenization approach	45	4.59

4. Conclusions

The modification of steel belt layer in shear band of airless tire model by homogenization approach with isotropic material provided more accurate result than model by REBAR element approach when compared to the experiment. The analysis time was also found to not significantly increase and can be neglected. The airless tire model with modified steel belt layers, which the rubber part was specified by hyperelastic material property, can be used to reduce the model complexity and analysis time while yields accurate results.

5. References

[1] Korunovic N, Fragassa C, Marinkovic D and Vitkovic N 2019 *J. Composite Structures.* **224** 111006.
[2] Kulikov G 1996 *J. Tire Science and Technology.* **24:1** 11-38.
[3] Krmela J and Krmelova V 2016 *J. Procedia Engineering.* **136** 132-136.
[4] Rugsaj R and Suvanjumrat C 2019 *J. Automotive Technology.* **20:4** 801-812.

Acknowledgement

This research was financial supported by Rubber Technology Research Center (RTEC), Mahidol University and the Thailand Research Fund (TRF) under the TRF Research Grant No. RDG60T0140.

Hydrothermal carbonization synthesis and KOH activation of porous carbons from waste marigold flowers

N Chaiammart[1], A Eiad-ua[2] and G Panomsuwan[1,*]

[1] Department of Materials Engineering, Faculty of Engineering, Kasetsart University, Bangkok 10900, Thailand.
[2] College of Nanotechnology, King Mongkut's Institute of Technology Ladkrabang, Bangkok 10520, Thailand.

*Corresponding author: gasidit.p@ku.ac.th

Abstract. Marigold flower-derived porous carbons were synthesized via hydrothermal carbonization (HTC) and KOH activation. The effect of HTC and KOH activation on the change in morphology, chemical functional group, and surface area were studied and discussed based on the results of scanning electron microscopy, Fourier transform infrared spectroscopy, and N_2 sorption analysis, respectively. Both HTC and KOH activation were found to play critical roles in changing morphology and enhancing surface area. Without HTC and KOH activation, carbons had low surface area and lacked porosity. In contrast, with both HTC and KOH activation, a sponge-like morphology with a large specific surface area of 1825 m^2/g was obtained. The results serve as a useful guideline for further development and synthesis of porous carbons in certain specific applications.

1. Introduction

Porous carbons with a high surface area and porosity have recently received considerable attention as potential materials for a broad range of applications, such as electrode in electrochemical energy devices [1], wastewater treatment adsorbents [2], and carbon dioxide captures [3]. Currently, several kinds of agricultural wastes are widely used as a carbon source precursor to prepare porous carbons, since their chemical structure typically consists of lignin, cellulose, and hemicellulose [4]. Over the past several years, most researchers reported the preparation of porous carbons from several agricultural waste products available in their local area, including rice husk [5], cornstalks [6], and palm empty fruit bunches [7]. In addition to agricultural waste products, a number of plant biomasses continue to require investigation and understanding. In our previous work, waste marigold flower was used as a carbon source precursor for the synthesis of porous carbon, since they are usually discarded as waste after religious ceremonies in sacred places and temples in Thailand. The synthesis was performed using hydrothermal carbonization (HTC) at different treatment times, followed by a pyrolysis process [8]. We found that the specific surface area of porous carbons derived from marigold flower increased in accordance with longer HTC time from 118 to 281 m^2/g, owing to the development of microporosity. However, the specific surface area and porosity were still low and insufficient for use as absorbents or electrodes in supercapacitors. Typically, carbon materials with relatively large surface areas can easily be prepared by either physical or chemical activation processes. Physical activation is usually achieved by heat treatment in an inert atmosphere under the presence of CO_2 or steam [9]. Meanwhile, chemical activation necessarily requires a chemical agent for activation, such as H_3PO_4, KOH, or NaOH [10].

Herein, we report the synthesis of porous carbons from marigold flower via HTC and KOH activation. The effects of HTC and KOH activation on the change in morphology, chemical functional group, and surface area of porous carbons were studied and discussed based on three characterization techniques, including scanning electron microscopy (SEM), Fourier transform infrared spectroscopy (FTIR), and N_2 adsorption–desorption isotherm analysis, respectively.

2. Experimental

2.1. Preparation of raw materials

Marigold flowers were collected from Three Musketeers' Kaset Monument at Kasetsart University. They were washed by deionized water followed drying at 80 °C for 24 h in a hot-air oven. Dried marigold flowers were cut into small pieces using a blender machine and then ground into a fine powder using a mortar and pestle, followed by sieving with a 0.79 mm-screen mesh.

2.2. Synthesis of porous carbon from marigold flowers

10 g of the fine powder and 50 mL of deionized water were mixed together and then added into a Teflon-lined stainless-steel autoclave. HTC was performed at 180 °C for 24 h in a hot-air oven (Memmert, UN55, Germany). The products obtained after HTC were washed with deionized water, followed by drying at 80 °C for 24 h. The dried products were mixed with KOH at a weight ratio of 1:1 and then heated in a ceramic tube furnace (Chavachote, Thailand) at 800 °C for 1 h at a heating rate of 10 °C/min under Ar gas flow (0.2 L/min). The samples after HTC with and without KOH activation were designated as MHAC and MHC, respectively. For comparison, marigold flowers were directly pyrolyzed under the same condition both with and without KOH activation, which were designated as MAC and MC, respectively. The sample names and the experimental conditions are summarized in Table 1.

Table 1. Sample names and the corresponding experimental conditions in this study.

Sample name	HTC temperature/time	Pyrolysis temperature/time	Hydrochar:KOH
MC	-	800 °C / 1 h	1:0
MHC	180 °C / 24 h	800 °C / 1 h	1:0
MAC	-	800 °C / 1 h	1:1
MHAC	180 °C / 24 h	800 °C / 1 h	1:1

2.3. Characterization

The morphology of the samples was examined using a field-emission scanning electron microscope (JEOL, JSM-7600F, Japan) at an acceleration voltage of 3 kV. Transmittance spectra were recorded using Fourier transform infrared spectrometer (Bruker, Alpha II, Germany). Textural properties were determined from the N_2 adsorption–desorption isotherms at 77 K, which were acquired with a surface area analyser (Micromeritics, 3Flex, USA).

3. Results and discussion

Figure 1 shows the SEM images of porous carbons prepared under different conditions. An clear morphological change is clearly observed for each sample, indicating the significant effect of HTC and KOH activation. As can be seen, MC exhibited smooth with wavy lines on the surface with remaining morphology similar to a raw marigold flower (Figure 1a). The MHC sample had a rougher and more ruptured surface. This implies that the HTC process resulted in the deterioration of lignocellulosic structure of marigold flower. In contrast, for the samples with KOH activation, MAC and MHAC showed a sponge-like morphology with a continuous pore-network structure. However, the MHAC structure was finer and more uniform than the MAC pore structure. This was likely due to a rougher and more ruptured surface of hydrochars which allowed a uniform interaction with KOH during activation. From the SEM observation, it can be concluded that HTC and KOH activation strongly affected the morphology of the carbons derived from marigold flowers.

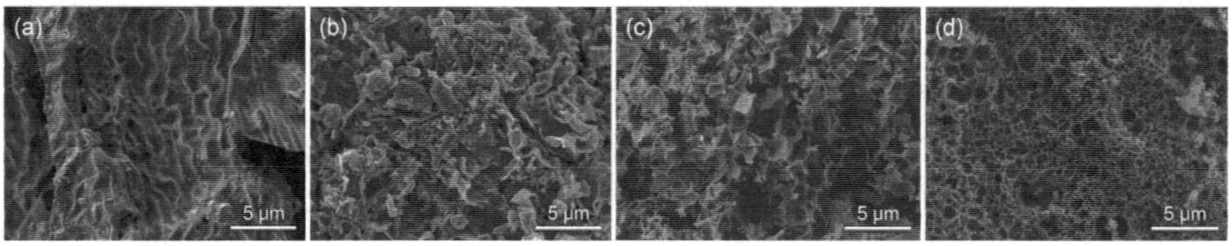

Figure 1. SEM images (5000×) of (a) MC, (b) MHTC, (c) MAC, and (d) MHTAC.

The FTIR spectra of all the samples are shown in Figure 2a. No obvious absorption peaks were detected for MC and MHC due to the decomposition of lignocellulosic structure during HTC and pyrolysis. For KOH activation, the absorption band of OH stretching was observed at about 3400 cm^{-1} for both MAC and MHAC. The presence of absorption peaks at 1404 and 1003 cm^{-1} of MAC were assigned to OH bending and C–O stretching, respectively. For MHAC, the absorption peaks at 1560 and 1080–1208 cm^{-1} were attributed to aromatic C=C stretching and C–O–C stretching, respectively [11]. The FTIR result suggests that the KOH activation at high temperatures created the oxidation on carbon surface, as reflected by the presence of oxygen functional groups for MAC and MHAC.

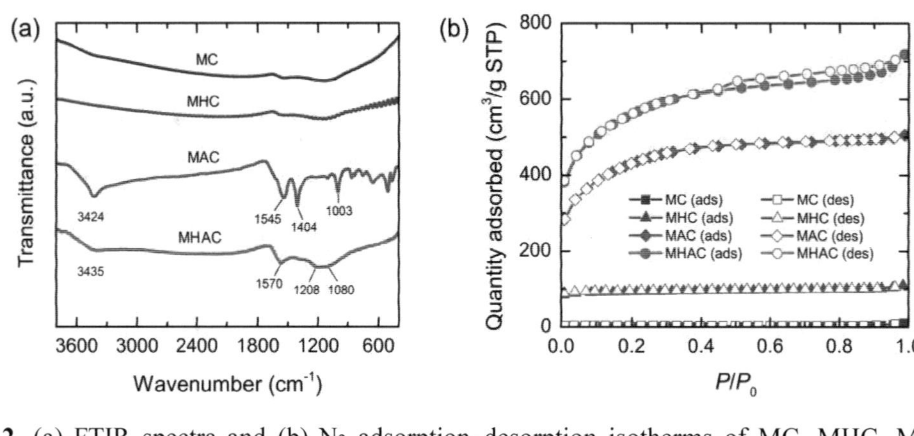

Figure 2. (a) FTIR spectra and (b) N_2 adsorption–desorption isotherms of MC, MHC, MAC, and MHAC.

Furthermore, the measurements of N_2 adsorption–desorption isotherms were used to elucidate the specific surface area and pore structure of carbons, as shown in Figure 2b. The specific surface area of MC, MHC, MAC, and MHAC determined using the Brunauer–Emmett–Teller (BET) method were 7, 281, 1416, and 1825 m^2/g, respectively. A small specific surface area of MC indicates a lack of porosity. This result is consistent with the morphology observed by SEM as previously discussed. For further information about the pore structure, the type of isotherm was analyzed based on the IUPAC classification. MHC and MAC were found to show type I isotherm with no difference between adsorption and desorption, which is the characteristic isotherm of micropore. However, a combination of type I and IV characteristics were observed for MHAC. The hysteresis loop at the relative pressure of 0.45–0.99 indicated the presence of mesopore. The *t*-plot was also used to determine the surface area contributed by micropores. The proportion of micropore in MHC, MAC, and MHAC was 83%, 38%, and 36%, respectively. The average pore size of MHC, MAC, and MHAC was determined from desorption isotherm using the Barrett-Joyner-Halenda (BJH) theory, and were about 2–3 nm. It can be concluded that HTC and KOH activation are important processes to enhance surface area through the formation of nanoporous formation. Moreover, KOH activation plays an important role in creating mesopores in the samples. Textural parameters of all the samples discussed above are listed in Table 2.

Table 2. Textural properties of MC, MHC, MAC, and MHAC.

Sample	BET surface area (m2/g)	Micropore area (m2/g)	External surface area (m2/g)	Pore volume (cm3/g)	Micropore volume (cm3/g)
MC	7	2 (29%)	5 (71%)	0.0108	0.0011
MHC	281	237 (84%)	44 (16%)	0.1629	0.1243
MAC	1416	550 (39%)	866 (61%)	0.7711	0.2843
MHAC	1825	675 (37%)	1150 (63%)	1.0801	0.3588

4. Conclusions

HTC and KOH activation were found to be necessary for the synthesis of porous carbon from marigold flowers with a large surface area. Direct pyrolysis without HTC and KOH activation resulted in MC having a very small surface area due to a lack of porosity. The MHAC with both HTC and KOH activation exhibited a sponge-like structure and had the largest surface area with a meso/macropore contribution of 63%. KOH activation also led to the surface oxidation of carbons. The preliminary results obtained in this study are highly useful for further maximizing surface area with controllable porosity by adjusting the activation temperature and KOH amount. The porous carbons derived from marigold flowers can be used as electrodes in supercapacitors or as adsorbents.

5. References

[1] Hu C, Xiao Y, Zou Y and Dia L 2018 *Electrochem. Energy Rev.* **1** 84.
[2] Wong S, Ngadi N, Inuwa IM and Hassan O 2018 *J. Cleaner Prod.* **20** 361.
[3] Creamer AE and Gao B 2016 *Environ. Sci. Technol.* **50** 7276.
[4] Deng J, Li M and Wang Y 2016 *Green Chem.* **18** 4824.
[5] Kim J, Yi Y, Peck DH, Yoon SH, Jung DW and Park HS 2019 *Environ. Sci.: Nano* **6** 916.
[6] Yu H, Zhang W, Li T, Dang L, Liu Z and Lei Z 2017 *RSC Adv.* **7** 1067.
[7] Guntagerng K, Panomsuwan G and Eiad-ua A 2018 *Walailak J. Sci. Technol.* **15** 779.
[8] Chaiammart N, Wongcharoen S, Eiad-ua A, Ishizaki T and Panomsuwan G 2019 *Key Eng. Mater.* **824** 23.
[9] Zhou J, Luo A and Zhao Y 2017 *J. Air Waste Manage. Assoc.* **68** 1269.
[10] Hunsom M and Autthanit C 2013 *Chem. Eng. J.* **229** 334.
[11] Siipola V, Tamminen T, Källi A, Lahti R, Romar H, Rasa K, Keskinen R, Hyväluoma J, Hannula M and Wikberg H 2018 *BioResources* **13** 1.

Acknowledgement

This work was supported by the Research Assistant Scholarship from the Faculty of Engineering, Kasetsart University (60/19/MATE/M.ENG).

Electrical and Mechanical Properties of PEDOT:PSS Strain Sensor based Microwave Plasma modified Pre-vulcanized Rubber Surface

K Thana[1], N Petchsang[2,3], and R Jaisutti[1*]

[1] Department of Physics, Faculty of Science and Technology, Thammasat University, Pathum Thani 12121, Thailand.
[2] Department of Materials Science, Faculty of Science, Kasetsart University, Bangkok 10900, Thailand.
[3] Specialized Center of Rubber and Polymer Materials for Agriculture and Industry (RPM), Faculty of Science, Kasetsart University, Bangkok 10900, Thailand

*Corresponding Author e-mail address: jrawat@tu.ac.th

Abstract. Poly(3,4- ethylenedioxythiophene) doped with poly(styrenesulfonate) (PEDOT:PSS) conductive polymer has been extensively used in various applications including flexible electrode and stretchable sensor. This study reports the electrical and mechanical strain behaviours of PEDOT:PSS deposited on pre-vulcanized rubber substrate, which is modified surface by microwave plasma. The effect of plasma treatment cycles on adhesion of PEDOT:PSS coated pre-vulcanized rubber and conductivity have been investigated. The results show a uniform PEDOT:PSS film on rubber surface after treating with microwave plasma. The electrical conductivity slightly increases by increasing treatment cycles. In addition, an electrical strain and mechanical strain increase up to 60 and 50%, respectively under 4 treatment cycles.

1. Introduction

Recently, flexible strain sensors have become highly increasing demand for their board applications in the field of health monitoring device, electronic skin and soft robot [1, 2]. Among the soft polymeric materials, poly(3,4- ethylenedioxythiophene) doped with poly(styrenesulfonate) (PEDOT: PSS) is widely applicable in flexible electronic devices [3]. By coating on the elastomeric substrate such as polymethyl methacrylate (PAMA), polyvinyl alcohol (PVA), poly(dimethylsiloxane) (PDMS) and rubber, it can be used as stretchable electrodes, pressure and strain sensors [3, 4]. As the substrate material of strain sensor, natural rubber that has a good elasticity, high strength and large elongation, giving a good recovery to original structure after releasing strain [5, 6]. To make conductive rubber, layer of PEDOT:PSS on rubber can be simply made by using spin coating technique. Plasma treatment to a surfaces cleaning and etching was used to modify a wettability of the surface before prepared the thin film on substrate.

In this study, we investigate the mechanical strain-responsive electrical behaviours of PEDOT:PSS films coated on natural rubber. The effect of plasma-treated rubber surface on electrical resistance was observed.

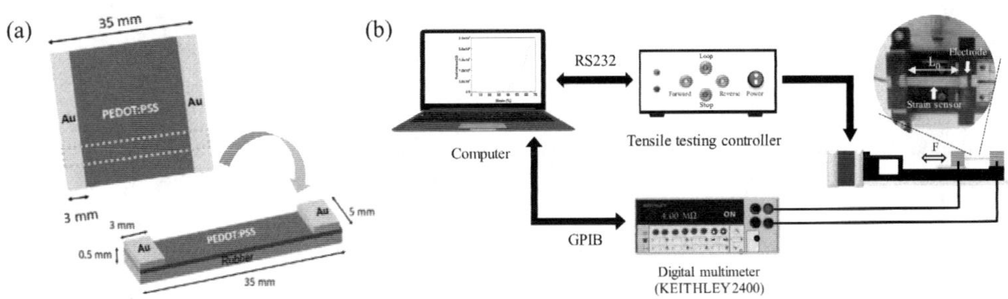

Figure 1. Schematic of (a) strain sensors fabrication and (b) their measurement system.

2. Experimental

The strain sensor was prepared by spin coating PEDOT:PSS on a pre-vulcanized rubber (received from S.K. Polymer, Thailand). Briefly, the rubber (0.5 mm thick) was attached on glass substrate, cleaned by rinsing with deionized water and dried with nitrogen gas. To remove an organic residues and increase the adhesion of PEDOT:PSS on the rubber surface, the pre-vulcanized rubber was treated under microwave plasma for 15 sec. The optimized condition was done by varying the treatment cycles ranging from 1 to 4 cycles. Then 1.1 wt% of PEDOT:PSS was spun on the substrate with spin speed of 4000 rpm for 30 sec and annealed on a hotplate at 60 $_o$C for 30 min. After that, the gold electrode was attached on the top of PEDOT:PSS to make an electrical contact. Finally, the device was peeled off from the glass substrate and cut into a fixed size of 5 x 35 mm$_2$, as shown in Figure 1(a).

The initial electrical conductivity of strain sensor was measured with Keithley 2400 source meter. Uniaxial stretching tests of the sensor were conducted using a homemade tensile system equipped with an electrical contactable jig for in-situ resistance measurements (Figure 1b). The change in resistance by uniaxial tension was measured at constant strain rate until it reaches breaking point. The relative resistance change ($S_R = \Delta R/R_0$) was calculated based on the initial resistance R_0 and the change in resistance $\Delta R = R_a - R_0$, where R_a the actual resistance at a given strain value.

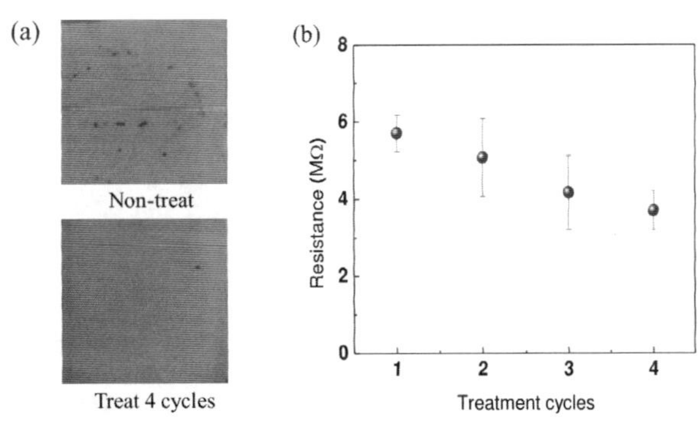

Figure 2. (a) Optical images of PEDOT:PSS coated on pre-vulcanized rubber without and with plasma treatment, and (b) electrical resistance of conductive rubber prepared under different plasma treatment cycles.

3. Results and discussion

Optical images of PEDOT:PSS coated on pre-vulcanized rubber without and with plasma treatment are showed in Figure. 2a. Due to its high hydrophobic of rubber surface [7], this made no adhesion of

PEDOT:PSS on the bare rubber. Under microwave plasma treatment, there is an evidence of a light blue surface of PEDOT:PSS appeared on the substrate. The electrical resistance slightly decreased by increasing the treatment cycles as demonstrated in Figure 2b. With a fixed dimension size of 5 x 35 mm2, the conductive rubber shows a measured resistance of 5.70 MΩ, 5.07 MΩ, 4.16 MΩ and 3.70 MΩ for 1, 2, 3 and 4 plasma treatment cycles, respectively. The plasma cleaning can be increased a wettability of the natural rubber surface and enhanced the on-surface coating. With increasing treatment cycles, a higher surface energy was observed as indicated higher adhesion of PEDOT:PSS on the rubber surface.

To investigate the variation of electrical properties of the strain sensor under mechanical stress, the sensing was tested in the stretching mode. Figure 3 shows the relative resistance change (S_R) as a function of tensile strain ($\varepsilon = \Delta L/L_0$, where ΔL is the change in the actual length at a given strain and L_0 is the initial length) for the different samples prepared with different plasma treatment cycles. The average resistance slightly decreases at the beginning, as seen in inset of Figure 3a, then it appears linearly increases and follows by drastic increases in the further elongation, finally the destruction of conductive network is occurred. The decrease in electrical resistance at the beginning of stretching may be due to a strain-induced the growth of conductive region of PEDOT:PSS by coalescence of conductive PEDOT-rich cores [8]. The increasing mechanical strain results in the loss of conductive path which causes difficult electron pass, and a crack generation lead to electrical breakdown at higher strain.

Figure 3b presents the maximum strain and gauge factor ($GF = S_R/\varepsilon$) as a function of plasma treatment cycles. A good stretchability and high sensitivity are observed for more repeating plasma treatment cycles. However, the natural rubber will be burned for further increase number of plasma treatment more than 4 cycles. Generally, the change in resistance of the PEDOT:PSS films on stretching elastic substrate depends on many factors such as the cracks generated by applied strain, the change in geometrical shape of the films and PEDOT:PSS-substrate interaction. Among all these factors, the PEDOT:PSS-rubber interaction is the most important parameter that influences sensing sensitivity of strain sensor after plasma treatment.

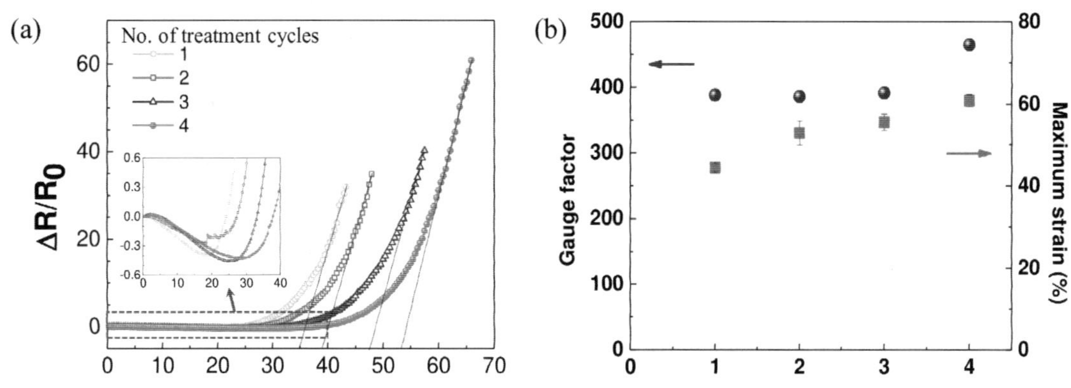

Figure 3. (a) Relative resistance change of PEDOT:PSS coated on rubber with different plasma treatment cycles. (b) Sensitivity and maximum mechanical strain of strain sensor based conductive rubber.

4. Conclusions

Flexible strain sensors were fabricated by spin coating PEDOT:PSS on pre-vulcanized rubber films. The microwave plasma treatment was used to enhance the hydrophilic of rubber surface which results in a homogeneous distribution of PEDOT:PSS on the surface of rubber. A good adhesive of PEDOT:PSS on natural rubber was observed after 4 treatment cycles (15 sec treatment interval time) with initial resistance of 2 MΩ. In the stretching mode strain sensor, the average resistance decreased and stretchibility of the strain sensor improved with an increase in plasma treatment cycles. The electrical and mechanical strain increase from 40 and 42% up to 60 and 65% for 1 and 4 treatment cycles, respectively. These results demonstrate that the conductive rubber can be used as a promising candidate for strain sensor or stretchable electrodes.

5. References

[1] Teo M Y, Kim N, Kee S, Kim B S, Kim G, Hong S, Jung S and Lee K 2017 *ACS Appl. Mater. Interfaces* **9** 819.
[2] Ko H U, Kim H C, Kim J W, Park J and Kim J 2018 *Mater. Lett.* **227** 276.
[3] Lipomi D J, Lee J A, Vosgueritchian, Tee C C-K, Bolander J A, and Bao Zhenan 2012 *Chem. Mater.* **24** 373.
[4] Fan X, Wang N, Wang J, Xu B and Yan F 2018 *Mater. Chem. Front.* **2** 355.
[5] Nattarajan T S, Eshwarn S B, Stöckelhuber K W, Wießner S, Pötschke P, Heinrich G and Das A 2017 *ACS Appl. Mater. Interfaces* **9** 4860.
[6] Zhng W, Yin B, Wang J, Mohamed A and Jia H 2019 *J. Alloys. Compd.* **785** 1001.
[7] Meesawan K, Thana T, Jaisutti R and Petchsang N 2019 *IOP Conf. Series: Mater. Sci. Eng.* **526** 012045.
[8] Lee Y-Y, Choi G E, Lim S-M, C J-Y, Choi I-S, Nam K T and Joo Y-C 2016 *Sci. Rep.* **6** 25332.

Acknowledgement

This work was financially supported by Thammasat University (Contract No.16/2561) and Thammasat University innovation scholar (Contract No. 2/2559). This work also was financially supported by Thailand Research Fund (TRF) (Grant No. RDG62T0004). We thank Scientific Equipment Center, Kasetsart University, for using their characterization equipment.

IOP Conf. Series: Materials Science and Engineering **773** (2020) 012050 doi:10.1088/1757-899X/773/1/012050

Development of a manual for rubber plantation owners and rubber wood consumers in Thailand for obtaining international forest management certification

K Duangsathaporn*, P Prasomsin, Y Omule, K Palakit, and P Lumyai,

Department of Forest Management, Faculty of Forestry, Kasetsart University, Bangkok, Thailand 10900

*Corresponding author: fforkcd@ku.ac.th

Abstract Currently, the export situation of rubber wood products in Thailand is at a high risk of price and market competition. All relevant organizations in Thailand hope to solve these problems sustainably. One of the important ways is to increase the value of rubber wood products and marketing channels for exporting to more countries besides China. However, Thailand still faces a significant obstacle to the export of rubber wood products, that is, most of the rubber plantations have not yet been certified for international forest management. Forest certification is one mechanism to prevent and manage global environmental problems including air, soil, water, biodiversity and the reduction of forest areas. Previous research [1] indicated that most rubber plantation owners and rubber wood consumers in Thailand did not fully understand the principles, criteria, and indicators of the Forest Stewardship Council (FSC), and that in Forest Management (FM) certification auditors need to process a request for Corrective Action Requests (CARs) or suspend certificates in some cases. Therefore, this research aimed to develop an appropriate manual for rubber plantation owners and rubber wood consumers in Thailand for obtaining the FSC standards. Secondary data of the forest management certification system at the international level, including organizations, certification process, and FSC certification standards, were analysed and related to data at the national level (i.e., laws, regulations and status of rubber plantations in economic, social and environmental dimensions). The consistency and limitations of FSC principles, criteria, and indicators in certificates issued in Thailand were also analysed. A manual, with recommendations for compliance with FSC standard, was developed. It consists of five chapters: 1) FSC definition; 2) FSC scoping; 3) FSC evaluation and surveillance; 4) FSC scoping assessment; and 5) implementations in accordance with FSC standards. This knowledge had been shared through workshops with the rubber plantation owners, rubber wood consumers, staff of the Rubber Authority of Thailand, and other organizations. In total, over 150 people attended the workshops in the following locations: Songkla, Surat Thani, Rayong, and Bangkok. The workshops aimed to develop the participant skills to request for FSC: FM certification.

1. Introduction

Thailand is currently having problems with the export of rubber wood products that are related to price and market competition, yet the rubber plantation area tends to increase all the time. All relevant government and private organizations in Thailand hope to solve these problems sustainably, for both wood and latex products. One of the important solutions is to increase the value of rubber wood products and marketing channels for exporting to more countries besides China, such as European countries, the United States and Japan. However, Thailand still faces a significant obstacle to the export of rubber wood products, that is, most of the rubber plantations have not yet been certified for international forest management, namely forest certification, which is one mechanism to prevent and manage global environmental problems including air, soil, water, biodiversity and the reduction of forest areas.

Forest certification can be divided into two categories, namely, the Forest Management (FM) and Chain of Custody (CoC), with the certifying organizations such as Forest Stewardship Council (FSC) and Pan European Forest Certification (PEFC). Based on FSC [2], only 11 private sector organizations and state enterprises were successfully certified FSC: FM, covering only 0.411 percent of rubber plantations in Thailand [2]. From the previous research [1], most rubber plantation owners and rubber wood consumers in Thailand did not understand enough about the principles, criteria, and indicators of FSC: FM certification, and that auditors need to process a request for Corrective Action Requests (CARs) or suspend certificates in some cases. Therefore, this research originated from the research plan titled "the Formulation of Policy Recommendations for the Potential Development of Rubber Plantation Owners and Rubber Wood Consumers in Thailand for Obtaining International Forest Management Certification", which was supported by the National Research Council of Thailand (NRCT) and the Thailand Research Fund (TRF) in the fiscal year 2016. The research aims to develop an appropriate manual for rubber plantation owners and rubber wood consumers in Thailand for achieving the Forest Stewardship Council (FSC) standards.

2. Methodology

This research included (1) the drafting of a manual of rubber plantation owners and rubber wood consumers in Thailand for obtaining the Forest Stewardship Council (FSC) standards, and (2) knowledge transfer to rubber plantation owners and rubber wood consumers in different regions of Thailand, with appropriate manual adjustment after training.

A draft manual was prepared based on secondary data on the structure of the international FM certification system, such as the relevant system organizations, the FSC certification process (consisting of Principles, Criteria, and Relevant Indicators), as well as compiling data on Thailand's FM certification associated with FSC certification standards, such as laws, regulations, and status of rubber plantations in terms of economic, social, and environmental dimensions. The manual was considered an important tool for internal audit by the rubber plantation owners and rubber wood consumers in Thailand for obtaining the FSC standards.

The draft manual was introduced through workshops to rubber plantation owners and rubber wood consumers in Thailand for obtaining the Forest Stewardship Council (FSC) standards. The workshops were held in four different regions of Thailand and involved a total of approximately 150 participants. Finally, the draft manual was then appropriately improved based on the experimental results of the training and the recommendations of the training participants.

3. Results and discussion

Forest Management (FSC-FM) Certification is a standard for sustainable forest management (SFM) in accordance with the guidelines of the Forest Stewardship Council (FSC). The FSC inspects the quality of forest management from forest plantation to production processes. Currently, the FSC standard version has changed from version 4 (FSC Principles and Criteria for Forest Stewardship; FSC-STD-01-001, V4-0 [3]) to version 5 (FSC Principles and Criteria for Forest Stewardship; FSC-STD-01-001, V5-2 [4]), which was released in July 2015. However, FSC standard version 5 has not yet been used for certification in Thailand, but it is preparing to use in the near future. Therefore, the analysis of the principles and criteria of the FSC certification system was conducted using Version 5 as the standard framework along with other relevant documents, including 1) recommendations, guidelines and other documents issued or certified by the FSC, 2) FSC standards, 3) standards for plant species, products and services, and 4) standards for various types of management units such as forest with small and low-intensity managed forests (SLIMF), forest with large and high-intensity managed forests, and forest with conservation or protected purpose as approved by FSC.

We analysed the scope of the FSC standard in Thailand based on the following issues: 1) covering all management activities related to both inside and outside the management unit (MU), whether by themselves or by a sub-contractor, 2) the geographic area of certification within the management unit that is requesting forest certification, but some principles and criteria are certified beyond the scope of the management unit, 3) principles and criteria can be applied to all types of forests and sizes, including natural forests, plantations and other plant species, 4) management units in an agricultural production system will not be subject to principles and criteria unless they are covered by the conditions specified in the management plan, 5) products and services cover the production of timber and non-timber forest products, conservation, protection, ecological services and other uses, and which ecological services include carbon sequestration according to guidelines for global warming action, and 6) compliance with international and national laws and regulations. Comparison of the FSC versions 4 and 5 showed that the FSC Principles and Criteria in version 5 contains 10 principles and 70 criteria, which is an increase of 14 criteria from version 4, and therefore requires a stricter process to certify than the previous version.

The draft manual consisted of five chapters: 1) FSC definition, 2) FSC scoping, 3) FSC evaluation and surveillance, 4) FSC scoping assessment, and 5) implementation in accordance with FSC standards (Figure 1). The FSC definition described the definition of the FSC, the establishment of the FSC and the FSC's commitment to promoting sustainable forest management, both socially, economically and environmentally. The FSC scoping was defined as the scopes of forest certification in six issues of the activities related to MU, the geographic area, types of forests and sizes, agricultural areas, products and services, and compliance with laws as described in the previous paragraph. The FSC evaluation and surveillance explained the procedures for requesting the forest certification and the timing of the main evaluation, surveillance, and Corrective Action Requests (CARs) improvements. The FSC scoping assessment composed of the name of the applicant, the geological area, types of forest certification and list of products and services. The last chapter of implementation in accordance with the FSC standard composed of desired characteristics, activities, and information, including lists of documents that must be prepared to support the FSC certification.

Figure 1. Manual of rubber plantation owners and rubber wood consumers in Thailand for obtaining international forest management certification.

The knowledge in the manual had been transferred to the rubber plantation owners, rubber wood consumers, staff of the Rubber Authority of Thailand, and other organizations in four workshops in Songkla, Surat Thani, Rayong, and Bangkok and involving over 150 participants. The aim of the workshops was to develop the participant skills on how to prepare to request for FSC: FM certification (Figure 2). The workshops not only provided knowledge and guidance on how to use the manual in the meeting room, but field trips to test the manual in the rubber plantation also helped workshop participants better understand how to use the manual. Questions, suggestions, and difficulty in understanding the draft manual derived from the workshop were used to improve the manual and make it more appropriate and clear. Examples of the improvements were changing the table which explained the appropriate personal protective equipment (PPE) in each of the activities to be images, for easy implementation, and increasing the font size to make it easier to read. All participants and the relevant organizations shall receive the revised manual by mail at the end of the research project.

Figure 2. Workshop and field trip to examine the draft manual.

4. Conclusions

In this research, we have developed a manual for rubber plantation owners and rubber wood consumers in Thailand for obtaining the Forest Stewardship Council (FSC) standards based on the principles and criteria of the FSC certification and other relevant documents. The draft manual has five chapters: 1) FSC definition, 2) FSC scoping, 3) FSC evaluation and surveillance, 4) FSC scoping assessment, and 5) implementations in accordance with FSC standards. The draft manual was shared with a total of over 150 participants including rubber plantation owners, rubber wood consumers, staffs of the Rubber Authority of Thailand, and other organizations in different regions of Thailand, and a manual adjustment was done after the training. At the end of the research project, the revised manual will be delivered to all participants and relevant organizations.

5. References

[1] Duangsathaporn K, Prasomsin P, Laemsak N, Omule Y, Palakit K, Lumyai P 2019 *IOP Conf. Series: Materials Science and Engineering* 526.
[2] https://info.fsc.org/certificate.php: November 2019.
[3] Forest Stewardship Council. 2016. *https://ic.fsc.org/en/certification/principles-and-criteria*: November 2016.
[4] Forest Stewardship Council. 2018. *https://ic.fsc.org/en/document-center/id/59*: June 2018.

Acknowledgement
This research was funded by the Thailand Science Research and Innovation (TSRI: RDG61T0006). We also thank the valuable staff from the Laboratory of Tropical Dendrochronology (LTD), Faculty of Forestry, Kasetsart University, for data preparation, collection and analysis.

IOP Conf. Series: Materials Science and Engineering **773** (2020) 012051 doi:10.1088/1757-899X/773/1/012051

Influence of ethylene glycol treatment on conductivity and stability of poly(3, 4-ethylenedioxythiophene) polystyrene sulfonate coated cotton yarn

K Pattanarat[1], N Petchsang[2,3] and R Jaisutti[1,*]

[1] Department of Physics, Faculty of Science and Technology, Thammasat University, Khlong Nueng, Khlong Laung, Pathumthani 12121, Thailand.
[2] Department of Materials Science, Faculty of Science, Kasetsart University, 50 Ngam Wong Wan Road, Lat Yao Chatuchack Bangkok 10900, Thailand.
[3] Specialized Center of Rubber and Polymer Materials for Agriculture and Industry (RPM), Faculty of Science, Kasetsart University, Bangkok 10900, Thailand

*Corresponding author: jrawat@tu.ac.th

Abstract. Recently electronic textiles have received much interest due to their board applications in wearable electronics such as solar energy, physical and chemical sensors. In this work, the conductive cotton is prepared by dip coating poly(3,4-ethylenedioxythiophene) polystyrene sulfonate (PEDOT:PSS) on commercial cotton yarn. The electrical conductivity of conductive cotton can be enhanced by secondary treated PEDOT:PSS with ethylene glycol (EG). The EG-treated conductive cotton exhibits lower resistance of 600 Ω/cm by comparison with the pristine PEDOT:PSS coated cotton yarn (110 kΩ/cm). The electrical resistances of the conductive cottons are also investigated as a function of temperature and humidity. The pristine conductive cotton increases in electrical resistance as the temperature increases, while there is no resistance change for EG-treated conductive cotton. SEM images show well distribute of PEDOT:PSS on the cotton yarn surface. FTIR analysis also confirms the residue of PEDOT:PSS coated on cotton yarn. In addition, the treated conductive cotton exhibit high stability under air as they show slightly change in electrical resistance after keeping for 30 days.

1. Introduction

Electric textiles with high performance and stretchability are of great interest owing to their potential applications in wearable electronics and portable devices [1, 2]. Generally, conductive textiles are relying on metal based technology which drawback in flexibility and gives weight burden to textiles. The coating of intrinsically conducting polymers on textile substrate is one of the suitable alternative method to fabricate conductive textiles. Among the different available conducting polymers, poly(3, 4-ethylenedioxythiophene) doped with poly(styrene sulfonate) (PEDOT:PSS) is one of the most used. It has attracted wide attention as electrode material in several applications, due to its excellent characteristics such as high transparency, good flexibility, and thermally stability [3]. However, intrinsic PEDOT:PSS films suffer from a low conductivity because the ambiguous phase segregation and the excessive components of insulating PSS in PEDOT:PSS matrix [4]. The conductivity of PEDOT:PSS has been reported to be increased by the addition of solvents, such as dimethyl sulphate (DMS) [5], dimethyl sulfoxide (DMSO) [6]. However, the use of DMSO may irritate human skin, and fabrics are vulnerable to acid, preventing the use of post-treatment with a strong acid. A several methods have been employed, for instance, doping the PEDOT:PSS aqueous solutions with secondary solvents of methanol, ethylene glycol (EG) [4, 7], and glycerol [8], or using other dopants (anionic and ionic surfactants) [3, 9]. However, EG-doping is considered as a simple and effective method for substantially boosting the conductivity of the PEDOT:PSS.

In this work, we report a simple method to fabricate conductive cotton by dip coating PEDOT:PSS on cotton yarn and enhance its conductivity by treating PEDOT:PSS with EG. The conductive cotton properties show higher conductivity and stability after treated with EG. They exhibited slightly change in electrical resistance under humidity and temperature increases.

2. Experimental

2.1. Conductive cotton preparation

Aqueous solution of PEDOT:PSS and EG were purchased from Sigma-Aldrich. All chemicals were used as received. The conductive cotton was prepared by dip coating process of undyed 100% cotton yarn (Venus Thread Co., Ltd.) in 1.1 wt% PEDOT:PSS solution. Firstly, cotton yarn was cleaned with acetone under ultra-sonication for 10 min to remove organic contaminants and dried in oven at 60°C for 30 min. Then, the cleaned cotton yarn was immersed in PEDOT:PSS for 5 min and dried in oven at 60°C for 30 min. For electrical enhancement of conductive yarn, the conductive cotton was dipped in EG and dried in oven at 60°C for 2 h.

2.2. Conductive cotton characterization

The surface morphologies of bare and conductive cottons were characterized by scanning electron microscope (SEM, Quanta 450 FEI). Fourier-transform infrared spectroscopy (FTIR, Nicolet iS50) with attenuated total reflectance (ATR) mode was used to determine the chemical structure of the bare and PEDOT:PSS coated cotton yarn. The electrical properties were characterized by using two probes digital multimeter (Agilent 34970A). The temperature-dependent electrical conductivity of conductive cotton was studied by a digital source meter (Keithley 2400) within the temperature ranging from 40°C to 80°C. In addition, the effect of humidity and electrical stability of conductive cotton were also investigated.

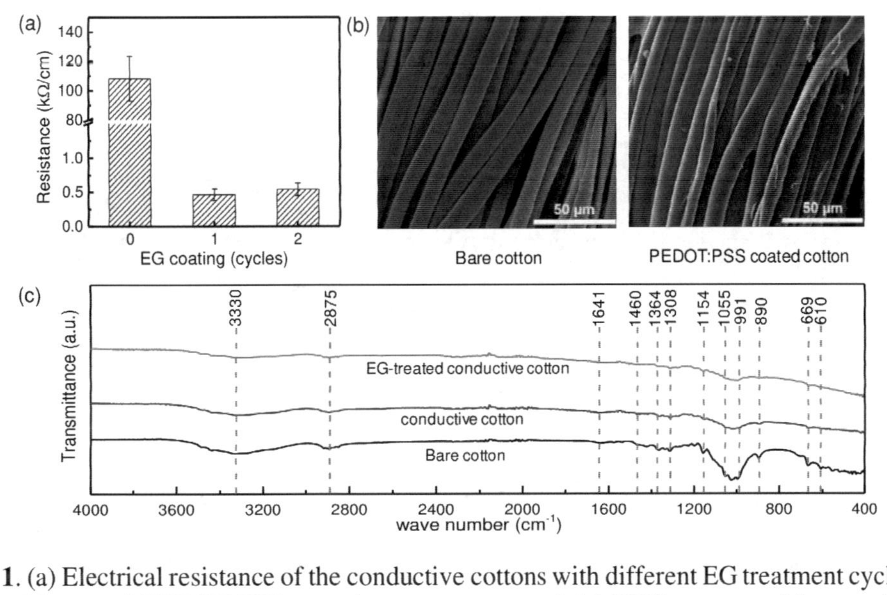

Figure 1. (a) Electrical resistance of the conductive cottons with different EG treatment cycles, (b) SEM images of bare and PEDOT:PSS coated cotton yarns, and (c) FTIR spectra of bare and conductive cottons with and without EG treatment.

IOP Conf. Series: Materials Science and Engineering **773** (2020) 012051 doi:10.1088/1757-899X/773/1/012051

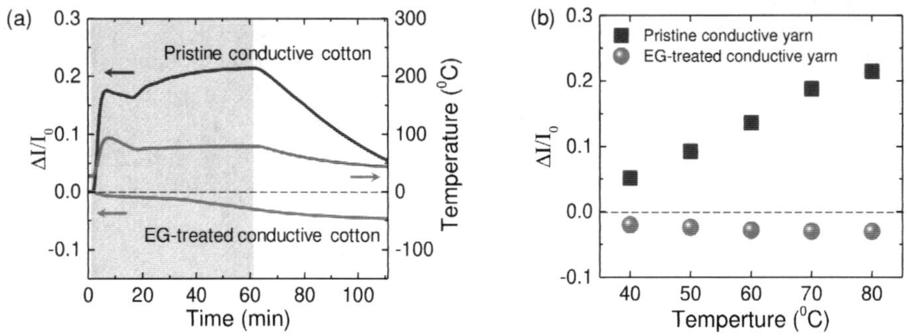

Figure 2. (a) Electrical characteristics of conductive cottons measured at temperature of 80 $_{o}$C, and (b) relative current change of pristine (square) and EG-treated conductive cotton (cycle) as a function of temperature.

3. Results and discussion

Figure 1a shows an electrical resistance of PEDOT:PSS coated cotton yarn with and without EG treatment. After coated with PEDOT:PSS, the cotton exhibits a resistance value of 110.6 kΩ/cm. This resistance value was dramatically decreased below 600 Ω/cm after EG-treated conductive cotton. However, there are no variation change under increase EG treatment cycles. The effect of EG on electrical conductivity of conductive cotton can be attributed to a relaxation distance between the polymer chains [7].

The surface morphology of bare and PEDOT:PSS coated cotton yarns are observed by SEM images as shown in Figure 1b. The bare cotton shows a smooth surface and a well distribution of PEDOT:PSS was observed on cotton yarn. To confirms the residue of PEDOT:PSS on cotton yarn, FTIR analysis was examined and results are showed in Figure 1c. The FTIR spectra exhibit almost the same peaks for both bare and conductive cottons. The FTIR spectrum of the pristine conductive cotton showed peaks at 1460, 1364, 1308, and 893 cm$_{-1}$ correspond to C=C, C-C, and C-S bonds, respectively, which originated from thiophene ring in PEDOT's monomer structure [3, 10]. The observed peak at 1054 cm$_{-1}$ is originated from to C-O-C bond stretching in the ethylenedioxy group in the PEDOT:PSS monomer [3, 10]. The peaks assigned at 1641 and 1154 cm$_{-1}$ correspond to C=C bond from aromatic ring of PSS and symmetrical vibration of SO_3H in PSS [3, 10].

To investigate the environment stability of the PEDOT:PSS coated cotton yarn, the effect of temperature on the electrical conductivity of conductive cotton was observed. A relative current change, $\Delta I = I - I_0$, where I_0 and I are the response current before and after applied heating temperature, is observed for pristine and EG-treated conductive cotton under a constant temperature of 80 $_{o}$C as shown in Figure 2a. The temperature profile (red solid line) of 80 $_{o}$C also revealed here. The pristine conductive cotton demonstrates a change in electrical current followed as increasing and decreasing of environment temperature. The relative current change and temperature is linearly increased as temperature increases (Figure 2b). In contrast, the EG-treated conductive cotton exhibit slightly decrease in electrical conductivity. A highly stable in temperature has been found for treated conductive cotton as indicated no-variation change in relative current in Figure 2b.

The effect of humidity on electrical characteristics of conductive cottons are investigated by applied voltage and measured current pass through the sample before and after soaking in a water drop (Figure 3). The current-voltage curve of the pristine conductive cotton shows a good ohmic behaviour. The decrease in conductivity and hysteresis of I-V characteristics was observed during soaking in a water drop. This is due to the surface effect between PEDOT:PSS and water molecules. In addition, some residues PEDOT:PSS can be dissolved in water. Whereas, the EG-treated conductive cotton does not change in conductivity under water drop.

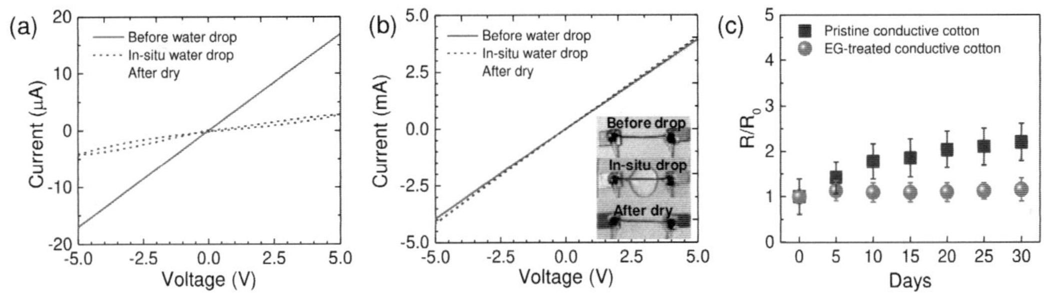

Figure 3. Electrical characteristics of (a) pristine and (b) EG-treated conductive cotton before, in-situ dropping and after drying of water. (c) Stability in electrical resistance of pristine and EG-treated conductive cotton measuring within 30 days. The inset of (b) show images of conductive cotton before and after testing under water dropping.

For the practical application of conductive cottons, a long-term stability was also examined for 30 days. As shown in Figure 3c, the pristine conductive cotton exhibits higher in resistance when keeping in air while there is no significant change in resistance for EG-treated conductive cotton. These results demonstrate that highly electrical conductivity and good stability of conductive cotton can be done by a simple EG treatment process.

4. Conclusions

In conclusions, the electrical conductivity of PEDOT:PSS can be enhanced by a simple method through secondary treated PEDOT:PSS with EG. After treatment, a decrease in electrical resistance was observed compared to pristine conductive cotton. The conductivity of EG-treated conductive cotton is not dependent upon the number of cycles in dipping EG solution. SEM images showed densely coating and perfectly connected of conducting polymer on cotton surface. In addition, the conductive cotton properties show no significantly change in electrical resistance under increasing of humidity and temperatures after treat with EG. The EG-treated conductive cotton are high stability for long-term used in the air. Considering the simplicity of the process, a low-cost EG-treated conductive cotton will contribute to realizing the mass production and utilizing for wearable electronic applications.

5. References

[1] Wu S, Liu P, Zhang Y, Zhang H and Qin X 2017 *Sens. Actuators,B.* **252** 697-705.
[2] Indarit N, Kim Y-H, Petchsang N and Jaisutti R 2019 *RSC Adv.* **9** 26773-26779.
[3] Yeon C, Kim G, Lim J W and Yun S J 2017 *RSC Adv.* **7** 5888-5897.
[4] Cui H-Q,Peng R-X, Song W, Zhang J-F, Huang J-M, Zhu L-Q and Ge Z-Y 2019 *Chin. J. Polym. Sci.* **37** 760-766
[5] Reyes-Reyes M, Cruz-Curz I and López-Sandoval R 2010 *J. Phys. Chem. C* **114** 20220-20224.
[6] Nagata R, Yanagi Y, Fujii S, Katauta H and Nishioka Y 2014 *IEICE. TRANS. ELECTRON.* 299-302.
[7] Lin Y-J, Ni W-S and Lee Y-J 2015 *J. Appl. Phys.* **117** 215501.
[8] Palumbiny C M, Heller C, Schaffer C J, Körstgens V, Santoro G, Stephan V R and Müller-Buschbaum P 2014 *J. Phys. Chem. C* **118** 13598-13606.
[9] Fan B, Mei X and Ouyang J 2008 *Macromolecules* **41** 5971-5973.
[10] Lee J and Choi W 2015 *J. Electrochem. Soc.* **162** A935-A939.

Acknowledgement
This work was financially supported by Thammasat University (Contract No.16/2561) and Thammasat University innovation scholar (Contract No. 2/2559). We thank Scientific Equipment Center, Kasetsart University, for using their characterization equipment.

IOP Conf. Series: Materials Science and Engineering **773** (2020) 012052 doi:10.1088/1757-899X/773/1/012052

Effect of Fe doping on the photocatalytic activity of TiO_2 hollow fibers under LED light irradiation

S Wongcharoen[1], C Chokradjaroen[2] and G Panomsuwan[1,*]

[1] Department of Materials Engineering, Faculty of Engineering, Kasetsart University, Bangkok 10900, Thailand.
[2] SIT Research Laboratories, Shibaura Institute of Technology, Tokyo 135-8548, Japan.

*Corresponding author: gasidit.p@ku.ac.th

Abstract. Fe-doped TiO_2 hollow fibers (Fe-THF) were prepared using a template replication method with kapok as the template. The Fe concentration was varied by adjusting the amount of iron (III) nitrate from 0 to 5 mol%. The effect of Fe doping on morphology, crystal structure, bandgap energy, and photocatalytic activity was investigated and discussed. Based on the characterization results, it was found that all samples possessed a hollow structure with anatase as the main crystalline phase. The rutile phase began to be observed with high-level doping. Bandgap energy decreased significantly from 3.02 to 2.12 eV as the Fe concentration increased, owing to the formation of intermediate energy level. The photocatalytic activity was probed using the degradation of methylene blue under light-emitting diode (LED) irradiation. We found that the photocatalytic activity of Fe-THF was inferior to undoped THF and decreased with increasing Fe concentration. This result indicated that Fe dopant could act as the recombination centers in the lattice system, thus inhibiting the photocatalytic activity of THF.

1. Introduction

Since 1972, TiO_2 has been found to be a promising photocatalyst material for the photo-electrochemical water splitting [1]. By absorbing the photons from a light source, the generation of electron–hole pairs in TiO_2 can accelerate chemical redox reactions. Owing to its unique optical properties and low cost, TiO_2 has significant potential for use in energy and environmental applications, such as dye-sensitized solar cells [2], wastewater treatment [3], and H_2 production [4]. However, the photocatalytic performance of TiO_2 is limited by its wide bandgap (3.0–3.2 eV) and high recombination rate of photogenerated electrons and holes. The enhancement of photocatalytic performance for TiO_2 is based on three strategies, as follows: (i) narrowing bandgap energy by creating impurity levels near the valence or conduction band through metal or non-metal doping [5], (ii) reduction of recombination rate by decorating the surface with noble-metal nanoparticles [6], and (ii) designing specific shapes and structures to enhance light scattering and absorb more incident photons [7]. Therefore, the rational design and integration of TiO_2 based on these strategies remains challenging and may provide the unexpected results.

In our previous work, we successfully synthesized TiO_2 hollow fibers (THF) using a template replication method with the use of kapok as biotemplate [8]. THF calcined at 450 °C showed the best photocatalytic activity on paraquat degradation under ultraviolet (UV) light exposure. In a further effort to improve the photocatalytic activity of THF, narrowing the bandgap by metal doping has been the focus of this work. Fe was selected as dopant since the ionic radius of Fe^{3+} ion (0.064 nm) is nearly similar to Ti^{4+} ion (0.068), which can allow easy incorporation into TiO_2 lattice [9]. The effect of Fe doping on morphology, crystal structure, and bandgap energy were investigated and discussed based on the following characterization techniques: scanning electron microscopy (SEM), X-ray diffraction (XRD), diffuse reflectance spectroscopy (DRS). Moreover, the photocatalytic activity of Fe-doped THF on the degradation of methylene blue (MB) was evaluated by measuring the change in absorption under light-emitting-diode (LED) irradiation.

2. Experimental

2.1. Synthesis of Fe-doped TiO_2 hollow fibers (Fe-THF)
A certain amount of $Fe(NO_3)_3 \cdot 9H_2O$ ranging (1, 3, and 5 mol%) was mixed with 150 mL of ethanol under vigorous stirring at room temperature for 1 h. Then, 3 mL of titanium butoxide (8.7×10^{-3} mol) was slowly dropped into the mixture by keeping vigorous stirring for 1 h. Dried kapok fibers (1 g) were immersed in the precursor for 1 h and dried 18 h at room temperature. The calcination was carried out at 450°C with a heating rate of 5°C/min under air atmosphere for 1 h in order to remove the kapok template and crystallization of TiO_2. White fluffy product was obtained for undoped THF. The products with Fe doping were light yellow, and became more yellow as the Fe concentration increased. THFs prepared by using $Fe(NO_3)_3 \cdot 9H_2O$ of 1, 3, and 5 mol% were named 1-Fe-THF, 3-Fe-THF, and 5-Fe-THF, respectively.

2.2. Characterization
Phase structures were identified using X-ray diffractometer (Rigaku, SmartLab) with Cu Kα radiation (λ = 1.54056 Å) at 40 kV and 40 mA. Morphology was observed with a scanning electron microscope (JEOL, JSM-7600F) operated at an accelerating voltage of 5 kV. Moreover, the elemental compositions were investigated by energy dispersive spectrometer (Oxford Instrument). Bandgap energy was estimated by UV-vis diffuse reflectance spectrometer (Shimadzu, UV-2550) and Tauc plot.

2.3. Photocatalytic activity measurement
5 ppm of MB solution was prepared and stirred in the dark for 24 h. 50 mg of the sample was suspended in 50 mL of MB solution and stirred under dark conditions for 1 h to obtain adsorption–desorption equilibrium. After that, the suspensions were stirred under white LED irradiation (20 W Phillips) for 6 h. 3 mL of suspension was collected hourly and centrifuged to collect only irradiated MB solutions. The MB degradation was evaluated with an UV-vis spectrometer (Shimadzu, UV-1700) through the change in absorption at 663.5 nm.

3. Results and discussion
SEM images of the THF and Fe-THF are illustrated in Figure 1. Clearly, all samples exhibited a hollow fiber structure with a diameter of about 20–30 μm. No obvious difference in morphology was observed regardless of Fe concentration. The presence of Fe dopant in THF was also confirmed by energy dispersive spectroscopy (EDS) measurement. It was found that the chemical composition of Fe-THFs comprised of only Ti, O, and Fe. The Fe concentration was found to be 0.22, 0.64, and 0.96 at% for 1-Fe-THF, 3-Fe-THF, and 5-Fe-THF, respectively.

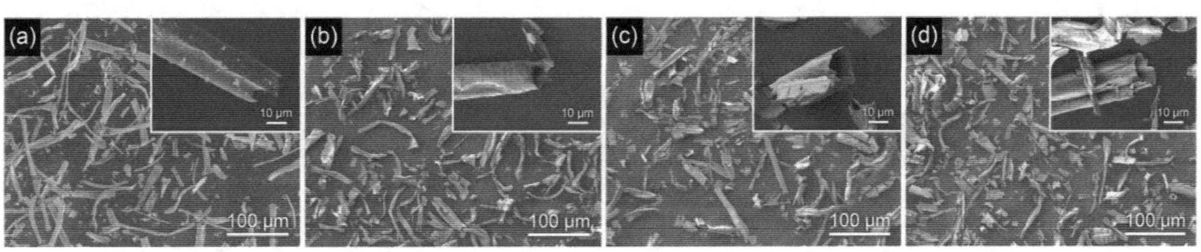

Figure 1. SEM images of (a) THF, (b) 1-Fe-THF, (c) 3-Fe-THF, and (d) 5-Fe-THF.

IOP Conf. Series: Materials Science and Engineering **773** (2020) 012052 doi:10.1088/1757-899X/773/1/012052

The XRD patterns of all samples are shown in Figure 2(a). All diffraction peaks of THF and 1-Fe-THF were associated with the anatase phase (JCPDS card no. 21-1272) without other crystalline phase. However, a small diffraction peak at 27.7° corresponding to the rutile phase (JCPDS card no. 21-1276) was observed for 3-Fe-THF (indicated with an arrow) and was stronger for 5-Fe-THF. The emergence of the rutile phase was likely due to the fact that Fe doping resulted in lower phase transformation temperature of TiO_2 to turn anatase into the rutile phase [10]. Moreover, there were no detectable diffraction peaks of the secondary phase (i.e., Fe_2O_3), even at high Fe concentration. This result is evidence that Fe^{3+} ion has been incorporated or replaced in the TiO_2 lattice due to the nearly similar ionic radii of Fe^{3+} (0.064 nm) and Ti^{4+} (0.068 nm).

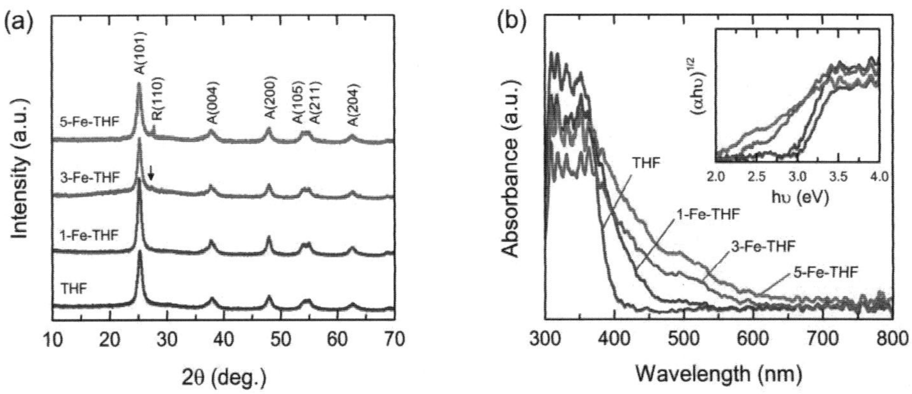

Figure 2. (a) XRD patterns and (b) absorption spectra of all samples. The corresponding Tuac plots between $(\alpha h\upsilon)^{1/2}$ versus $h\upsilon$ are shown in the inset of (b).

UV-vis absorption spectra of all samples in the wavelength range of 300–800 nm are demonstrated in Figure 2(b). It was seen that the absorption band edge shifted significantly toward longer wavelengths with increasing Fe concentration, indicating the redshift phenomenon. In other words, the absorption of Fe-THFs was extended from UV to the visible-light region. Furthermore, the bandgap energy (E_g) can be calculated using the Tauc plot of $(\alpha h\upsilon)^{1/2}$ for indirect allowed transition versus energy ($h\upsilon$), as shown in the inset of Figure 2(b). The bandgap energy of the samples was obtained by extrapolation of the linear line to intercept the $h\upsilon$-axis. The bandgap energy was estimated to be 3.02, 2.92, 2.49, and 2.12 eV for THF, 1-Fe-THF, 3-Fe-THF, and 5-Fe-THF, respectively. The reduction of bandgap energy at higher Fe concentration is attributable to the formation of an impurity state below the conduction band, which can thus enable visible-light absorption.

The photocatalytic activity of all samples was further evaluated by the degradation of MB under LED irradiation. Figure 3a presents the change in the absorption spectrum of MB solution in the presence of 1-Fe-THF at different irradiation times. It was evident that the absorption of MB solution decreased continuously with prolonged irradiation time, indicating its photocatalytic degradation. Figure 3b displays the relative concentration (C_t/C_0) as a function of irradiation time for all samples. Note that C_t and C_0 are the concentrations of MB at the given reaction and initial times, respectively. There was no noticeable reduction in the absorption of MB in the absence of samples, suggesting that MB was not self-degraded by LED. Under dark conditions for 1 h, MB adsorption was observed about 12% on THF and decreased to about 1% for 5-Fe-THF. Further adsorption on the sample surface did not occur after 1 h. The decrease in MB adsorption on the sample surface may be due to the decrease in surface area and a difference in surface chemistry at high Fe concentration. After LED irradiation for 6 h, MB was degraded by 67%, 66%, 55%, and 49% for THF, 1-Fe-THF, 3-Fe-THF, and 5-Fe-THF, respectively. From these results, although Fe doping could narrow bandgap energy of THF to be responsible in the visible light region, the photocatalytic activity of Fe-THF was not improved. Rather, the opposite occurred. This finding was similar to the previous works reported by Li et al [9] and Moradi et al [10] showing that excess Fe doping led to poor photocatalytic activity. The adverse effect of Fe doping on photocatalytic activity can be ascribed to the fact that overly high levels of Fe dopants can serve as a recombination center for electrons and holes, instead of electron scavenger sites, thus inhibiting photocatalytic activity [9,10].

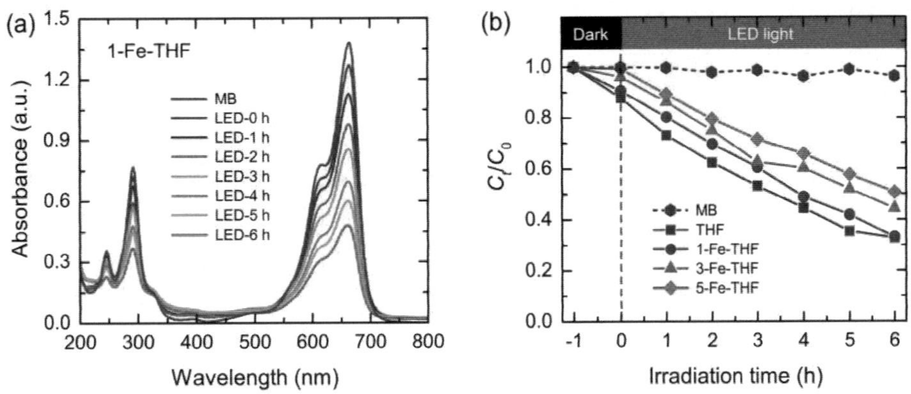

Figure 3 (a) UV–vis absorption spectra of MB solution at the presence of 1-Fe-THF at different irradiation times. (b) Plot of C_t/C_0 versus irradiation time for all samples.

4. Conclusions

In summary, Fe-THFs with different Fe concentrations (0.22–0.96 at%) have been successfully synthesized via the biotemplate method. Anatase was found as the main crystalline phase in Fe-THFs, while rutile phase was formed at high Fe concentrations. The Fe-THFs revealed a redshift phenomenon, which extended their absorption to visible light region, resulting in a significant reduction of bandgap energy. Despite narrow bandgap energy, the photocatalytic activity of Fe-THFs was inferior to undoped THF and became worse at higher Fe concentration. Poor photocatalytic activity of Fe-THFs in this work could be attributed to too much Fe dopant in the lattice structure, which in turn led the formation of recombination centers, thereby inhibiting electron–hole pair separation. The results obtained in this work suggested that more detailed investigations are necessary to seek the optimum Fe concentration for enhancing the photocatalytic activity of THF. Surface area and chemical bonding state of dopant must also be considered for comparison in future work.

5. References

[1] Fujishima A and Honda K 1972 *Nature* **238**, 37.
[2] Roose B, Pathak S and Steiner U 2015 *Chem. Soc. Rev.* **44** 8326.
[3] Lee SY and Park SJ 2013 *J. Ind. Eng. Chem.* **19** 1761.
[4] Chiarello GL, Dozzi MV and Selli E 2017 *J. Energy Chem.* **26** 250.
[5] Girish Kumar S and Gomathi Devi L 2011 *J. Phys. Chem. A* **115** 13211.
[6] Primo A, Corma A and Garcia H 2011 *Phys. Chem. Chem. Phys.* **13** 886.
[7] Lee K, Yoon H, Ahn C, Park J and Jeon S 2019 *Nanoscale* **11** 7025.
[8] Wongcharoen S and Panomsuwan G 2018 *Mater. Lett.* **228** 482.
[9] Li Z, Shen W, He W and Zu X 2007 *J. Hazard. Mater.* **155** 590.
[10] Moradi V, Ahmed F, Jun MBG, Blackburn A and Herring R 2019 *J. Environ. Sci.* **83** 183.

Acknowledgement
This work was supported by the Innovation Development Scholarship for Master Student from the Faculty of Engineering, Kasetsart University (06-61-MET-Innovation).

Study on processability and mechanical properties of parawood-powder filled PLA for 3D printing material

K Chansoda, C Suwanjamrat and W Chookaew*

Material and Manufacturing Innovation Research Group, Department of Mechanical Engineering, Faculty of Engineering, Mahidol University, Nakhon Pathom 73170, Thailand.

*Email: watcharapong.chk@mahidol.ac.th

Abstract. Recently, there has been an increasing interest in the development of 3-Dimensional (3D) printing technologies. Among those, material development play an important role in helping the growth of 3D printing technique, which led to more supporting of wide range of applications and user requirements. In that regards, advanced materials are continually developed, i.e. filament filled reinforcing fiber and plastic containing metal granulate or wood powder. Special material supporting 3D printer normally showed an expensive cost. On the other hand, it is well known that the processes involving natural rubber from Para rubber trees produced a large amount of wood a year, which resulted in releasing of an abundant waste from wood processing. Therefore, this work focused on the utilization of parawood powder derived from the furniture industry. The processability of polylactic acid (PLA) filament containing wood powder was investigated. Twin and single-screw extrusion machines were used to produce the composite filaments. Coupling agent and treated method were also studied. The results of this work showed that the maleic anhydride (MAH) and sodium hydroxide (NaOH) did not have significantly effects in improving the compatibility and processability. In addition, wood contents were found to influence the quality of filament and mechanical properties. Moreover, it was also found that parawood powder can be filled into PLA with approximately 10 wt%.

1. Introduction

3-Dimensional (3D) printer is gaining popularity in conventional customer due to the lower machine cost, easy to use and various types of material selection. Moreover, it can quickly respond and adapt to the changing needs of consumers. Definitely, the special materials are enhancing considerable attention in terms of reinforcing material, biomaterial [1-3], renewable or degradable material [4,5]. Thermoplastic composites play an important role 3D printing material, i.e. acrylonitrile butadiene styrene (ABS), polypropylene (PP) [6] polyethylene terephthalate (PET) and polylactic acid (PLA) etc. Conventional material using for 3D printer is distributed as a roll of filament, which passed extrusion machine. A price of plastic roll is normally higher than a granulate material. Moreover, special roll supporting 3D printed material is more expensive, about 30 times than that of pellet thermoplastic. Further problem of wood by-product, the waste from parawood industrial was produced a large amount such as sawdust, polishing wood powder in any stage of processing. Therefore, the utilization of residue material in manufacturing of 3D printing technique can lead to the value added of filament and also eliminated the environmental concern. Hence, the objective of this works was to investigate the possibility of producing the parawood filled into PLA. Processability by twin and single screw extruder was also tested. Processability of wood filament by 3D printer was compared to commercial filament and imported wood powder.

2. Experimental

Different wood powders were used in this study, i.e. commercial grad of wood powder, and blasting powder and residue wood from the furniture industry as shown in Table 1. In Fig x(a)-(b), the degradable polymer, (Poly lactic acid, PLA) was blended with wood powder by twin and single screw extruders, respectively. PLA and wood powder were mixed using the twin-screw extruder at the temperature of 200oC with the rotational speed of 140 rev/min. Each batch of composite was rolled to produce a filament of approximately 1.50 mm in diameter. The powder content in ranges of 5-20 wt% was varying to analysis the processability. For the comparison of mechanical strength, a material test piece in accordance with the standard tensile test ASTM D638 was formed by the 3D printer as illustrated in Fig x. The tensile strength was characterized by using the Universal Testing Machine (UTM, COMETECH Model LRK-20) at a crosshead speed of 500 mm/min.

Table 1. Formulas of wood composites in percentage by weight.

Mixture	Content (wt%)		
	Wood powder	Poly lactic acid (PLA)	Maleic anhydride (MAH)
CW#1	5	95	-
CW#2	10	90	-
CW#3	15	85	-
CW#4	20	80	-
CW#5	5	90	5
CW#6	10	85	5
CW#7	15	80	5
CW#8	20	75	5
PW#1	5	95	-
PW#2	10	90	-
PW#3	15	85	-
PW#4	20	80	-
RW#1	5	95	-
RW#2	7.5	92.5	-
RW#3	10	90	-
RW#5	15	85	-
RW#4	20	80	-

3. Results and discussions

3.1. Effect of wood powder content on the processability by screw extruder

Commercial grade of wood powder (CW) was mixed together with PLA in content of 0-20 wt% as illustrated in Table 1 (CW#1- CW#4). It was found that all formulation varying CW content was able to compound by twin screw machine. The primary mixtures were then dumped to single screw extruder in order to produce a filament. Figure 1 depicts the effect of wood content on the appearance of the composites. A perfectly smooth surface of extrudate was presented in the neat PLA. This was due to a mobility of polymer molecules without any obstacle from wood element. Further increasing the wood content, slight irregularities were observed on the surface of composite extrudate. This result presented that highest contents of filling wood power was 10 wt%.

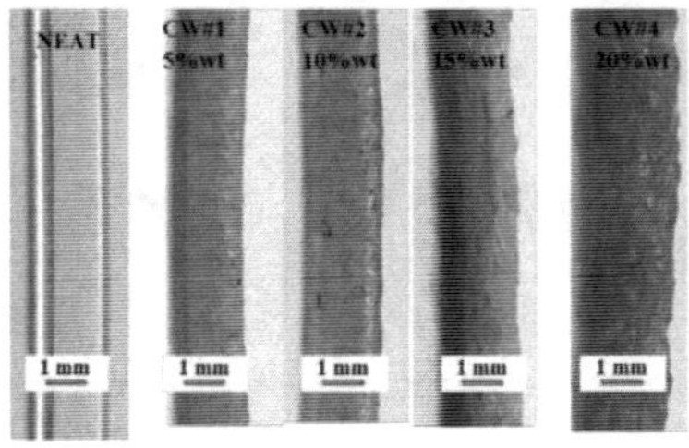

Figure 1. Appearance extrudate of composite varying wood content.

3.2. Effect of maleic anhydride (MAH) on the compatibility of composite
In order to improve the compatibility of composite, the filling of coupling agent was investigated. This study used MAH as a coupling agent in a content of 5 wt%. It was found that all ingredients are able to blend by using screw extruder. However, the primary mixtures could not be made a filament when it applied to single screw extruder. Figure 2 shows the low viscosity of composites. The result was expected from the decomposition of MAH during the extruding process. MAH at the high processing temperature was transformed into gas, which are released from die as in Figure 1. Other gases were still composed in the mixture, which made compound lower viscosity and aggregated at external machine.

Figure 2. Composites filled MAH passing single screw machine.

3.3. Effect type of preparing process of natural rubber wood.
Former testing indicated that MAH did not attribute the enhancing incorporation of wood content. In this case, wood powder prepared by sand paper polishing (PW) was compound without coupling agent (PW#1-PW#4, RW#1-RW#5). In the Figure 3, PW could not be prepared an accepted extrudate, which presented the non-uniform distribution of particle. This observation was due to the large differential density of mixing composition, which derived from particle size of wood powder. Residue wood powder (RW) introduced the good filament in ranges of 5-15 wt%. They can be printed by conventional 3D printer.

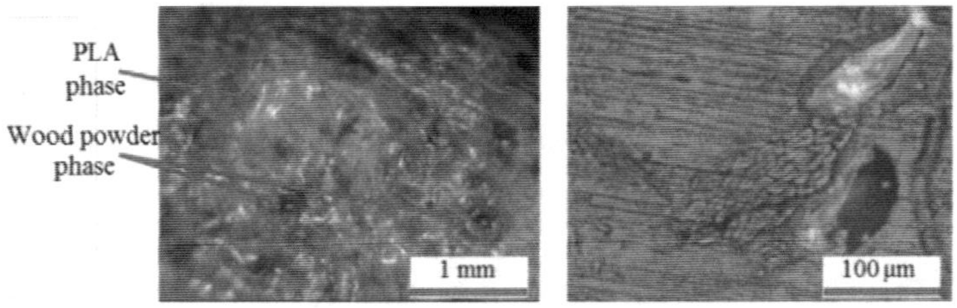

Figure 3. Non-uniform distribution of wood particle prepared by sand paper polishing.

3.4. Effect of wood powder types on the mechanical properties

In this work, the different types of wood filler were loaded into the PLA passing screw extruder. The highest mechanical strength of neat PLA was found. Two types of commercial filament were used to compare with prepared composites. Figure 4 indicated that the tensile strength of commercial filament were lower than those of producing extrudates. This observation was due to the higher content of wood particle in commercial product. When considering the properties of two types of wood powder, natural rubber wood from grinding process was not much different from the imported wood powder.

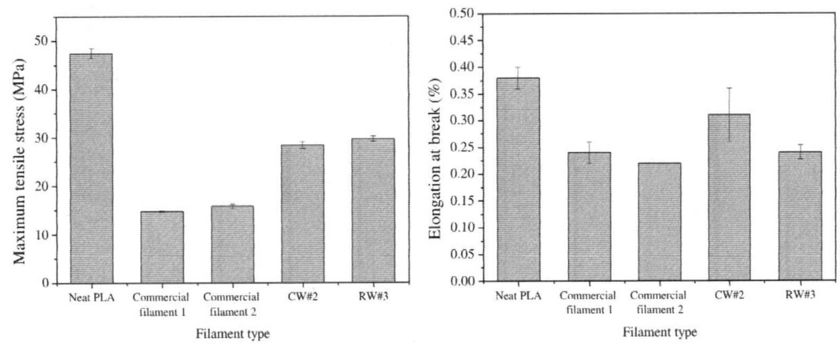

(a) (b)

Figure 4. Mechanical properties of wood filament (a) maximum tensile strength (b) elongation at break

4. Conclusions

The study result was found that all formulations can compounded in the twin screw extruder. While, the maximum loading of wood powder producing with single screw was 15 wt% (CW#1-CW#3, RW#1-RW#5). Composites containing wood content of 10 wt% can be printed by the 3D printer (CW#1, CW#2, RW#1-RW#4). In the results of preparing filament, the commercial powder and residue wood from furniture industry showed non-significant difference in mechanical strength. From the corresponding result, it should be concluded that the parawood filament is able to produce the product as a filament type for 3D printer. However, an adding more content of powder will be continually investigated in the further work.

5. References

[1] Jang et al., 2018 T.S. Jang, H.-D. Jung, H.M. Pan, W.T. Han, S. Chen, J. Song. 3D printing of hydrogel composite systems: recent advances in technology for 4. Int. J. of Bioprinting, 4 (1) (2018).
[2] Diana Gregor-Svetec, Mirjam Leskovšek, Urška Vrabič Brodnjak, Urška Stankovič Elesini, Deja Muck, Raša Urbas, Characteristics of HDPE/cardboard dust 3D printable composite filaments., Journal of Materials Processing Tech., (2020).
[3] A.J. Capel, R.P. Rimington, M.P. Lewis, S.D. Christie. 3D printing for chemical, pharmaceutical and biological applications. Int. Rev. Chem. Eng., 1 (2018).
[4] F. Brites, C. Malça, F. Gaspar, J.F. Horta, M.C. Franco, S. Biscaia, A. Mateus. Cork plastic composite optimization for 3D printing applications. Procedia Manuf., 12 (2017), pp. 156-165.
[5] Le Duigou et al., 2016 A. Le Duigou, M. Castro, R. Bevan, N. Martin. 3D printing of wood fibre biocomposites: from mechanical to actuation functionality, Mater. Design., 96 (2016), pp. 106-114.
[6] O.S. Carneiro, A.F. Silva, R. Gomes. Fused deposition modelling with polypropylene. Mater. Design., 83 (2015), pp. 768-776.

Acknowledgement

Highest acknowledgement is given to Thailand Research Fund (TRF) under contract number RDG60T136 for providing the financial support throughout this work.

IOP Conf. Series: Materials Science and Engineering **773** (2020) 012054 doi:10.1088/1757-899X/773/1/012054

On the mechanical properties of para rubber-oil palm ash derived geosynthetic clay liner

P Promputthangkoon[1],*, A Rungvichaniwat [2], N Kaewthai Andrei [3] and T Kuasakul [1]

[1] Faculty of Engineering, Rajamangala University of Technology Srivijaya, Songkhla, Thailand 90000
[2] Faculty of Science and Technology, Prince of Songkhla University, Pattani Campus, Thailand 94000
[3] Faculty of Agro-Industry, Rajamangala University of Technology Srivijaya, Nakhon si thammarat, Thailand 80240

*Corresponding author: panu_pptk@hotmail.com

Abstract. A mixture of kaolinite and oil palm ash with latex binder were from to make a core of the geosynthetic clay liner. Then, it was covered by latex-coated calico. The GCL developed had the width and length of 100 cm and 200 cm, with the total thickness of about 7 mm. It is to be used as liner at the bottom of a landfill in order to prevent the leachate to have a contact with the groundwater that could harm people who make use of that contaminated water. Several laboratory tests were carried out to evaluate whether it has the same properties as commercial ones; and, it can be employed in field. Only punching shear resistance and tensile strength are reported in this paper. It was found that both tests yield results that are in accordance with the industrial standard, indicating that it can be really employed in the field.

1. Introduction

Para rubber and oil palm are ones of the most important crops in Thailand, especially in the South. The price for the former has progressively decreased over the years, resulting in the economic problem countrywide. The latter has been in a similar situation, but in a lesser extent. The associated problem is that, however, ashes left from manufacturing processes have caused air pollution that requires a solution to improve the environment. The production of natural para rubber in Thailand has steadily increased from 3.12 million tonnes in 2006 to 4.47 million tonnes in 2015 [1]. Of the total production of 4.47 million tonnes, 3.75 million tonnes were exported, while only 0.60 million tonnes were domestically consumed. From these figures, no doubt that why it is very important to Thailand in terms of economy. The oil palm production in Thailand has as well gradually increased from 3.3 million tonnes in 2000 to 12.5 million tonnes in 2014; and, it has been forecast to be 19.2 million tonnes in 2022 [2]. It can be seen that the crop is so vital for the economy of the country, as high as for the para rubber. As such, when the price for both crops goes down, it is important for policy makers to improve the situation. Researchers are one of the key players that could find solutions in order that make them value added. In Thailand, dumping waste in unprotected hole is not abnormal. This practice leads to environmental

problems somewhat [3]. For instance, the leachate could have a contact with the groundwater thereby causing it contaminated. People who have utilised that water will surely suffer.
As a result, this study attempted to develop a geosynthetic clay liner (GCL) derived from para rubber and oil palm ash. Note that the GCL is employed at the bottom and sides of a landfill in order to prevent the leachate to flow through and contaminate the groundwater as a consequence. The GCL was made according to figure 1: it comprises top and bottom layers made from calico coated with latex compound and the internal core made from the mixture of artificial clay, oil palm ash, and latex compound. This creates an almost impermeable layer that could prevent the flowing of leachate generated from the waste.

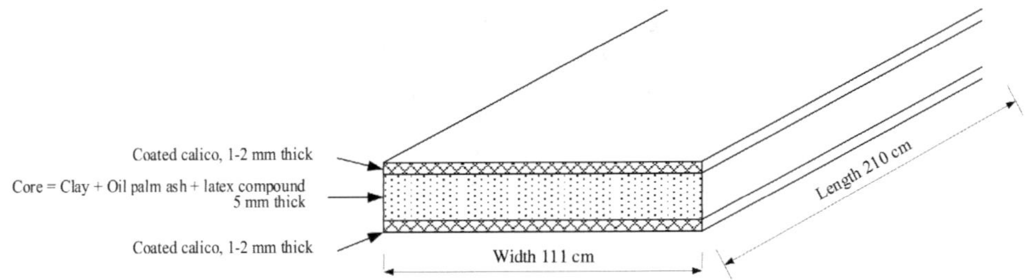

Figure 1. Proposed GCL made from clay, oil palm ash, and para rubber.

2. Experimental

Normally, bentonite is used as core of the GCL. In this case, kaolinite was chosen instead. This is because bentonite must be imported, while kaolinite can be obtained from a mine in Southern Thailand. Please be noted that bentonite is normally employed because it has lower permeability. The oil palm ash in this study was obtained from Krabi. The latex compound was prepared using the following ingredients: 60% HA latex, 10% KOH, 20% K-oleate, 50% ZDEC, 50% Wing stay L, 50% sulphur, 75% $CaCO_3$, 15% carbon black, and 50% ZnO. Figure 2 displays the kaolinite and oil palm ash employed; figure 3 shows how the latex compound was developed.

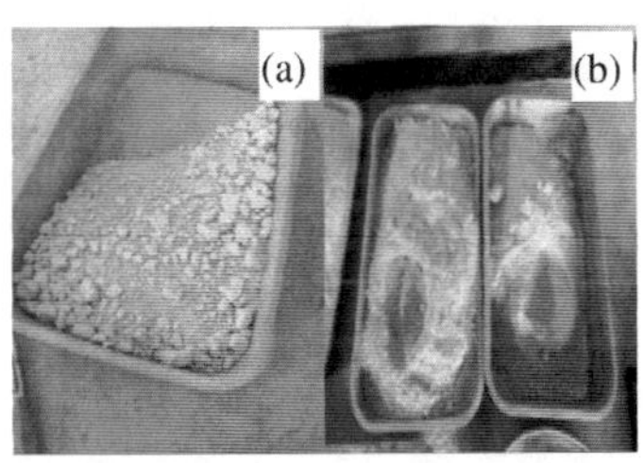

Figure 2. Kaolinite (a) Oil palm ash (b)

Figure 3. Preparation of latex compound (a) and (b), adding carbon black to the compound (c)

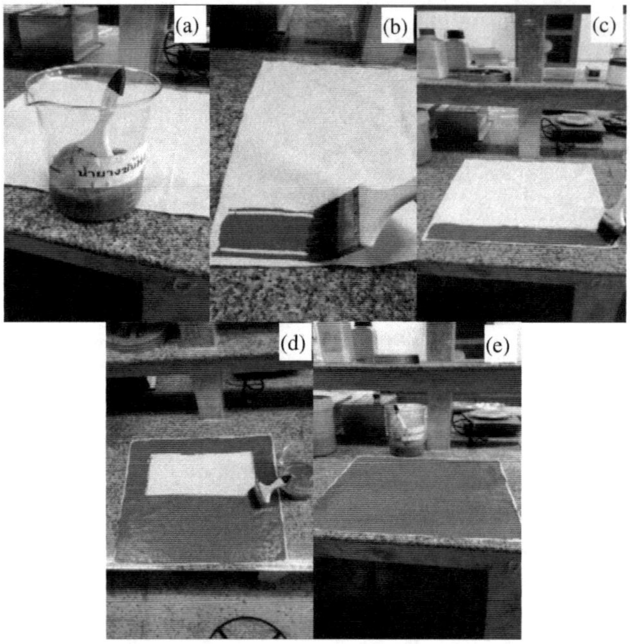

Figure 4. Preparation for the coated calico

To develop the GCL, first the coated calico was made according to the procedures shown in Figure 4. The bottom sheet then was laid on a flat floor; the compound latex was sprayed over. Next, the mixture of clay and OPA was laid over; and, the compound latex was sprayed over. After the top sheet was placed to cover, making a mixture sandwiched by the two sheets. To make it moveable, the GCL was manually sewed. To be able to use the GCL many tests were carried out. These include (1) permeability of the core, (2) punching shear resistance of the sheet, and (3) tensile strength of the sheet. In this paper, only the strengths were reported and discussed.

3. Results and discussion

The important properties for GCL are an ability to withstand both punching shear and axial shear forces. It should be noted that calico and geotextile (normally used to cover the GCL) are very different in terms of texture and manufacturing processes. The latter is developed specifically to withstand those forces; while the former is made for totally different purposes. That is why the geotextile is employed to assembly the GCL.

A total of 10 samples were employed for both punching shear and axial shear tests in order to obtain an average value. Table 1 displays the results obtained from punching shear test. The average value showed that it can withstand up to 2235 N (equivalent to 228 kg force). According to the standard set for the GCL, it states that the minimum required punching shear is 600 N. This result shows that the material used in this study is more than enough. In other words, thinner calico may be employed thereby reducing the cost of manufacturing. From the tensile test, it was found that the averaged value from 10 testes is 2832 N. Comparing this figure to the value set by the industrial standard of 2000 N, it can be seen that, again, the material has more than enough tensile strength to withstand the force during construction.

Table 1. Punching shear resistance of the coated calico

Sample no.	Area (sq.m)	Weight (g)	Punching force (N)
1	0.0625	33.55	2100
2	0.0625	35.64	1950
3	0.0625	33.55	2800
4	0.0625	35.64	2650
5	0.0625	33.55	1850
6	0.0625	35.64	2750
7	0.0625	33.55	1900
8	0.0625	35.64	2350
9	0.0625	33.55	1950
10	0.0625	35.64	2050
	Average punching force		2235

Table 2. Tensile strength of the coated calio

Sample no.	Max. tensile force (N)	Max. elongation (mm)
1	2433	34.07
2	2373	33.77
3	2439	33.68
4	2673	38.84
5	2708	33.47
6	3068	41.26
7	3151	41.86
8	3451	47.85
9	2303	32.27
10	3722	49.48
Average	2832	39

4. Conclusions

This study attempted to find a new means for making use of para rubber and oil palm ash. The development of GCL was chosen to achieve the determination. Artificial clay was mixed with the OPA and latex binder to form a core of the GCL. The, it was covered with latex-coated calico, both sides. Several laboratory testes were carried out in order to evaluate its properties. In this paper, it was found that both punching shear and tensile strength of the GCL developed is in accordance with the standard, indicating that it can be employed in the field.

5. References

[1] Win HE 2017 Economic importance of rubber in Thailand, FFTC Agricultural Policy Platform, Food & Fertilizer Technology Center.
[2] Kumnerdpetch K 2018, A future scenario of oil palm throughout Thailand and adjustment management guidelines for oil palm farmers: a case study of global warming, J. of Community Development Research (Humanities and Sciences), 11(4), 92 – 105.
[3] Lee GF, and Jones-Lee, A 2005, Municipal solid waste landfills – Water quality issues, Water Encyclopaedia: Water Quality and Resource Development, Wiley, 163 – 169.

Acknowledgement

This study was successfully carried out because of the research grant number RDG60T0117 provided by Thailand Research Fund, now Thailand Science Research and Innovation, of which the authors are greatly in debt. In addition, during the research, help and support provided by the people at Para rubber Office Coordination, Kasetsart University, Thailand, are kindly acknowledge; without those support the project would not be successful.

IOP Conf. Series: Materials Science and Engineering **773** (2020) 012055 doi:10.1088/1757-899X/773/1/012055

Performance of lightweight cement board using coconut coir fiber and expanded polystyrene foam waste

C Pusri[1], B Ksapabutr[1,2], N Chaiyut[1] and M Panapoy[1,2,*]

[1] Department of Materials Science and Engineering, Faculty of Engineering and Industrial Technology, Silpakorn University, Nakhon Pathom 73000, Thailand.
[2] Center of Excellence on Petrochemical and Materials Technology, Chulalongkorn University, Bangkok 10330, Thailand.

*Corresponding author: mpanapoy@gmail.com, panapoy_m@su.ac.th

Abstract. Lightweight cement hybrid materials based on expanded polystyrene (EPS) foam waste and coconut coir fiber (CF) have been developed as an effective method for industrial and agricultural wastes management. The purpose of this study is to investigate the effects of CF and EPS foam as cement replacements at various proportions of EPS foam and CF at various lengths of CF on the properties of cement composites. The mixtures containing partial replacement of cement by 1, 2 and 3% CF with three different CF lengths (1, 2 and 3 cm) and 1, 2 and 3% EPS foam were prepared. The physical and mechanical properties of the specimens were determined after 28 days of hydration. The flexural strength and impact strength of the composites increased with CF content. Meanwhile, the bulk density clearly decreased along with the addition of EPS foam. The cement composites containing 3% CF and 1% EPS as well as 2% CF and 2% EPS at 2 cm fiber length showed acceptable strengths and density for structural lightweight cement. The results indicated that EPS foam and CF can be used as replacements for cement and asbestos in the production of fiber board cement to obtain inexpensive, lightweight and strong product.

1. Introduction

Nowadays, cement board is one of the most widely used construction materials. The most common type of fiber cement is asbestos cement. The major attractions of asbestos cement products to the users are durability and cost-effectiveness. However, asbestos fibers are carcinogenic. Alternatively, there are other types of organic or inorganic fibers, which are safer to human health [1]. The rapid growth of urbanization and industrialization has generated huge amounts of industrial and agricultural wastes. A circular economy in recycling and waste management is a systematic approach to minimize the problem. In this regard, the use of industrial and agricultural wastes should be considered in attempt to develop new cement board composites. Many researchers have shown several advantages in using natural fibers in cement composites: increased impact strength, increased flexural strength, and decreased bulk density [2-3]. Coconut fiber has one of the highest lignin contents, making it stronger than other natural fibers [2]. For industrial wastes, EPS is mainly used as an insulating or packaging material in various industrial fields. A large number of EPS is consumed, and is disposed as a waste, whilst the availability of mechanical test results will extend EPS geofoam to geotechnical applications, contributing to decreasing the amount of EPS ending up in landfills and more sustainable infrastructure systems [4]. The natural fiber reinforced concrete and EPS concrete have been separately studied in previous works; but until now, the study is carried out with combination of CF and EPS in cement board. Therefore, the current study presented herein is an attempt to develop lightweight EPS/CF/cement composite with considerable flexural and impact strengths. The influence of EPS foam from waste material in packaging industry and CF as cement and asbestos replacements at different weight contents of EPS foam and CF at different CF lengths on the physical and mechanical properties of fiber cement board was investigated.

2. Experimental

2.1. Materials and specimen preparation
Type I Portland cement satisfying the requirements of ASTM C150 was supplied by Siam Cement Group Company. In this work, CF was brown fiber obtained from Suphan Buri, Thailand. The raw materials were washed with distilled water to remove impurities and then dried at 60°C for 24 h. The cleaned CF with 0.14 mm diameter were then cut into 1, 2 and 3 cm length. Recycled EPS waste from electronics packaging was used. The cleaned EPS waste was ground to reduce the particle size to < 3 mm. All specimens test setup was developed with acrylic mold dimension of 12 cm in length, 5 cm in width and 5 mm in depth. The specimens were removed from the mold after 24 h and then kept in water for 28 days before the test date. In this study, various mixtures were produced by substituting cement with 1, 2 and 3% CF with three different fiber lengths (1, 2 and 3 cm) and/or 1, 2 and 3% EPS foam. The sample name was abbreviated as $C_{xy}E_z$, where the wt% of CF, the fiber length of CF (cm), and the wt% of EPS were denoted as x, y, z, respectively. Meanwhile, CF and EPS were given notation as C and E in the sample name.

2.2 Characterization and testing method
The tests were undertaken in accordance with BS EN 12467:2012 (Fibre-cement flat sheets-Product specification and test methods). The bending test was performed to investigate flexural behaviour of the specimens using the Instron 5965 50kN universal testing machine by the three-point bending tests, whose displacement control rate and span were 2 mm/min and 75 mm, respectively. An Instron CEAST 9050 impact pendulum machine was used to evaluate the impact behaviour of the specimens. The bulk density of the specimens was measured using a density kit (Mettler Toledo) mounted on a precision balance. The morphology of CF, EPS raw materials and the composites was characterized using a Hitachi TM3030 SEM.

3. Results and discussion
The SEM images of EPS and CF wastes are presented in Figures 1(a) and (b), respectively. The morphology of EPS revealed hollow and closed-cellular foam structure, that plays an important role in its properties. CF showed a rough surface morphology with irregular shape. Figures 1(c)-(f) demonstrate the SEM images of the fractured surface for EPS cement and CF cement composites after impact test. It can be seen that there was a small gap between EPS and cement matrix which means a weak adhesion for 3 wt% EPS. The gap formed was probably caused by incomplete wettability between cement and EPS during the composite fabrication (Figures 1(c)-(d)). This was also reported by other literatures [5]. Meanwhile, a better interfacial bonding is observed between CF and cement for the CF cement composite and thereby found relatively little fiber pull-out with a rough interface and more effectively bridge microcracks within the CF cement composite (Figures 1(e)-(f)). Probably because physical interlocking and hydrogen bonds could form between fiber and cement matrix during the hydration of cement [6]. The results of the physical and mechanical tests of EPS cement and CF cement composites are shown in Figure 2. The bulk density of plain cement ranged from 2.07 to 2.09 g/cm^3. Both the composites, particularly EPS cement, have lower density than plain cement (Figure 2(a)). The decrease in bulk density of the composites was mainly due to the substitution of heavier material (cement) by lighter one (CF and EPS). Depending on replacement by EPS or CF, the bulk densities of the composites were decreased by 66% when specimen was dosed with 3 wt% EPS. The EPS cements have the bulk densities ranged between 0.71 to 1.43 g/cm^3. Whist a smaller decrease (~11%) in bulk density was

observed after adding 3 wt% CF at all the fiber lengths. The bulk densities of CF cements ranged between 1.85 to 1.95 g/cm_3.

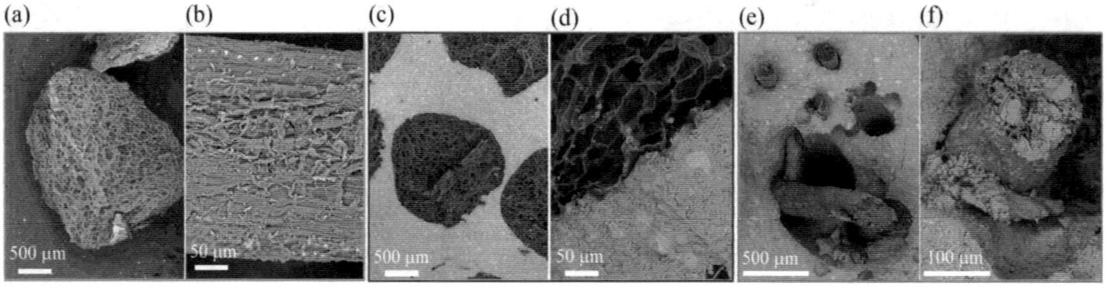

Figure 1. SEM images of (a) EPS, (b) CF and SEM images of the fractured surface of the composites (c-d) EPS cement and (e-f) CF cement.

The effect of EPS content on the flexural and impact strengths is shown in Figures 2(b) and (c). The flexural and impact strengths of EPS cement decreased as EPS content increased. Moreover, it can be seen that after EPS content exceeding 1 wt%, its effect on the impact strength of EPS cement become insignificant. The results showed that higher EPS content resulted in lower flexural strength, impact strength and density due to higher air-void content and looser structure in EPS composite with lower volume of cement. For the CF cements, the replacement of cement by CF notably improved flexural and impact strengths compared with plain cement (Figures 2(b) and (c)). The improvement rate increased with increasing cement replacement (flexural strength improved by 60-70% for all the CF lengths at 3% replacement). On replacing 3% cement by CF, at the CF lengths of 1, 2 and 3 cm, impact strength increased by 407, 668 and 860%, respectively. This may be because the better interaction between CF and cement and the fracture behavior of high strength composite contains crack bridging and some fiber pull-out (as shown in Figures 1(e)-(f))) that are responsible to resist crack propagation and improve the composite strength [6-8].

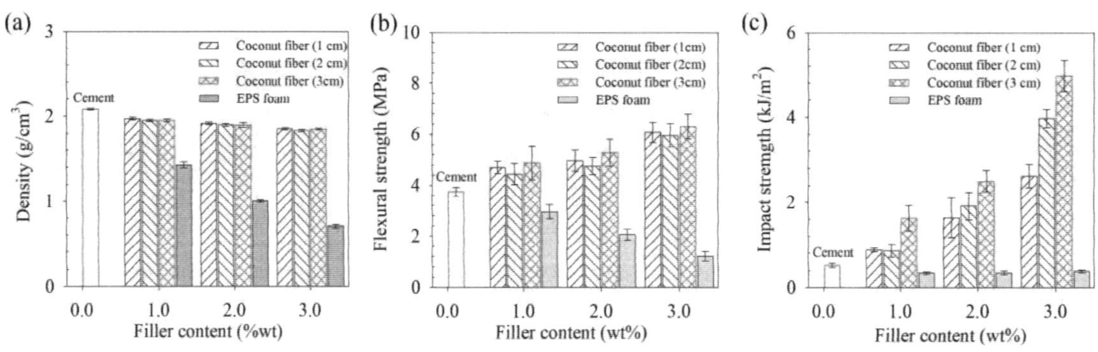

Figure 2. Effect of CF or EPS content at various CF lengths on (a) density, (b) flexural strength and (c) impact strength.

The combination of CF and EPS in cement replacement seems to improve their flexural and impact strengths and decrease the bulk density for structural lightweight cement board (Figures 3(a)-(c)). The increased flexural and impact strengths resulted from the addition of CF, whereas the existence of EPS is responsible for the reduction of bulk density. It seems that somewhat adverse effect of EPS beads in flexural and impact strengths could be compensated by adding CF to the EPS cement composite. Cement mixed with both CF and EPS showed improvements on increased flexural strength (24%), increased impact strength (547%) and reduced density (41%) using 3% CF for 3 cm fiber length and 1% EPS). The CF/EPS/cement composite containing 3% CF at 3 cm fiber length and 1% EPS ($C_{33}E_1$) as well as 2% CF at 3 cm fiber length and 2% EPS ($C_{23}E_2$) showed acceptable flexural strength, impact strength and density for structural lightweight cement. According to $C_{33}E_1$, the structural lightweight cement has the average flexural strength of 4.62 MPa, impact strength of 3.35 kJ/m^2, bulk density of 1.23 g/cm^3, and consists entirely of lightweight cement.

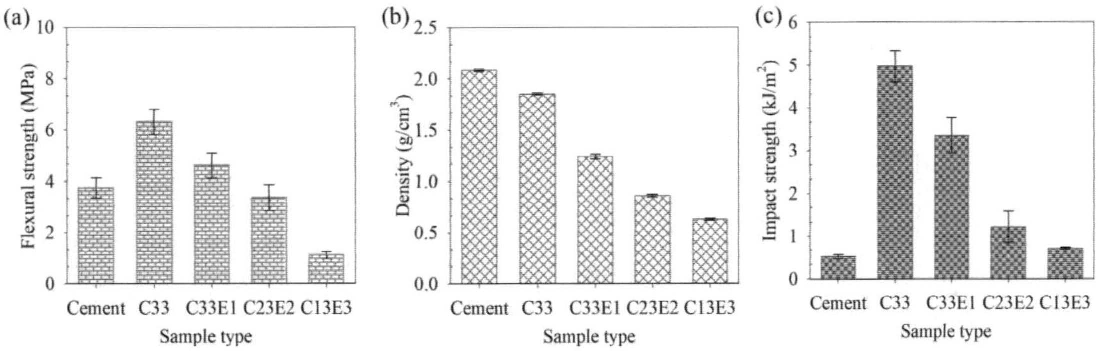

Figure 3. Effect of replacement level by the mixture of CF and EPS at 3 cm CF length on (a) flexural strength, (b) density and impact strength.

4. Conclusions

In this study, CF and EPS wastes can be used as a reinforcement and replacement of cement in the development of lightweight cement board. Application of CF improved the EPS /CF /cement composite properties including increased flexural and impact strengths and decreased bulk density. Adding EPS into cementitious mixtures significantly decreased the bulk densities of composites. Application of EPS also decreased the strength properties; however, the results showed the potential for producing different strength and lightweight cement composites; in particular satisfactory cement composite can be produced by replacing cement with 3% CF and 1% EPS as well as replacing with 2% CF and 2% EPS for the CF length of 3 cm. In this regard, to some adverse effect of EPS beads in strengths could be compensated by adding CF to the EPS cement composite. The findings of the present study supported acceptable performance in cement based composite materials for low-cost housing applications.

5. References

[1] Ingham JP 2013 Geomaterials Under the Microscope, *Academic Press,* 192 pages.
[2] Lertwattanaruk P and Suntijitto 2015 *Constr. Build. Mater.* **94** 664.
[3] Wang W and Chouw N 2018 *Constr. Build. Mater.* **171** 546.
[4] Herki BA and Khatib JM 2017 *Eur. J. Environ. Civ. En.* **21** 1384.
[5] Dong X, Wang S, Gong C and Lu L 2014 *Constr. Build. Mater.* **773** 255.
[6] Pehanich JL, Blankenhorn PR and Silsbee MR 2004 *Cement Concrete Res.* **24** 59.
[7] Hwang CL, Tran VA, Hong JW and Hsieh YC 2016 *Constr. Build. Mater.* **127** 984.
[8] Anggraini V, Asadi A, Syamsir A and Huat BBK 2017 *Measurement* **111** 158.

Acknowledgement
The authors would like to thank the Department of Materials Science and Engineering, Faculty of Engineering and Industrial Technology, Silpakorn University, and the Center of Excellence on Petrochemical and Materials Technology, Chulalongkorn University.

IOP Conf. Series: Materials Science and Engineering **773** (2020) 012056 doi:10.1088/1757-899X/773/1/012056

Efficient removal of methylene blue by low-cost and biodegradable highly effective adsorbents based on biomass in the fixed bed column

D Sasaengta[1,2], M Panapoy[1,2], N Chaiyut[1] and B Ksapabutr[1,2,*]

[1] Department of Materials Science and Engineering, Faculty of Engineering and Industrial Technology, Silpakorn University, Nakhon Pathom 73000, Thailand.
[2] Center of Excellence on Petrochemical and Materials Technology, Chulalongkorn University, Bangkok 10330, Thailand.

*Corresponding author: kbussarin@gmail.com, ksapabutr_b@su.ac.th

Abstract. The adsorptive ability of sustainable biomass (spent coffee ground (SC) and water hyacinth (WH)) and activated biochar derived from WH to remove methylene blue (MB) from aqueous solution was evaluated under continuous fixed-bed column. Morphological structure and functional groups of WH and SC were determined by SEM and FTIR, respectively. The SEM showed the presence of porous structure of WH, whilst FTIR confirmed the presence of more hydroxyl groups at the WH surface, resulting in higher MB adsorption. A series of column experiments were performed with varying bed height and initial MB concentration. To determine the breakthrough curves and characteristic parameters for process design, Yoon-Nelson, Thomas and modified Dose-Response models were applied to experimental breakthrough data. The MB adsorption was dependent on the bed height and initial MB concentration. An increase in bed height resulted in improved adsorption capacity. With increasing initial MB concentration, the adsorption capacity decreased. The maximum adsorption capacity (Q_0) of ~391 mg/g by WH (at 0.75 cm bed height and 300 mg/L initial MB concentration) and ~393 mg/g by its activated biochar (at 2 cm bed height and 200 mg/L initial MB concentration) was observed. The results showed that WH gave comparable MB adsorption capacity with a greater cost-effectiveness.

1. Introduction

Dyes are extensively used for several industries, e.g., textile. Methylene blue (MB) is a cationic dye that is most commonly utilized in textile industries. Approximately 10-15% of the dyes are released into the environment, causing significant health and environmental hazards. A lot of biological, physical and chemical techniques have been adopted for dye removal from wastewater. Among these methods, adsorption is regarded as an attractive and effective approach for water treatment due to its remarkable advantages, e.g., simple operation, low cost and no secondary pollution [1]. Generally, continuous flow adsorption process offers several advantages over batch process. The application is economic and practical as the operation is conducted continuously and the process is controllable [2]. Various types of adsorbents have been reported (e.g. metal oxide, activated carbon). Particularly, biomass has attracted increasing attention due to its cheap source and great potential to be modified as a highly efficient adsorbent [1]. In this study, WH and SC were selected for their abundance and economic significance. They are carbonaceous materials containing cellulose, lignin and some functional groups (hydroxyl, carboxyl groups, etc) that make the adsorption process feasible [3]. Moreover, to avoid any environmental related issues for the poor WH management, WH was converted into activated biochar for using as an efficient and sustainable adsorbent material. Hence, this work focused on performing continuous column study for the MB removal and investigating the performance of WH and SC biosorbents and activated biochar under variation of bed height and initial MB concentration. Thomas, Yoon-Nelson, and modified Dose-Response models were applied to experimental data for evaluating the column adsorption performance and for predicting the scale-up of a unit plant for the MB removal.

2. Experimental

2.1. Materials and methods

WH was collected from Tha Chin River at Nakhon Pathom, Thailand. SC was obtained from coffee shop in Silpakorn University, Thailand. The raw materials were washed with distilled water to remove impurities and then dried at 70°C for 24 h. The dried WH and SC were ground and sieved to desired size (150-250 μm). The activated biochar was prepared from WH via a KOH activation and was denoted as ABW. Dried WH and KOH mixture with mass ratio of 1:1 was used. After mixing, the mixture was heated at 800°C for 1 h under N_2 flow, followed by washing to reach pH 7 and drying at 80°C for 6 h. The resulting ABW was crushed and sieved to 75-106 μm. The BET surface area, total pore volume and average pore diameter of ABW were 911 m^2/g, 0.50 cm^2/g and 2.16 nm, respectively. The morphology was characterized using a Hitachi TM3030 SEM. To determine the functional groups, biomass powders were analyzed by a Bruker Vertex 70 FTIR spectrometer. For continuous column experiment, the lab scale fixed-bed parameters were applied for the actual large scale fixed-bed reactor design. A series of fixed-bed column tests were conducted in a plastic column (2 cm inside diameter, 10 cm long) for the simulation of adsorbents for MB adsorption.

2.2 Application of empirical models for breakthrough curves

The breakthrough curves were analyzed using Thomas, Yoon-Nelson, and modified Dose-Response models. These models are given, respectively, by the following equations [4]:

$$\frac{C_t}{C_o} = \left(1 + exp\left(\frac{k_{Th}Q_o m}{Q} - k_{Th}C_o t\right)\right)^{-1} \quad (1), \quad \frac{C_t}{C_o} = \left(1 + exp(k_{YN}\tau - k_{YN}t)\right)^{-1} \quad (2), \quad \frac{C_t}{C_o} = 1 - \left(1 + \left(\frac{C_o Q}{Q_o m}\right)^{a}\right)^{-1} \quad (3)$$

where k_{Th} is the Thomas kinetic coefficient (mL/min mg), k_{YN} is the Yoon-Nelson rate of constant (1/min), C_o is the concentration of initial time (mg/L), C_t is the concentration at arbitrary time t (mg/L), Q_o is the maximum adsorption capacity of adsorbent (mg/g), τ is the time required for 50% adsorbate breakthrough (min), Q is the flow rate (mL/min), m is the mass of adsorbent (g), and a is the constant.

3. Results and discussion

The morphological structure of SC and WH biosorbents is analyzed by the SEM analysis (Figure 1(a)-(b)). SC particles presented denser morphology than WH particles. The sponge-like rough surface morphology with high porosity was observed in WH. This surface roughness may possibly be useful in MB adsorption. The FTIR spectra of WH and SC are shown in Figure 1(c). The strong band at 3400 cm^{-1} indicates the stretching vibrations of OH groups. The band at 2920 cm^{-1} corresponds to the C-H stretching. The peaks at 1631 and 1731 cm^{-1} are related to the adsorbed water in WH and either acetyl or uronic ether linkages of carboxylic group in the ferulic and p-coumeric acids of lignin [5]. The peaks of SC at 1744, 1654, and 1543 cm^{-1} correspond to the C=O stretching of hemicellulose and chlorogenic acids, and stretching of caffeine CN bonds [6]. This suggests that functional groups, particularly hydroxyl groups, may potentially act as the active sites for the binding of MB. To predict the breakthrough curves and determine the characteristic column parameters, Yoon-Nelson, Thomas and modified Dose-Response models were compared with the experimental points and exhibited in Figure 1(d)-(f). All the models predicted the column performance parameters by including the bed height (0.5 and 0.75 cm) and initial MB concentration (200 and 300 mg/L) of the column system reported in Table

1. The experimental data were well described by three models in different column conditions ($R_2 > 0.95$), suggesting their suitability to be used for the design and scale-up. On studying the effect of bed height, it was observed that at a fixed initial MB concentration (300 mg/L), the maximum adsorption capacity (Q_0) and breakthrough time (t_b at $C/C_0 \approx 0.1$) increased with bed height of WH (Table 1). At higher bed height, the amount of adsorbent in the column is increased, therefore there is an increased number of biosorbent active sites. This leads to increased adsorption capacity at higher bed height, and the breakthrough times are longer as well since more active sites available would become slowly saturated. For the effect of initial MB concentration, it was found that the t_b and Q_0 values decreased with increasing adsorbate concentration as the binding sites became more quickly saturated. At higher concentration, more dye molecules are left unadsorbed in the solution, probably because of the saturation of binding sites resulting in decreased adsorption capacity. Similar trend has also been observed by other researchers [1,7]. The highest Q_0 of ~391 mg/g was obtained at 0.75 cm bed height and 300 mg/L initial MB concentration. At the same bed height (0.5 cm) and initial concentration (300 mg/L), the Q_0 value of WH was higher than that of SC by 51%. This may be due to the presence of more hydroxyl groups at the surface of WH, resulting in the formation of surface hydrogen bonds between the hydroxyl groups on the WH surface and the nitrogen atoms of MB [8].

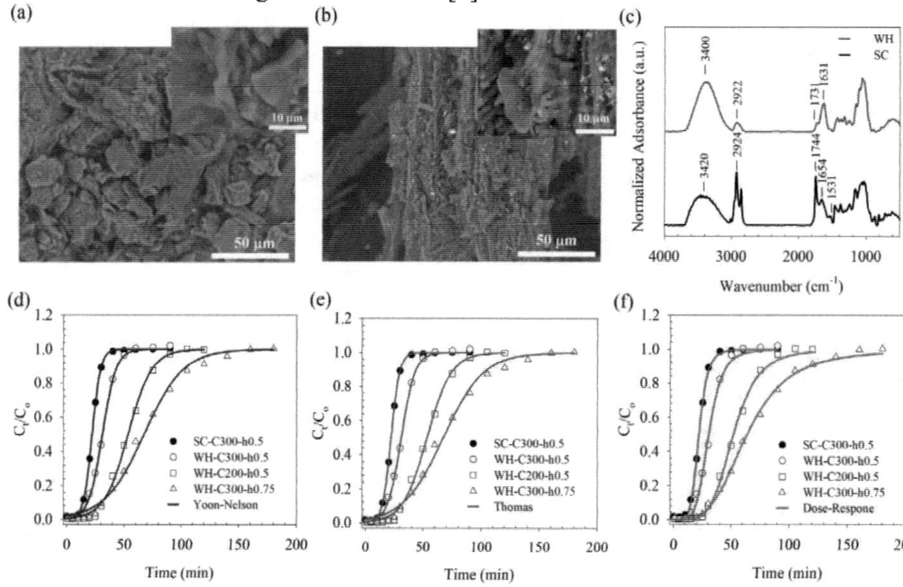

Figure 1. SEM images of (a) SC and (b) WH, (c) FTIR spectra of SC and WH, (d) Yoon-Nelson, (e) Thomas and (f) modified Dose-Response non-linear breakthrough modeling.

Table 1. Parameters of three models for SC and WH in fixed-bed column (flow rate: 10 mL/min).

Sample	C_0 (mg/L)	h (cm)	t_b (min)	Yoon–Nelson k_{YN} (1/min)	Yoon–Nelson τ (min)	Yoon–Nelson R_2	Thomas k_{TH} (mL/mg·min)	Thomas Q_0 (mg/g)	Thomas R_2	Dose-Response Q_0 (mg/g)	Dose-Response a	Dose-Response R_2
SC	300	0.5	15	0.278	22.171	0.999	0.911	198.993	0.999	197.007	5.639	0.998
	200	0.5	30	0.097	53.374	0.998	0.460	362.720	0.998	352.551	4.889	0.998
WH	300	0.5	20	0.171	31.119	0.999	0.573	304.931	0.999	297.516	5.245	0.996
	300	0.75	30	0.058	67.267	0.994	0.179	413.322	0.994	391.101	3.571	0.998

For a preliminary study of activated biochar from WH (Table 2), the analysis of the applicability of three models for predicting MB adsorption on ABW revealed that the correlation coefficient (R_2) value was between 0.996-0.998 for modified Dose-Response as compared with 0.995-0.9996 for Yoon-Nelson and Thomas model data fittings. The MB adsorption on ABW can best be predicted by modified Dose-Response. The Q_0 values increased with the increase in bed height and the decrease in initial MB concentration. The same observation has been reported by previous researchers [9]. The highest Q_0 of ~393 mg/g was found at 2 cm bed height and 200 mg/L initial MB concentration. Column tests of other activated biochars will be reported in future.

Table 2. Three model parameters for ABW in fixed bed column (flow rate: 10 mL/min).

Sample	C_o (mg/L)	h (cm)	t_b (min)	Yoon–Nelson k_{YN} (1/min)	τ (min)	R_2	Thomas k_{TH} (mL/mg·min)	Q_o (mg/g)	R_2	Dose-Response Q_o (mg/g)	a	R_2
AC-800	300	1.0	25	0.109	41.389	0.996	0.359	308.238	0.996	298.291	4.376	0.998
	300	2.0	50	0.041	94.518	0.985	0.136	354.633	0.985	336.603	3.395	0.995
	300	4.0	110	0.021	204.508	0.987	0.068	383.663	0.987	362.155	3.893	0.993
	200	2.0	90	0.028	161.267	0.992	0.133	412.481	0.992	393.269	3.985	0.995
	400	2.0	35	0.065	67.557	0.995	0.160	335.540	0.995	319.999	4.0437	0.996

4. Conclusions

The effectiveness of WH and SC as biosorbents and biochar obtained from WH was tested in the MB removal from aqueous solution in continuous fixed-bed column system by varying bed height and initial MB concentration. The breakthrough times were increased with increasing bed height and decreasing initial MB concentration. The adsorption capacity of MB on WH was higher than that of SC by 51%. In terms of adsorption capacity and economic efficiency, the use of a WH-packed column has demonstrated potential for the remediation of MB contaminated wastewater. For both biosorbents, all three models provided a satisfactory correlation of experimental column data with $R_2 > 0.95$, suggesting its suitability for the design and large-scale process. The modified Dose-Response model was found best fitted for biochar and could be applied to scale up the process to an industrial operation.

5. References

[1] Afroze S, Sen TK and Ang HM 2016 *Res. Chem. Intermed.* **42** 2343.
[2] Dichiara AB, Weinstein SJ and Rogers RE 2015 *Ind. Eng. Chem. Res.* **54** 8579.
[3] Correia IKS, Santos PF, Santana CS, Neris JB, Luzardo FHM and Velasco FG 2018 *J. Environ. Chem. Eng.* **6** 2319.
[4] Zhang Y, Jin F, Shen Z, Lynch R and Al-TA 2018 *J. Hazard. Mater.* **5** 461.
[5] Das S, Bhattacharya A, Haldar S, Ganguly A, Gud S, Ting YP and Chatterjee PK 2015 *SM&T* **3** 17.
[6] Mendes JF, Martins JT, Manrich A, Neto ARS, Pinheiro ACM, Mattoso LHC and Martins MA 2019 *Carbohydr. Polym.* **210** 92.
[7] Chen J, Liu M, Pu Y, Wang C, Han J, Jiang M and Liu K 2019 *J. Environ. Manage.* **245** 105.
[8] Ahmad T, Danish M, Rafatullah M, Ghazali A, Sulaiman O, Hashim R and Ibrahim MNM 2012 *Environ. Sci. Pollut. Res.* **19** 1464.
[9] Fan, Z, Zhang Q, Gao B, Li M, Liu C and Qiu Y 2019 *Chemosphere* **217** 85.

Acknowledgement

The authors would like to thank the Department of Materials Science and Engineering, Faculty of Engineering and Industrial Technology, Silpakorn University, and the Center of Excellence on Petrochemical and Materials Technology, Chulalongkorn University.

IOP Conf. Series: Materials Science and Engineering **773** (2020) 012057 doi:10.1088/1757-899X/773/1/012057

Enhanced mechanical performance of cement board composite reinforced with coconut coir fiber and tire rubber waste

C Thepthong[1], B Ksapabutr[1,2], N Chaiyut[1] and M Panapoy[1,2,*]

[1] Department of Materials Science and Engineering, Faculty of Engineering and Industrial Technology, Silpakorn University, Nakhon Pathom 73000, Thailand.
[2] Center of Excellence on Petrochemical and Materials Technology, Chulalongkorn University, Bangkok 10330, Thailand.

*Corresponding author: mpanapoy@gmail.com, panapoy_m@su.ac.th

Abstract. This work aimed to study the use of tire rubber waste (TW) and coconut coir fiber (CF) as the recycled material and partial replacement of cement and asbestos to produce cement board composite and obtain optimal replacement level. The amounts of TW and CF were the variables in this experimental study. Various mixtures were produced by partial replacement of cement with TW at different percentages (2.5, 5.0, 7.5 and 10%) and with CF at different percentages (1, 2 and 4%) using a fiber length of 3 cm. The mechanical and physical properties of the cement boards were evaluated. Both the flexural and impact strengths of the cement boards increased with an increase in the content of TW and CF. The mechanical properties of the cement composites based on the combination of TW and CF were significantly better due to the synergy of TW and CF additions. Moreover, the results also showed that the density decreased with increasing in TW and CF. Therefore, TW and CF can be utilized as a good replacement of cement for cheap, lightweight and eco-friendly reinforced cement boards.

1. Introduction

Composite materials with both economic and environmental benefits are being considered for applications in the building and construction industries. After banning asbestos fibers because of its hazardous impacts on human health, developing low-cost, strong, and durable cement board remains one of the greatest challenges [1]. In present, there are many waste materials for cement board production to achieve desired properties or reduce the cost of the final products [2]. Management of TW has caused an environmental crisis due to the large volume of waste generated from automobile industry and increasing use of car. In considering the reuse of recycled rubber in concrete, extensive researches have been carried out on tire-based concretes. These results have revealed that rubberized concrete mixtures provide a decreased density, higher impact resistance, lower compressive and tensile strength, and better sound insulation [3]. However, some observations have suggested that the loss in strength might be minimized by incorporating natural fibers [4]. CF was found to be one of the most effective fibers and widely available. Large quantities of coconut waste are generated from agroindustry that when discarded or burnt can cause environmental problems. Therefore, it is imperative to reduce the volume of TW and CF waste burned or dumped to landfill through modern recycling processes, without harming the environment. Using TW and CF in cement as a partial replacement of some components is one of the possible recycling approaches currently being explored. In order to solve the above problem, this study replaced part of cement and asbestos with TW powder and CF, which was then mixed into cement composites. We tested the physical and mechanical properties (density, impact and flexural strengths) of cement composites based on different weight percentages of added TW (2.5, 5, 7.5 and 10%) and CF (1, 2 and 4%) to find out the optimal replacement level.

2. Experimental

2.1. Materials and composite preparation

Type I Portland cement conformed to ASTM C150 specifications and two types of replaced cementitious materials; TW and CF; were used in this work. Type I Portland cement was purchased from the Siam Cement Group Company. Brown CF was obtained from Suphan Buri (Thailand). Prior to use, CF was washed with distilled water and then dried for 24 h at 70°C, followed by cutting into 3 cm length. The TW powder was produced by the KKI Recycle factory. The TW powder is sized less than 1 mm. The mixtures were cast in acrylic molds (5 mm in height, 12 cm in length and 5 cm in width). After casting, the specimens were cured at room temperature for 24 h. Then the samples were demolded and soaked in water for 28 days before testing. In this study, the mixtures include partial replacement of cement weight with 2.5, 5.0, 7.5 and 10% TW and/or 1, 2 and 4% CF with 3 cm fiber length. The material name was designated as TW_xCF_y, where x and y indicate the wt% of TW and CF in cement, respectively.

2.2. Experimental method

The fractured surface morphology of the composites was studied by SEM using a Hitachi TM3030 electron microscope. All the mechanical properties were performed in accordance with BS EN 12467:2012 (Fibre-cement flat sheets Product specification and test methods) to determine the optimal TW and CF contents of fiber cement composites. Flexural behaviour of the specimens was analyzed under a three-point load system. The three-point bending tests were done under displacement control (cross-head feed rate of 2 mm/min and span of 75 mm) using a universal testing machine (Instron 5965), with a 50 kN load cell. The impact resistance test was carried out using an Instron CEAST 9050 impact pendulum machine and a pendulum of 1 J. The specimen density was measured using a top-loading electronic balance, equipped with a Mettler Toledo accessory density kit.

3. Results and discussion

The mechanical properties and densities of cement composite containing different TW contents are shown in Figure 1. The flexural strength of TW cement composites was not significantly altered with the addition of up to 10% TW (Figure 1(a)). However, the impact strength of TW composites significantly depended on TW content (Figure 1(b). The results showed that the addition of 2.5, 5, 7.5 and 10% TW resulted in 3.3, 16.3, 23.0 and 34.6% increment in impact strength compared with the reference (plain cement), respectively. This may be because the higher the TW content, the higher tendency to absorb impact energy attributing to higher toughness and ductility of TW particles that is resistant to failure. Similar results were also reported in earlier studies [5]. The effect of TW content on bulk density of the specimens is demonstrated in Figure 1(c). The addition of TW affected the bulk density of cement and as a consequence the TW composites at 2.5, 5, 7.5 and 10% TW replacement revealed about 4.6, 6.4, 8.1 and 9.9% reduction, respectively. A decrease in bulk density resulted from the decreased amount of cement paste required to fill the interstitial space between TW particles. For the CF composites, the addition of CF in TW composites leads to the increase in flexural strength and impact strength compared with the reference because of a synergistic effect of TW and CF (Figure 2(a) and (b)). In TW/CF/cement composites, the addition of 1, 2 and 4% CF (TW5CF1, TW5CF2, TW5CF4) in 5% TW composites presented an increase of 90.9, 106.6 and 126.8% in flexural strength, followed by the mixtures with 1, 2 and 4% CF (TW7.5CF1, TW7.5CF2, TW7.5CF4) in 7.5% TW composites were 68.0, 86.2 and 98.3%. The physical-mechanical properties of composites are directly related to the adhesion between CF and cement. The specimens with lower TW content (TW_5CF_1-TW_5CF_4) presented higher flexural strength than those with higher TW content ($TW_{7.5}CF_1$-$TW_{7.5}CF_4$). This fact may have occurred due to higher TW content ($TW_{7.5}CF_1$-$TW_{7.5}CF_4$) with lower amount of cement matrix, leading to a decreasing tendency of the adhesion between CF and cement.

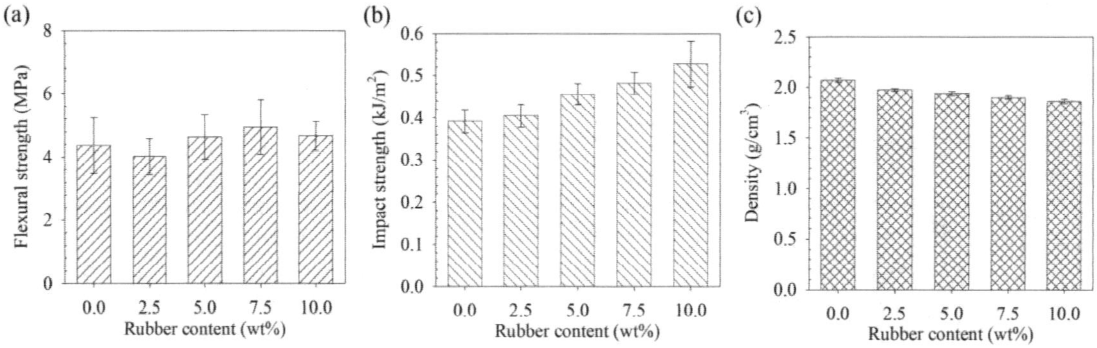

Figure 1. Effect of TW content on (a) flexural strength, (b) impact strength and density.

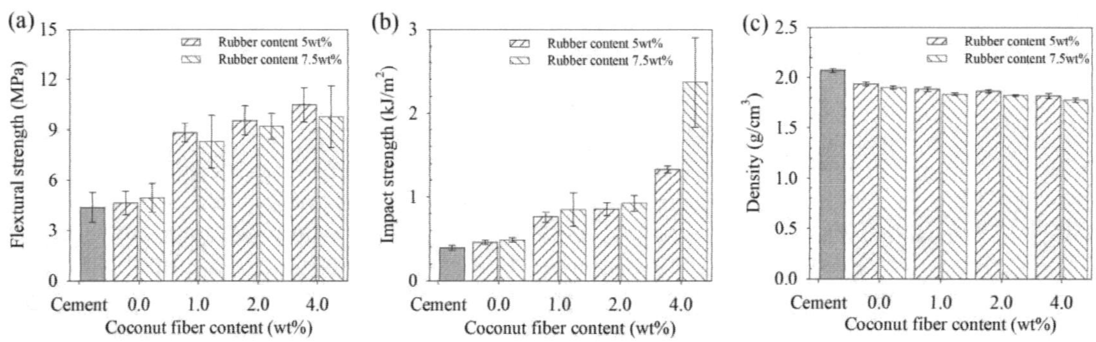

Figure 2. Effect of adding TW and CF on (a) flexural strength, (b) impact strength and (c) density.

Figure 3(b) shows the existence of microcracks (blue circle) in the interface between TW particles and matrix. These cracks are probably because of high elasticity, deformability and also softness of TW with respect to matrix, which consequently results in the loss in strength. Moreover, the irregular shape of TW particles (Figure 3(a)) may cause the presence of microcavities in the matrix (Figure 3(b) (red circle)), resulting in weak bonding between TW particles and matrix. This weak interfacial zone may act as microcrack, leading to crack initiation at the interface between materials accelerating the matrix breakdown. For the addition of CF, the microcrack could be transferred to CF through the interfacial bonding between CF and cement (Figure 3(c)). CF can also restrain the crack propagation and traverse across the cracks to transfer internal force, and CF and cement matrix would bear the load [6-7].

Figure 3. SEM images of (a) tire rubber waste (TW), (b) $TW_{7.5}$ and (c) $TW_{7.5}CF_4$.

IOP Conf. Series: Materials Science and Engineering **773** (2020) 012057 doi:10.1088/1757-899X/773/1/012057

The effect of CF content on impact strength is demonstrated in Figure 2(b). The impact strength increased from 0.46 kJ/m^2 for TW_5 to 0.76, 0.86 and 1.33 kJ/m^2 for TW_5CF_1, TW_5CF_2 and TW_5CF_4, respectively. Whilst the impact strength increased from 0.48 kJ/m^2 for $TW_{7.5}$ to 0.85, 0.93, and 2.37 kJ/m^2 for $TW_{7.5}CF_1$, $TW_{7.5}CF_2$ and $TW_{7.5}CF_4$, respectively. It is clear that the increased impact strength resulted from adding CF. $TW_{7.5}CF_4$ showed higher impact strength than TW_5CF_4 (1.8 times) probably because TW particles have low stiffness which affected the matrix by increasing its flexibility, ability to absorb and transfer energy to CF (Figure 4). $TW_{7.5}CF_4$ has lower density than cement and TW_5CF_4 about 14% and 2.3% respectively (Figure 2(c)). Therefore, according to $TW_{7.5}CF_4$, the lightweight cement composite has the flexural strength ranged from 7.98-11.63 MPa, impact strength of 1.83-2.91 kJ/m^2 and bulk density of 1.75-1.80 g/cm^3.

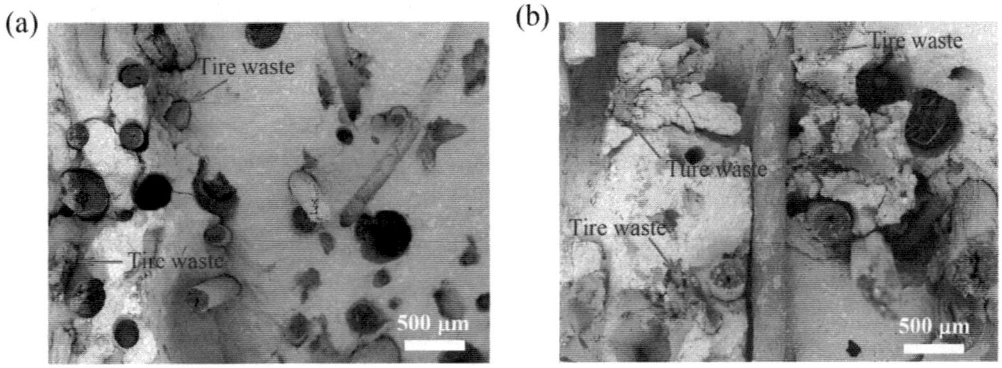

Figure 4. SEM images of (a) TW_5CF_4 and (b) $TW_{7.5}CF_4$.

4. Conclusions

The use of CF and TW showed potential as reinforcement of cement-based composites. The drastic improvement in the flexural strength of the TW/CF/cement composites was observed with adding CF only 1%. According to the mechanical and physical tests on the flexural strength, impact strength, and bulk density, it is verified that the replacement of cement by the combination of TW and CF worked synergistically to reinforce and modify the cement composites. The cement composites reinforced with 4% CF and 7.5% TW ($TW_{7.5}CF_4$) presented the optimal average values for flexural strength, impact strength and bulk density. The results of this research proved that the application of TW and CF in cement continues to be an essential part in the process of reducing waste and a promising alternative construction material.

5. References

[1] Park S-H 2018 *SH@W.* **9** 360.
[2] Asasutjarit C, Hirunlabh J, Khedari J, Charoenvai S, Zeghmati B and Cheul SU 2007 *Constr.Build. Mater.* **21** 277.
[3] Ghizdăveţ Z, Ştefan BM, Nastac D, Vasile O and Bratu M 2016 *Constr.Build.Mater.* **124** 755.
[4] Chaikaew C, Sukontasukkul P, Chaisakulkiet U, Sata V and Chindaprasirt P 2019 *Case Stud. Constr. Mater.* **11** e00304.
[5] Anggraini V, Asadi A, Syamsir A and Huat BKB 2017 *Measurement* **111** 158.
[6] Youssf O, Hassanli R and Mills EJ 2017 *J. Build. Eng.* **11** 115.
[7] Abdelmonem A, El-Feky MS, Nasr ARES and Kohail M 2019 *Constr.Build.Mater* **227** 116660.

Acknowledgement

The authors would like to thank the Department of Materials Science and Engineering, Faculty of Engineering and Industrial Technology, Silpakorn University, and the Center of Excellence on Petrochemical and Materials Technology, Chulalongkorn University.

Role of Sn-based and Ti-based catalysts on melt copolymerization of PLA-Polyols

J Luangkachao and K Sirisinha*

Department of Chemistry, Faculty of Science, Mahidol University, Bangkok, 10400

*Corresponding author: kalyanee.sir@mahidol.ac.th

Abstract. Aiming at improving the ductility of PLA, two polyols: polyethylene glycol (PEG), and polypropylene glycol (PPG) were introduced individually to the PLA structure as a soft component. Copolymerization of PLA and polyols was performed in a melt-mixer via transesterification in the presence of a catalyst. $_1$H-NMR and GPC results revealed that PLA-PEG and PLA-PPG copolymers with different molecular weights were successfully formed. The copolymer formation and characteristics depended strongly on the structure of polyol and type of catalyst. Titanium butoxide (TBT) was found to be a more effective catalyst than dibutyltin dilaurate (DBTL). The effects of catalyst content and reaction time on the characteristics of the copolymers are also reported.

1. Introduction

Interest in polylactic acid (PLA)-based material has grown continuously in various industries. Copolymerization of PLA with soft segment could turn a brittle PLA into a ductile product. A traditional method for producing PLA copolymer is ring-opening copolymerization of lactide monomer with soft components as polyethylene glycol (PEG), polypropylene glycol (PPG), polycaprolactone (PCL), etc. [1-3]. The reaction is commonly performed under reduced pressure for 4 - 8 h and a large amount of solvent is required in the process [1, 5]. In this study, melt-copolymerization of PLA polymer with two types of polyols: PEG and PPG, was carried out via transesterification reaction in a melt-mixer in the absence of solvent. The effects of catalyst types (i.e. titanium butoxide (TBT) and dibutyltin dilaurate (DBTL)), content, and time of reaction on the characteristics of the copolymers are reported.

2. Experimental

2.1. Materials

PLA 2003D was from NatureWorks LLC. PEG (Mw ~ 4000 g/mol) was from Ajax Finechem Pty Ltd. PPG (Mw ~ 2000 g/mol) was from Acros Organics. TBT and DBTL catalysts were from Sigma-Aldrich.

2.2. Melt copolymerization of PLA and polyols in a mixer

PLA-PEG and PLA-PPG copolymers were prepared via melt-transesterification method in a Haake internal mixer at 170°C. The detail of this method has been reported elsewhere [4]. The contents of PEG and PPG were kept constant at 20 w/w%. TBT and DBTL catalysts were used with various amounts of 0, 0.5, 0.75 and 1.5 phr. The reaction time was varied at 5, 10, 20 and 30 min.

2.3. Characterizations

Chemical structure of the obtained PLA-PEG and PLA-PPG copolymers was investigated using $_1$H-NMR spectroscopy (Bruker 500 MHz, USA). Molecular weights of the copolymers were evaluated by Gel permeation chromatography (GPC) (Waters 1515, USA) using polystyrene as a standard. Mechanical properties of the copolymers were characterized by Instron 5566 tensile testing machine.

3. Results and discussion

3.1. Chemical structure of PLA-PEG and PLA-PPG copolymer

Figures 1a and 1b demonstrate the chemical structures of PLA-PEG and PLA-PPG copolymers, as characterized by $_1$H-NMR. Methyl protons (-CH_3, a) and methine protons (-CH, b) of the PLA unit are clearly seen at δ of 1.60 ppm and 5.20 ppm, respectively. The PEG unit exhibits a characteristic signal of methylene protons (-CH_2, c) at δ of 3.60 ppm [1, 4] whereas the PPG unit shows a signal of methyl protons (-CH_3, d), methylene protons (-CH_2, e) and methine protons (-CH, f) at δ of 1.14, 3.55 and 3.40 ppm, respectively [2]. The signal at δ of 4.20 - 4.30 ppm belonging to methylene protons (-CH_2, g) of PEG that attached to the PLA unit, and the signal at δ of 4.15 ppm belonging to methylene protons (-CH_2, h) of PPG that attached to the PLA unit confirm that PLA-PEG and PLA-PPG copolymers are successfully prepared by a melt-transesterification approach.

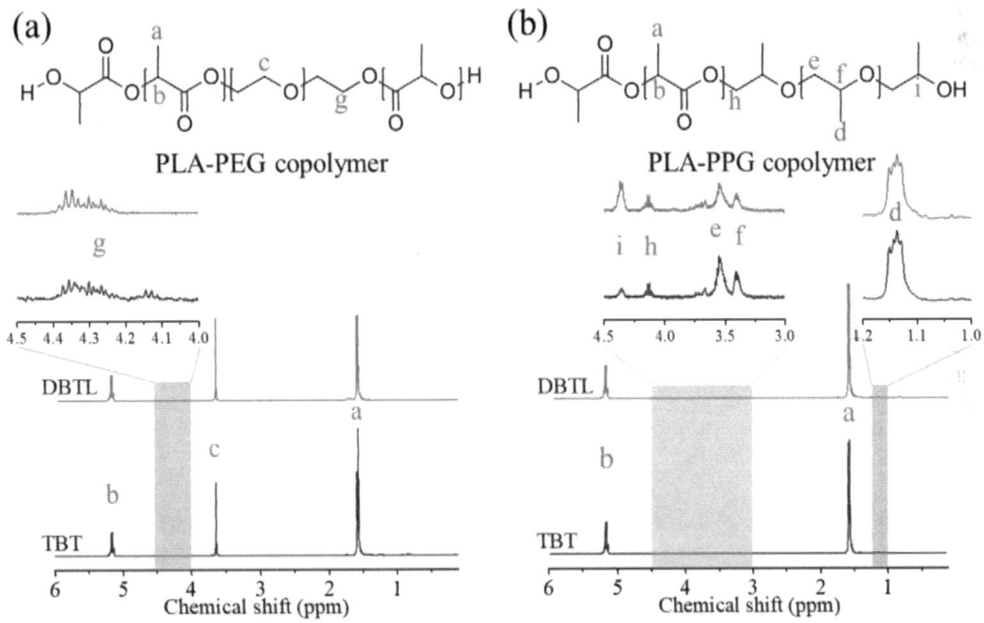

Figure 1 $_1$H-NMR spectra of a.) PLA-PEG copolymer and b.) PLA-PPG copolymer.

3.2. Effect of catalyst type, content, and reaction time

Figure 2 shows the GPC curves of the copolymers prepared by using different catalysts. Individual peaks of the PLA and PEG component are clearly seen in the sample prepared without adding catalyst (PLA/PEG blend) (Figure 2a), indicating that no copolymer is formed under this condition. In the presence of catalyst, copolymerization between PLA and PEG segments occurs as evidenced by a single-modal curve centered at a higher molecular weight region than that of PEG. Compared between Sn-based and Ti-based catalyst, the results of Figure 2a show that TBT is more effective than DBTL in catalyzing the copolymerization of PLA and PEG. In the case of PLA-PPG system, both catalysts exhibit similar effects and two separate peaks corresponding to the individual species are still observed even after a longer period of reaction (Figure 2c). The bi-modal peaks of PLA-PPG are observed in the sample prepared by using a higher loading of TBT catalyst (Figure 2d). The difficulties of copolymer formation in the PLA-PPG system is mainly due to the immiscibility between the molten PLA and PPG components. Additionally, the reactivity of the hydroxyl-end groups of PPG is rather different to that of the PEG since the PPG structure consists of reactive primary hydroxyl group on the one side and secondary hydroxyl-end group which is less effective, on the other side. The ^{1}H-NMR signal at δ of 4.35 ppm (-CH, i) of figure 1b is corresponding to the attached secondary hydroxyl protons of PPG.

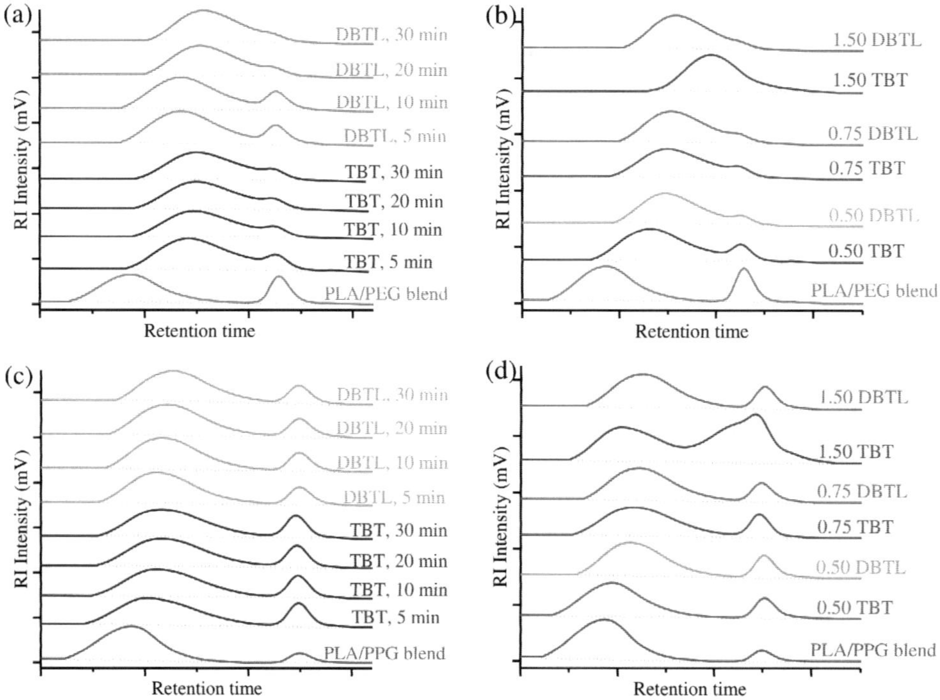

Figure 2 Effect of TBT and DBTL catalysts, reaction time, and content of catalyst on the GPC chromatograms of various copolymers (a, b) PLA-PEG, and (c, d) PLA-PPG system.

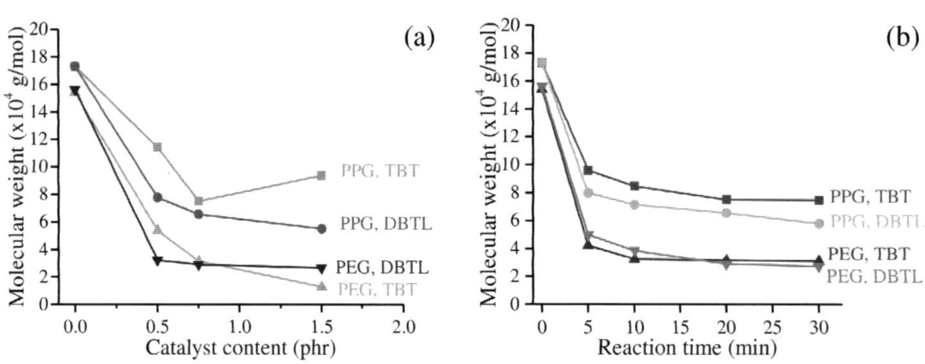

Figure 3 Evolution of molecular weights of PLA-PEG and PLA-PPG copolymer as affected by type and (a) content of catalyst, and (b) reaction time.

The effects of catalyst type, content, and reaction time on the molecular weight of PLA copolymers are shown in Figure 3. Comparing to the starting PLA (~ 1.7×10^5 g/mol), all copolymers prepared exhibit lower molecular weights. Type of catalyst plays a strong role in determining the molecular weight of the copolymers whereas catalyst content and reaction time in the studied range show minor effects on the molecular weight of the copolymer. In this work, the PLA-PEG copolymers obtained have molecular weight in a range of 5.0 x 10^{4} – 1.1 x 10^{4} g/mol when TBT was used as a catalyst and 3.3 x 10^{4} – 2.6 x 10^{4} g/mol in case of using DBTL catalyst. The molecular weight of PLA-PPG copolymers is in a range of 1.1 x 10^{5} – 7.5 x 10^{4} g/mol in the case of using TBT catalyst and 7.8 x 10^{4} – 5.5 x 10^{4} g/mol for the system with DBTL catalyst.

3.3. Mechanical properties

Figure 4 shows the tensile stress-strain curves of PLA and its copolymers. In order to obtain the materials with good mechanical performance, chain extender was added with the objective of linking the short-chain copolymers formed via melt-transesterification to form into the long-chain copolymers. From Figure 4, the PLA sample exhibits a high tensile strength of 64 MPa with a very low elongation at break of ~ 5%. As expected, the PLA copolymers with a soft component in the structure show much higher ductility than the PLA. The PLA-PEG copolymers prepared in this study have a strength of 25 MPa with a high elongation of 600%. The properties of the copolymers vary depending on the type of soft segment, type of catalyst, and also reaction conditions.

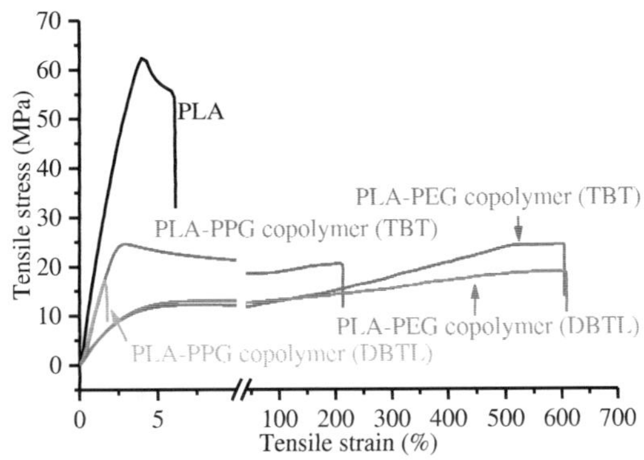

Figure 4 stress-strain curves of PLA, PLA-PEG, and PLA-PPG copolymers

4. Conclusions

The PLA-PEG and PLA-PPG copolymers were prepared using a transesterification reaction in melt state in the presence of either TBT or DBTL catalyst. The effects of catalyst type, content, and reaction time on the characteristics of the copolymers have been investigated. ^{1}H-NMR results confirmed the formation of these copolymers. The copolymers with different structures and molecular weights could be prepared by controlling the type of polyol used as a soft component, the type and content of catalyst, as well as the time of reaction. The long-chain copolymers obtained in this study exhibited a remarkable high elongation and their tensile properties are comparable to those of the petroleum-based polyolefins.

5. References

[1] Han L, Yu C, Zhou J, Shan G, Bao Y, Yun X, Dong T and Pan P 2016 *Polymer* **103** 376
[2] Pholharn D, Cheerarot O and Baimark Y 2017 *Chinese J. Polym. Sci* **35** 1391.
[3] Cohn D and Salomon A 2005 *Biomaterials* **26** 2297.
[4] Koosomsuan W, Phinyocheep P and Sirisinha K 2018 *Polym. Degrad. Stabil.* **157** 160.
[5] Kimura Y, Matsuzaki Y, Yamane H and Kitao T 1989 *Polymer* **30** 1342.

Acknowledgement

This work was financially supported by the National Research Council of Thailand (NRCT) under the Research University Network Initiative. The authors would like to express gratitude to the Central Instrument Facility (CIF) and the Rubber Technology Research Centre (RTEC) at Faculty of Science, Mahidol University for providing the supports in terms of research facilities.

IOP Conf. Series: Materials Science and Engineering **773** (2020) 012059 doi:10.1088/1757-899X/773/1/012059

Biodegradable poly(butylene adipate-co-terephthalate)/wheat gluten blends: Effect of PBAT modification on morphological, mechanical and water adsorption properties

K Saraphat[1], R Phaorung[1], U Saengsuk[1] and S Hemsri[1,*]

[1] Department of Materials Science and Engineering, Faculty of Engineering and Industrial Technology, Silpakorn University, Nakhon Pathom 73000, Thailand

*Corresponding author: sudsirihemsri@yahoo.com

Abstract. Biodegradable polymer blends between poly(butylene adipate-co-terephthalate) (PBAT) and wheat gluten (WG) were prepared in various ratios of 70:30, 60:40 and 50:50 by weight. Modified PBAT with maleic anhydride (PBAT-g-MA) was used as a compatibilizer in this work. Morphological, mechanical and impact properties as well as water adsorption of the PBAT/WG blends with and without the compatibilizer were investigated. Scanning electron microscopic (SEM) analysis revealed that the PBAT/WG blends with PBAT-g-MA showed a better compatibility between PBAT and WG phases. Moreover, addition of the compatibilizer into the blends significantly improved tensile strength, elongation at break and impact strength compared to the PBAT/WG blends without compatibilizer. Furthermore, water adsorption of the blends with compatibilizer is lower than that of uncompatibilized PBAT/WG blends.

1. Introduction

Poly(butylene adipate-co-terephthalate) (PBAT) is a biodegradable aromatic-aliphatic copolyester with higher flexibility and elongation than PLA and PBS [1] and suitable for various packaging and agricultural applications [2]. However, a major drawback of PBAT is its high cost. In order to overcome the drawback, blending of PBAT with low cost polymers has been an alternative approach. In this work, wheat gluten (WG) was selected to blend with PBAT because of its large availability, low price and biodegradability [3]. Nevertheless, polarity of PBAT and WG is different leading to immiscible blend between PBAT and WG. Thus, in this work, maleic anhydride grafted PBAT, which was synthesized by reactive grafting process using Luperox® 101 as an initiator, was used as a compatibilizer between PBAT and WG phases. Morphological, mechanical and water adsorption properties of PBAT/WG with and without PBAT-g-MA compatibilizer were studies.

2. Experimental

2.1. Materials

PBAT (Ecoflex© F Blend C1200) with MFI of 3.8 g/10 min was purchased from BASF Corporation. WG was supplied from Anhui Ante Food Co., Ltd, China. Maleic anhydride (MA) and 2,5-Bis(*tert*-butylperoxy)-2,5-dimethylhexane (Luperox® 101) were provided from Sigma-Aldrich, Inc. Glycerol was obtained from Ajax Finechem PTY Ltd.

IOP Conf. Series: Materials Science and Engineering **773** (2020) 012059 doi:10.1088/1757-899X/773/1/012059

2.2. Preparation of PBAT-g-MA

PBAT-g-MA was prepared in a twin-screw extruder at 180°C with a screw speed of 80 rpm. MA and Luperox as an initiator were used at constant amount of 5 and 0.5 phr for grafting MA on PBAT chains.

2.3. Preparation of PBAT/WG blends

WG was plasticized by glycerol at content of 20 phr before mixed with PBAT. In the case of the compatibilized PBAT/WG blends, the PBAT-g-MA with amount of 10 wt% of PBAT was used. The PBAT/WG and PBAT/WG/PBAT-g-MA blends were blended in the twin screw extruder at 150°C with a screw speed of 60 rpm. The PBAT/WG ratios were varied at 70:30, 60:40 and 50:50 by weight. All specimens of the blends were heat-pressed by compression molding.

2.4. Characterization Methods

Morphologies of cryo-fractured surface of PBAT/WG blends with and without the PBAT-g-MA compatibilizer were observed by a scanning Electron microscope (TM3030, Hitachi, Japan). Tensile properties including maximum tensile strength and elongation at break of the PBAT, WG, PBAT/WG and PBAT/WG containing the compatibilizer were investigated by Universal testing machine (Instron 5969, Instron, USA.) according to ASTM D638 type IV. Izod impact testing was performed by Pendulum impact tester (CEAST 9050, Instron, USA.), following the procedure as described in ASTM D256-10. Water adsorption of PBAT, WG, PBAT/WG and PBAT/WG/PBAT-g-MA was determined by soaking the specimens in deionized water at a set time interval until 7 days. The percentage of weight change (WC) was calculated as followed equation: WC (%) = [(Ws-Wo)/Wo] x 100 where Ws and Wo are the weight of a specimen after and before soaked in the deionized water, respectively.

3. Results and discussion

3.1. Morphology characterization

The SEM was used in order to investigate the effect of the PBAT-g-MA compatibilizer on the phase morphology of the PBAT/WG blends. Figure 1 showed SEM micrographs of the cryogenically fractured surface of the PBAT/WG and PBAT/WG/PBAT-g-MA blends at ratios of 70/30, 60/40 and 50/50. Regards to the SEM images of the uncompatibilized PBAT/WG (Fig.1 (a)-(c)), the dispersion of WG was non-uniformed in the PBAT matrix with the large particle size of the WG as indicated in the SEM images. Moreover, the interfacial de-bonding was observed in all uncompatibilized blends. These evidences indicated poor adhesion between PBAT and WG phases. In the case of the PBAT/WG/PBAT-g-MA blends (Fig.1 (d)-(f)), the dispersion of WG phase became uniform with a smooth surface and without a presence of WG particles in PBAT matrix. This result could be attributed to improvement of interfacial adhesion between PBAT matrix and WG by addition of the PBAT-g-MA resulting in the better compatibility between PBAT and WG phases.

PBAT/WG	**70:30**	**60:40**	**50:50**
No added compatibilizer			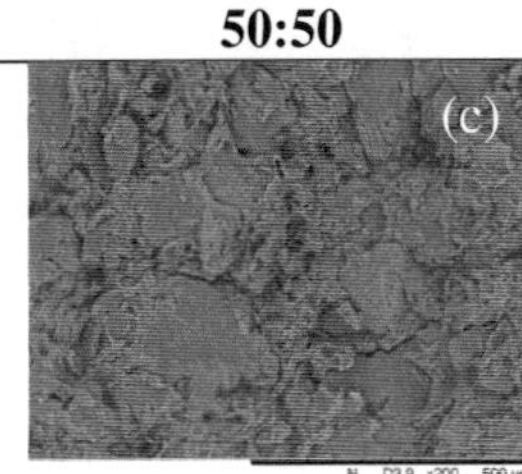

PBAT/WG	**70:30**	**60:40**	**50:50**
Added compatibilizer	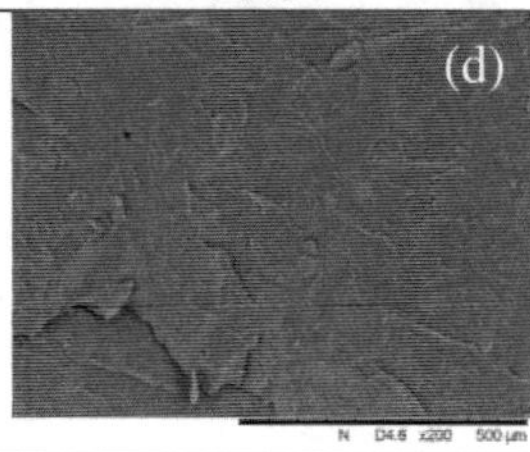	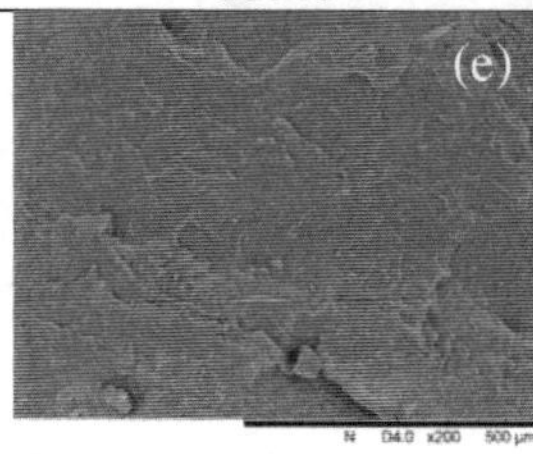	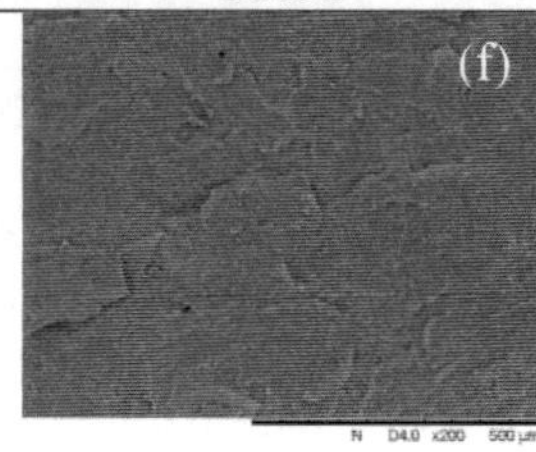

Figure 1. SEM images of the PBAT/WG and PBAT/WG/PBAT-g-MA blends at the ratios of 70/30, 60/40 and 50/50 at magnification x 200.

3.2. Mechanical and impact properties

The mechanical properties including maximum tensile strength and elongation at break and impact properties of PBAT, WG, PBAT/WG blends with and without the PBAT-g-MA were represented in Figure 2 (a)-(d), respectively. Regardless of adding the PBAT-g-MA compatibilizer, elongation at break and impact strength decreased as increasing WG content in the blends. As seen in Fig.2 (a)-(c), plasticized WG had high modulus and low elongation at break compared with PBAT, These results indicated that WG showed more brittleness behaviour and PBAT possessed more ductile characteristic. Thus, incorporation of stiffness WG polymer into PBAT could reduce elongation at break and impact strength in the PBAT/WG blends with and without the compatibilizer. It was found that the mechanical and impact properties, the compatibilized PBAT/WG blends were higher than those of the uncompatibilized blends. This could be attributed to reduced interfacial tension between PBAT and WG polymers, resulting from reaction of MA on PBAT-g-MA with hydroxyl and carboxyl groups of PBAT and interaction of MA of the compatibilizer with amino and hydroxyl groups of WG.

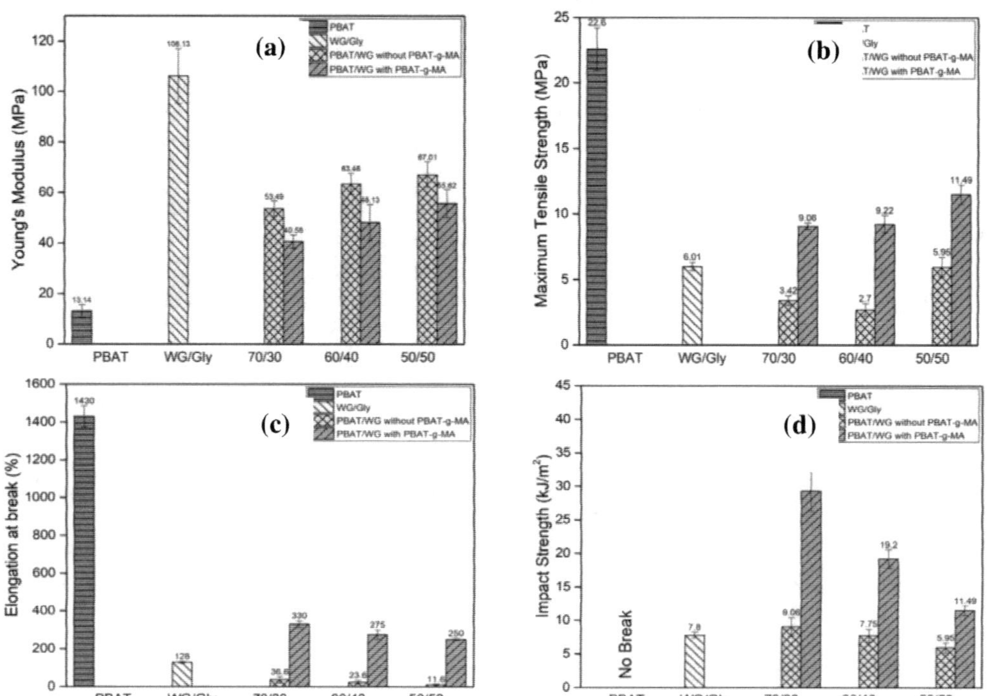

Figure 2. Mechanical and Impact properties of PBAT, WG, uncompatibilized PBAT/WG and compatibilized PBAT/WG blends at various ratios: (a) Maximum tensile strength, (b) Elongation at break and (c) Impact strength.

3.3. Water adsorption

Figure 3 represents percentage of weight change of plasticized WG, PBAT, PBAT/WG blends with and without PPBAT-g-MA after immersion in DI water at a set time interval. Among all specimens, the plasticized WG showed the highest %weight change after soaked in water, indicating poor water resistance of the plasticized WG due to a presence of hydrophilic characteristic of polar peptides (amino and carboxyl groups) in WG responsible for interacting with water via hydrogen bonding [4] and a presence of glycerol plasticizer added in WG. On the other hand, the hydrophobic PBAT provided the lowest water adsorption. The water weight change increased as an increase in polar plasticized WG content. Compared to the uncompatibilized PBAT/WG blends, the blends containing PBAT-g-MA compatibilizer gave the lower water weight change, attributed to the higher water resistance in the compatibilized PBAT/WG blends. This is because the polar groups such as amino and hydroxyl groups of WG reacted with MA in PBAT-g-MA compatibilizer, resulting in a reduction of the hydrophilic groups in the compatibilized blends.

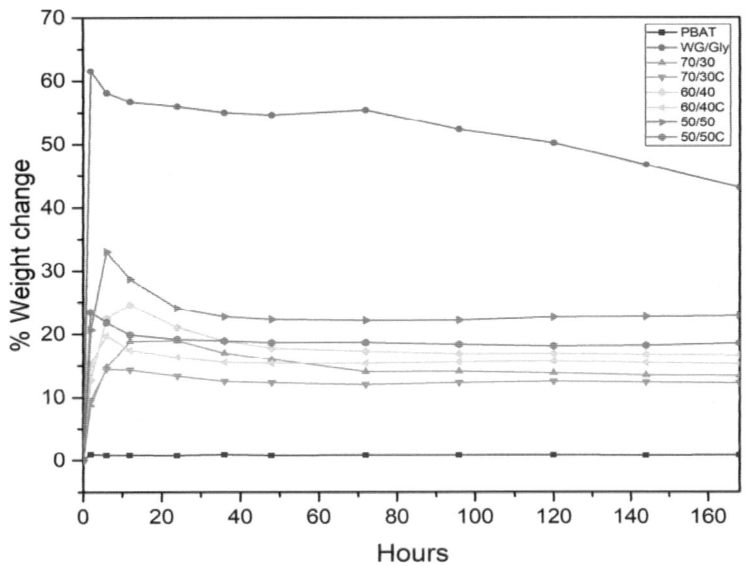

Figure 3. % weight change of plasticized WG, PBAT, PBAT/WG blends and PBAT/WG/PBAT-g-MA blends at ratios of 70/30, 60/40 and 50/50.

4. Conclusions

Incorporating PBAT-g-MA as a reactive compatibilizer in the PBAT/WG blends had great effect on phase morphology, mechanical and impact properties as well as water adsorption of the blends. Addition of the compatibilizer into the PBAT/WG provides the better interfacial adhesion between two polymers leading to an enhancement of tensile, impact and water adsorption properties with respect to uncompatibilized PBAT/WG blends.

5. References

[1] Ray S S and Okamoto M 2003 *Macromolecules.* **36** 2355.
[2] Averous L 2004 *J. Macromol. Sci. Part C Polym. Rev.* **4** 231.
[3] Gontard N, Guilbert S and Cuq J 1992 *J. Food Sci.* **57** 190.
[4] Wang C, Carriere C J, Willett J L. 2002 *J. Polym. Sci. Part B: Polym. Phys.* **40** 2324.

Acknowledgement

The authors are thankful for the financial support from Department of Materials Science and Engineering, Faculty of Engineering and Industrial Technology, Silpakorn University.

A 3D-RISM study of water and potassium ion adsorption in Montmorillonite nanoclay

N Ruankaew[1], A Kristinawati [1,2], N Yoshida[3], and S Phongphanphanee[1,4,5,*]

[1] Department of Materials Science, Faculty of Science, Kasetsart University, Bangkok 10900, Thailand.
[2] Department of Physics, Institut Pertanian Bogor, Bogor 16680, Indonesia.
[3] Department of Chemistry, Graduate School of Science, Kyushu University, Fukuoka 819-0395, Japan.
[4] Computational Biomodelling Laboratory for Agricultural Science and Technology (CBLAST), Faculty of Science, Kasetsart University, Bangkok 10900, Thailand.
[5] Specialized Center of Rubber and Polymer Materials in Agriculture and Industry (RPM), Faculty of Science, Kasetsart University, Bangkok 10900, Thailand.

*E-mail: fscisrph@ku.ac.th

Abstract. In this study, we investigated the intercalation of water and K^+ to interlayer of Montmorillonite (MMT) nanoclay by using the statistical mechanics theory of molecular liquids or three-dimensional reference interaction site model (3D-RISM). The results from 3D-RISM show that the distribution of water and K^+ bound to the MMT's surfaces with different sites. Moreover, the results from 3D-RISM show the amount of water corresponds to the d-basal spacing of the clay, which are in good agreements with other simulation methods as well as experimental.

1. Introduction

Montmorillonite is in the smectite class of clay, which is an important smectite clay mineral and found in many places around the world. Structurally, MMT consists of two silica tetrahedral sheets and one aluminum octahedral sheet, as well as its isomorphous replacement of silica by aluminum can be occur as shown in figure 1a. It has been received attention for application in many materials due to its great swelling, adsorption, and selective to specific cation. For these useful properties made MMT has been widely used in agriculture, industries, engineering, food and drug [1]. Recently, Fafad *et al.* (2013) and Hlushak *et al.* (2016) successfully used the 3D-RISM theory with Kovalenko-Hirata (KH) closure to investigate the adsorption preference of Kaolinite in aqueous electrolyte and non-aqueous solution, respectively [2,3]. Their studies provided very insight conclusion such as the adsorption configuration of indole on kaolinite surface, the desolvation free energies calculation, and the interactions of clay particles in solution. However, MMT structure is different from Kaolinite; MMT is a 2:1 mineral whereas Kaolinite is a 1:1 mineral. Kaolinite, by itself, has poor swelling and low cation adsorption abilities when compared to MMT. In this work, we studied the water and K^+ distribution between swelling able interlayer of MMT by using 3DRISM-KH and CLAYFF force field.

IOP Conf. Series: Materials Science and Engineering **773** (2020) 012060 doi:10.1088/1757-899X/773/1/012060

2. Computational Method

A bi-layer of MMT (figure 1a), which each layer contained $\{7 \times 4 \times 1\}$ unit cells in $\{a \times b \times c\}$ direction (2540 atoms) was considered as the solute in this study. The solute structure was derived from the X-ray studied by C.E. Marshall (1935) [4]. The d-basal spacing of clay in our calculation was vary from 9.6 Å to 24 Å with gradually increased at 0.1 Å. The CLAYFF force field parameters were applied for all solute atom species as well as the ions in KCl electrolyte solution. For water, we adopt the SPC/E water model [5,6]. The condition for the solvent–solvent correlation functions in DRISM and solute-solvent correlation functions in 3D-RISM are the same the condition as used in references [7,8]. We also replaced one silica with aluminum atom on our clay's surface for investigating the surface charge effect on the adsorption process.

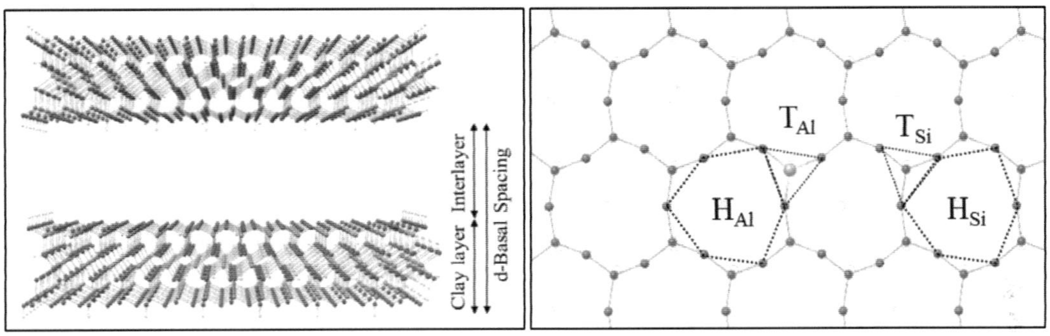

(a) (b)

Figure 1. (a) A structure of bi-layer MMT with terms. (b) A most top surface of MMT with the isomorphous replacement of silica by aluminum. Atomic legends were colored highlight as silica (light blue), oxygen (red), hydrogen (white) and aluminum (orange). Dotted-triangles, T_{Si} and T_{Al}, represent the triangular site of silica and aluminum, respectively. Dotted-hexagons, H_{Si} and H_{Al}, represent the hexagonal site next to the triangular site of silica and aluminum, respectively.

3. Results and Discussion

3.1 Water and K^+ adsorption

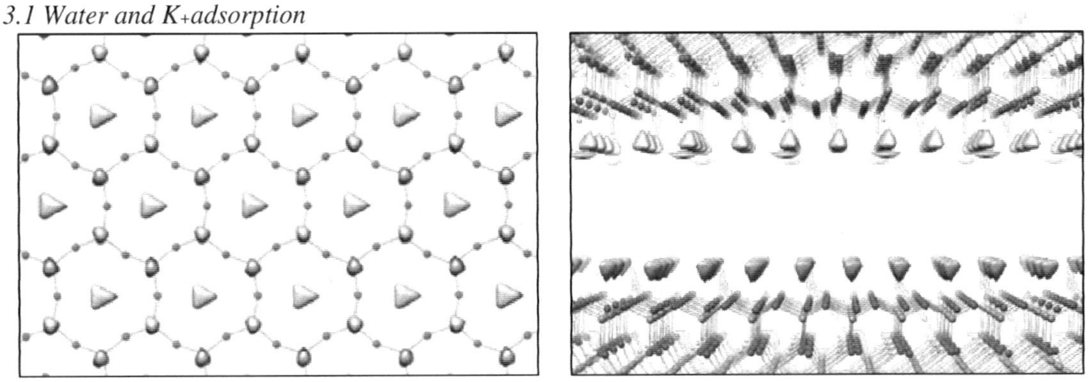

(a) (b)

Figure 2. Isosurface plots of 3D-DF of O_w (gray) at $g^{his} = 4$ (a) on MMT's inner-surface, and (b) cross-section view.

Figure 2a illustrates the isosurface plot of three-dimensional distribution function (3D-DF) for oxygen of water (O_w) at $g^{his} = 4$ at the interlayer space of MMT. The term $g_{Ow}^{his} = 4$ in the context means that the probability of finding the O_w at position **r** is 4 times greater than that of the bulk; this term also

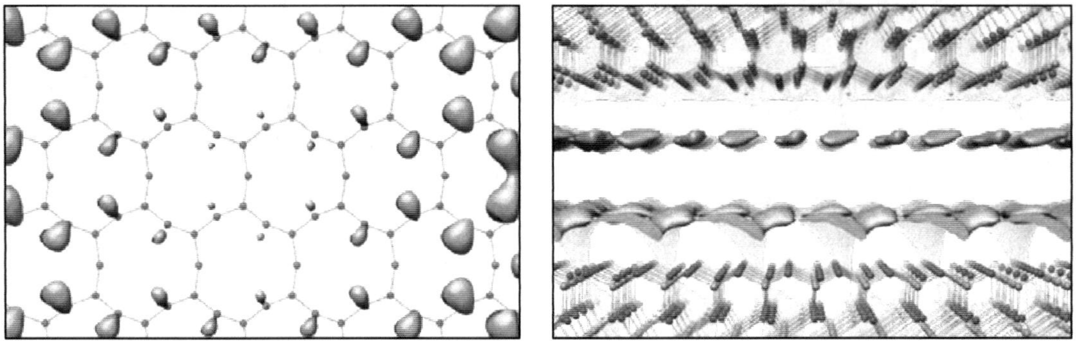

(a) (b)

Figure 3. Isosurface plots of 3D-DF of K_+ (green) at ghis = 5 (a) on MMT's inner-surface, and (b) cross-section view.

applies to the K_+ species in further discussion. The 3D-DF of O_w in cross-section view is presented in figure 2b. The results from 3D-DFs of O_w demonstrate the adsorption sites of water are at the T_{Si} and his sites (higher distribution) on the clay surfaces, the sites are defined in figure 1b. We also examined the amount of water in the interlayer space from d-basal spacing at 10 Å to 24 Å, as show in Figure 4a. Holmboe and Bourg (2014) suggested that when MMT adsorbed 5 water molecules per unit cell, one water layer (called 1W) will form at d-basal spacing is 12.5 Å. And the 2W, 3W and 4W form when 10, 15, and 20 water molecules per unit cell are absorbed at d-basal spacing are 15.7, 18.9 and 22 Å, respectively. As shown in figure 4a, the plot of number of water integrated from the 3D-DF and d-basal spacing are in good agreement of the previous experiments and simulations [9,10]. In the case of the adsorption site of K_+, the highest peak of 3D-DF of K_+ presented closely to the oxygen atoms on the his site; as shown in Figure 3a and 3b. These peaks correspond to the binding site of K_+ after absorbed to MMT's interlayer space. Moreover, we calculated the density profiles of all atom species in KCl (show only at 2W, Figure 4b) along the z-axis, the result confirmed the water layer formed in our studies.

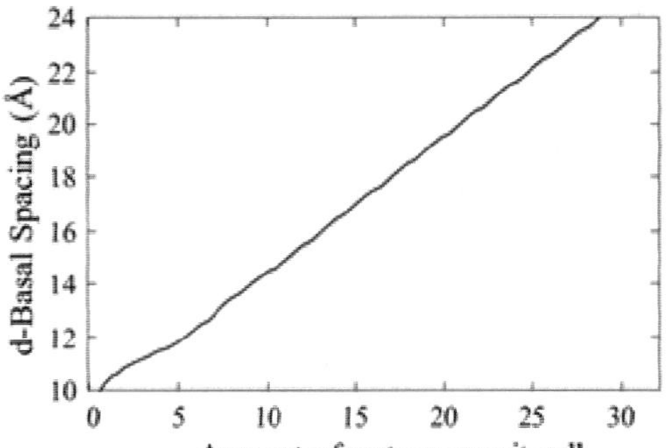

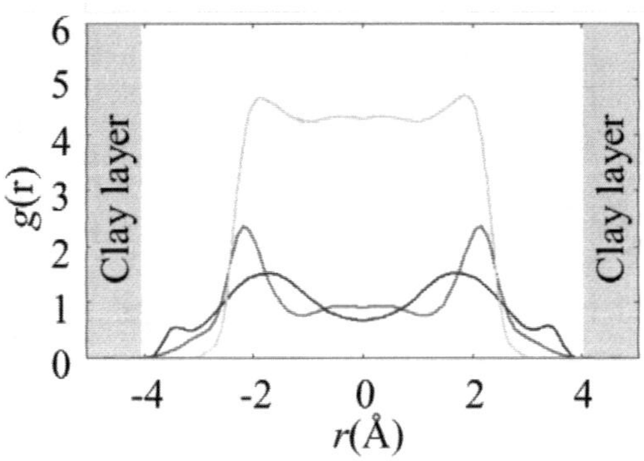

(a) (b)

Figure 4. (a) A plot between the amount of water and clay's d-basal spacing. (b) The density profiles along z-axis of KCl at 2W, blue line for O_w and red for hydrogen, and green for K_+.

3.2 Surface charges effect

The isomorphous structure of MMT are generally occur in nature by replacement of silica with aluminum atom. This replacement is considered to be one of important factors which can affect to the adsorption ability of MMT. From the CLAYFF parameters, after occurring an isomorphous replacement,

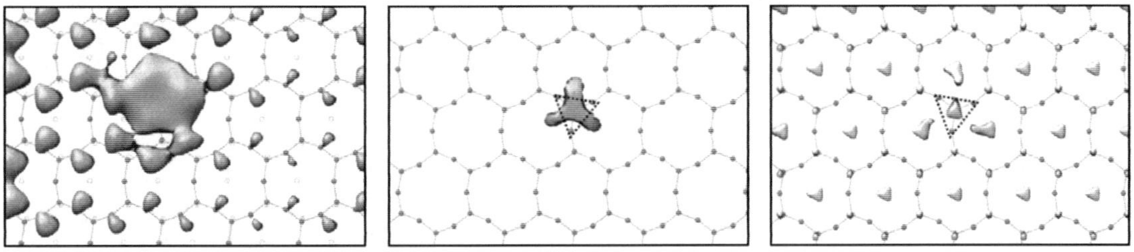

(a) (b) (c)

Figure 5. Isosurface plots of 3D-DF of K_+ (green) on MMT's inner-surface (a) at g(**r**) = 5 (b) g(**r**) = 10 and (c) an isosurface plot of 3D-DF of O_w (gray) at g(**r**) = 4. Dotted-triangles are the T_{Al} site.

the surface charges will become more negative due to the atomic charges of tetrahedral silica (2.100e) is reduced to 1.575e (tetrahedral aluminum). Likewise, the oxygen bonded with the replaced aluminum (-1.168e) is more negative than that of the oxygen bonded with the silica (-1.050e) [5]. In this case, our results support the hypotheses from the reports of Liu *et al.* (2007) and Peng *et al.* (2019) that the surface charges of MMT can induce clay to adsorb more cation and water to MMT's interlayer [11,12]. The 3D-DFs of O_w and K_+ are at the H_{Al} and T_{Al} sites, shown in figure 5a to 5c, are clearly demonstrated the higher peaks of O_w and K_+ are binding to these sites.

4. Conclusion

We have studied the distribution of water and K_+ inside MMT's interlayer. The results of 3D-DFs show that the adsorbed water amount of MMT depends on the d-basal spacing of the clay and give a good agreement with the hypothesis that water can form 1-4 water layers in the interlayer space. Our results reveal the adsorption sites of water are at the both triangular and hexagonal site of silica and aluminum. However, the adsorption site of the K_+ is closely to the oxygen atom of hexagonal site of silica, H_{Si}, and T_{Al} in the isomorphous replacement structure, respectively. Moreover, we also confirmed that the surface charges of MMT is able to enhance the adsorption ability of the clay.

5. References

[1] Yamamoto K, Shiono T, Yoshimura R, Matsui Y and Yoneda M 2018 *Adsorpt. Sci. Technol.* **36** 967-81
[2] Fafard J, Lyubimova O, Stoyanov S R, Dedzo G K, Gusarov S, Kovalenko A and Detellier C 2013 *J. Phys. Chem. C* **117** 18556–66
[3] Hlushak S, Stoyanov S R and Kovalenko A 2016 *J. Phys. Chem. C* **120** 21344–57
[4] Marshall C.E. 1935 *Zeitschrift für Krist. - Cryst. Mater.* **91** 433
[5] Cygan R T, Liang J-J and Kalinichev A G 2004 *J. Phys. Chem. B* **108** 1255–66
[6] Whitley H D and Smith D E 2004 *J. Chem. Phys.* **120** 5387–95
[7] Ruankaew N, Yoshida N, Watanabe Y, Nakayama A, Nakano H and Phongphanphanee S 2019 *Phys. Chem. Chem. Phys.* **21** 22569–76
[8] Maruyama Y, Yoshida N, Tadano H, Takahashi D, Sato M and Hirata F 2014 *J. Comput. Chem.* **35** 1347–55
[9] Ngouana W. B F and Kalinichev A G 2014 *J. Phys. Chem. C* **118** 12758–73
[10] Holmboe M and Bourg I C 2014 *J. Phys. Chem. C* **118** 1001–13
[11] Liu X, Lu X, Wang R and Zhou H 2008 e *Geochim. Cosmochim. Acta* **72** 1837–47
[12] Peng J, Yi H, Song S, Zhan W and Zhao Y 2019 *Results Phys.* **12** 113–7

Acknowledgement

This work was supported by the Kasetsart University Research and Development Institute (KURDI).

IOP Conf. Series: Materials Science and Engineering **773** (2020) 012061 doi:10.1088/1757-899X/773/1/012061

Applicability of density functional and wave function theories combined with the three-dimensional reference interaction site model self-consistent field method to the d–d transitions of a transition metal aqua complex

C Yang, Y Watanabe, N Yoshida, and H Nakano*

Department of Chemistry, Graduate School of Science, Kyushu University, 744 Motooka, Nishi-ku, Fukuoka 819-0395, Japan

Email: nakano@chem.kyushu-univ.jp

Abstract. The applicability of density functional theory (DFT) and wave function theory combined with the three-dimensional reference interaction site model self-consistent field method to the d–d transitions of transition metal aqua complexes was examined using $[Cr(H_2O)_6]^{3+}$ in aqueous solution as an example. DFTs with hybrid functionals, multiconfigurational self-consistent field followed by perturbation theory, and coupled-cluster singles and doubles (CCSD) followed by the equation of motion CCSD, gave reasonable d–d transition energies.

1. Introduction

Transition metal compounds are widely known to exhibit a variety of colors. Therefore, transition metal salts and oxides have been used as colorants in glass and ceramics, and as pigments in paintings. Transition metal complexes in solution also exhibit a variety of colors. These properties are due to the fact that the central element of the metal complex is a transition element, and the d or f orbitals are not completely filled with electrons.

The d–d transition is an electronic transition between the split d orbitals. The d–d transition energy is in the range of 1–4 eV. This corresponds to the wavelength range of 300–1200 nm, which covers the visible light wavelength range of 360–800 nm. To elucidate the optical properties of transition metal complexes, a number of experimental and theoretical studies have been conducted.

Previously, we reported on the molecular and solvation structures and excitation spectra of $[Cr(H_2O)_6]^{3+}$ and $[Cu(H_2O)_{4–6}]^{2+}$ in aqueous solution using the three-dimensional reference interaction site model self-consistent field (3D-RISM-SCF) method.[1,2] These two systems are of a rather different nature in terms of electronic structure. Since $[Cr(H_2O)_6]^{3+}$ is a d^3 system, the ground state is nondegenerate and the octahedral complex is stable. On the other hand, $[Cu(H_2O)_{4–6}]^{2+}$ are d^9 systems, and therefore the ground state is quasidegenerate in octahedral symmetry, resulting in distorted geometries due to the Jahn–Teller effect. In both cases, density functional theory (DFT) with global hybrid generalized gradient approximation (the B3LYP and PBE0 functionals) worked well. However, the applicability of other exchange–correlation functionals and wave function theory (WFT) has not been clarified. In the present study, we examine their applicability to the d–d transitions using the $[Cr(H_2O)_6]^{3+}$ system as an example.

IOP Conf. Series: Materials Science and Engineering **773** (2020) 012061 doi:10.1088/1757-899X/773/1/012061

2. Method

2.1. 3D-RISM-SCF method

The 3D-RISM-SCF method is a hybrid method of the electronic structure theory and integral equation theory of liquids (3D-RISM theory).[3,4] In this method, solute molecules are treated quantum-chemically and solvent molecules are treated statistical-mechanically; thereby, the electronic wave function of the solute and the distribution of the solvent (spatial distribution function (SDF), $g_s(\mathbf{r})$; s indicates the solvent site) around the solute are determined simultaneously.

The Helmholtz energy of a system (A) is defined as the sum of the solute electronic energy (E_{solute}) and the excess chemical potential ($\Delta\mu$)

$$A = E_{\mathrm{solute}} + \Delta\mu. \quad (1)$$

E_{solute} is calculated from electronic structure (density functional orwave function) calculation; hence, it is a functional of an electron density $E_{\mathrm{solute}}[n]$ or a wave function $E_{\mathrm{solute}}[\Psi]$. $\Delta\mu$ is calculated from the equation

$$\Delta\mu = \rho k_B T \sum_{s}^{\text{solvent-sites}} \int d\mathbf{r} \left[\frac{1}{2}\left(h_s(\mathbf{r})\right)^2 \theta(-h_s(\mathbf{r})) - c_s(\mathbf{r}) - \frac{1}{2} h_s(\mathbf{r}) c_s(\mathbf{r})\right] \quad (2)$$

where k_B and T indicate the Boltzmann constant and the absolute temperature, respectively, and ρ is the number density of the solvent. The functions $h_s(\mathbf{r}) = g_s(\mathbf{r}) - 1$ and $c_s(\mathbf{r})$ are the 3D total and direct correlation functions of solvent site γ, respectively, and $\theta(x)$ is the Heaviside step function. The Kovalenko–Hirata (KH) closure[3,5] is used to derive equation (2). 3D correlation functions can be obtained by solving the 3D-RISM equation

$$h_s(\mathbf{r}) = c_t(\mathbf{r}) * \left(\omega_{ts}^{\mathrm{vv}}(r) + \rho h_{ts}^{\mathrm{vv}}(r)\right) \quad (3)$$

coupled with the KH closure, where $\omega_{st}^{\mathrm{vv}}(r)$ and $h_{st}^{\mathrm{vv}}(r)$ are the site–site intramolecular and total correlation functions of the solvent, respectively (superscript v denotes the solvent). The symbol * indicates convolution in direct space and summation over repeated site indices. Before carrying out the 3D-RISM calculation, we need to solve the (1D) RISM equation between the solvent molecules

$$h_{st}^{\mathrm{vv}}(r) = \omega_{ss'}^{\mathrm{vv}}(r) * c_{s't'}^{\mathrm{vv}}(r) * \left(\omega_{t't}^{\mathrm{vv}}(r) + \rho h_{t't}^{\mathrm{vv}}(r)\right) \quad (4)$$

and obtain $h_{st}^{\mathrm{vv}}(r)$ in equation (3), where $c_{st}^{\mathrm{vv}}(r)$ indicates the site–site direct correlation function of the solvent molecules.

2.2. Computational details

We performed the 3D-RISM-SCF calculations for the ground state in aqueous solution and excited state calculations for the d–d transitions of $[Cr(H_2O)_6]^{3+}$ using the solvation structure of the 3D-RISM-SCF method. The molecular structure of the ion was optimized using the solvation model based on density (SMD), proposed by Marenich et al.[6] The basis sets used were the 6-311G(d,p) basis set[7] for H and O, and the Wachters–Hay all-electron basis set [8,9] supplemented with one f-function (exponent: 0.87) for Cr. The electronic structure theories used in the 3D-RISM-SCF method were DFT with several exchange–correlation functionals [BLYP, PBE, and OLYP (pure generalized gradient approximation (GGA)); B3LYP, PBE0, and X3LYP (global hybrid (GH) GGA); TPSS and revTPSS (pure meta-GGA (mGGA)); TPSSh (GH-mGGA); and CAM-B3LYP (range-separated hybrid (RSH) GGA)],[10] and also the Hartree–Fock (HF), complete active space self-consistent field (CASSCF),[11] and coupled-cluster singles and doubles (CCSD)[12] methods. For excited state calculations, the time-dependent (TD) HF, TDDFT, state-averaged (SA) CASSCF, multiconfigurational quasidegenerate perturbation theory (MC-QDPT),[13] and equation of motion (EOM) CCSD[12,14,15] were also used. In the CASSCF method, the active space was constructed by distributing three electrons in 10 orbitals (3d- and 4d-like orbitals). The parameters in the 3D-RISM calculations were the same as reported previously.[1]

Table 1. First and second d–d transition energies of $[Cr(H_2O)_6]^{3+}$ in aqueous solution (eV).

Method[a]	Optimized geometry			PBE0 geometry	
	$^4T_{1g}$	$^4T_{2g}$	d(Cr–O) / Å	$^4T_{1g}$	$^4T_{2g}$
HF/TDHF	1.53	3.08	2.080	1.78	3.32
DFT/TDDFT					
BLYP	2.06	2.48	2.045	2.06	2.48
PBE	2.13	2.49	2.026	2.16	2.53
OLYP	1.86	2.17	2.057	2.06	2.38
B3LYP	2.10	2.73	2.036	2.09	2.73
PBE0	2.15	2.80	2.020	(2.15)	(2.80)
X3LYP	2.09	2.76	2.034	2.16	2.82
TPSS	2.39	2.87	2.022	m	2.88
revTPSS	2.46	2.96	2.018	2.46	2.96
TPSSh	2.39	2.98	2.020	2.40	2.98
CAM-B3LYP	2.29	2.85	2.020	2.18	2.85
CASSCF/SA-CASSCF	1.81	2.85	2.000	1.72	2.74
CASSCF/MC-QDPT	1.97	2.87	2.000	1.88	2.90
CCSD/EOM-CCSD	1.81	2.97	2.048	1.95	3.12
Exp.[b]	2.11	2.98	1.96–2.03[c]		

[a] The notation A/B indicates that the 3D-RISM-SCF and excitation energy calculations were performed using the A and B methods, respectively. [b] Reference 16. [c] Reference 17.

3. Results and discussion

Table 1 shows the transition energies of two d–d transitions ($^4T_{1g}(F)\leftarrow{}^4A_{2g}(F)$ and $^4T_{2g}(F)\leftarrow{}^4A_{2g}(F)$). In the table, the Cr–O distances are also listed. All the geometries optimized were in T_h symmetry; hence the six Cr–O distances are all equal.

The method not including electron correlation, i.e., the HF method, gives a rather long Cr–O bond distance of 2.080 Å, compared with the experimental value range of 1.96–2.03 Å. The transition energies by HF/TDHF at the optimized geometry are 1.53 and 3.08 eV. The deviation of the lower transition energy from the experimental value is not small (−0.58 eV). The DFTs with various exchange–correlation functionals give somewhat longer Cr–O distances. The average distance is 2.03 Å, and four of 10 results are over the experimental value range of 1.96–2.03 Å. Many of the DFT/TDDFT give improved transition energies compared with the HF/TDHF. In particular, the results of the hybrid functionals (GH-GGA, GH-mGGA, and RSH-GGA) are in good agreement with experimental results. The average deviations from the experimental values are 0.11 and 0.16 eV for $^4T_{1g}$ and $^4T_{2g}$, respectively. In contrast, the deviations of pure functionals (pure GGA and mGGA) are 0.19 and 0.39 eV. The CASSCF Cr–O distance is 2.000 Å. The transition energies by CASSCF/SA-CASSCF are 1.81 and 2.85 eV, which are slightly underestimated. The CASSCF method takes account of only static electron correlation, not dynamic electron correlation. By also taking account of dynamic electron correlation, MC-QDPT (CASSCF/MC-QDPT) improves the transition energies, giving 1.97 and 2.87 eV. The Cr–O distance by CCSD is somewhat longer (2.048 Å) compared with that by CASSCF. The transition energies of CCSD/EOM-CCSD are 1.81 and 2.97 eV (the former is slightly underestimated).

Table 1 also shows the transition energies at the PBE0 optimized geometry; the trend of each theoretical method can be seen by excluding the geometry effect. The results show that the DFTs tend to underestimate the $^4T_{2g}$ energy (average 2.74 eV), while the wave function theories (CASSCF, MC-QDPT, and EOM-CCSD) tend to underestimate the $^4T_{1g}$ energy (average 1.85 eV).

Figure 1 shows the radial distribution functions (RDFs) of O and H atoms around the Cr atom, which are obtained by angular averaging of the SDFs by 3D-RISM-SCF. Each method gives very similar RDFs. The peak positions are 4.0 and 4.9 Å, corresponding to the experimental values of 4.0 and 4.5 Å,[17] respectively. These peaks indicate the second hydration shell around the Cr^{3+} ion, which is mainly composed of solvent water molecules that form hydrogen bonds with the ligand water molecules.

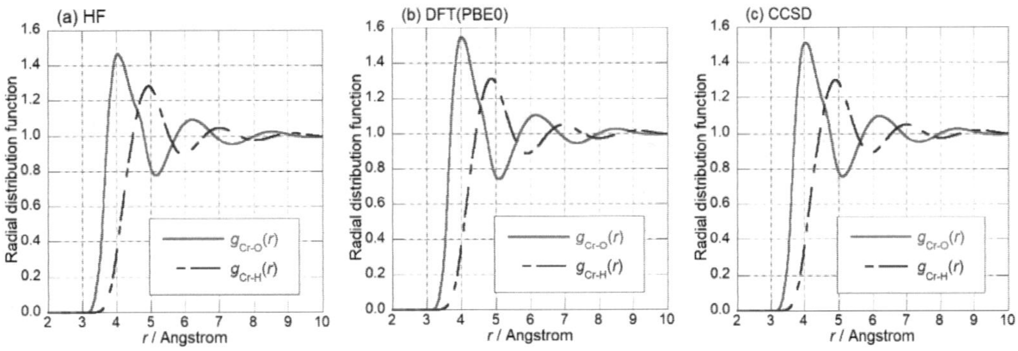

Figure 1. Radial distribution functions calculated by angular averaging of spatial distribution functions of the 3D-RISM-SCF with (a) HF, (b) DFT(PBE0), and (c) CCSD.

4. Summary

We have examined the geometry, d–d transitions, and solvent distribution functions of $[Cr(H_2O)_6]^{3+}$ in aqueous solution, which is an example of a transition metal complex with a nondegenerate ground state. The DFTs with hybrid functionals work well for the d–d transitions. The wave function theories also give reasonable results. The distribution functions of solvent obtained by 3D-RISM-SCF are similar, which are almost independent of the electronic structure methods.

5. References

[1] Fujishige S, Kawashima Y, Yoshida N and Nakano H 2013 *J. Phys. Chem. A* **117** 8314–22
[2] Yang C, Watanabe Y, Yoshida N and Nakano H 2019 *J. Phys. Chem. A* **123** 3344–54
[3] Kovalenko A and Hirata F 1999 *J. Chem. Phys.* **110** 10095
[4] Sato H, Kovalenko A and Hirata F 2000 *J. Chem. Phys.* **112** 9463–68
[5] Kovalenko A and Hirata F 1998 *Chem. Phys. Lett.* **290** 237–44
[6] Marenich A V., Cramer C J and Truhlar D G 2009 *J. Phys. Chem. B* **113** 6378–96
[7] Krishnan R, Binkley J S, Seeger R and Pople J A 1980 *J. Chem. Phys.* **72** 650–4
[8] Wachters A J H 1970 *J. Chem. Phys.* **52** 1033–6
[9] Hay P J 1977 *J. Chem. Phys.* **66** 4377–84
[10] Mardirossian N and Head-Gordon M 2017 *Mol. Phys.* **115** 2315–72 and references therein
[11] Roos B O 1987 *Ab Initio Methods in Quantum Chemistry - II* (New York: Wiley) pp. 399–445
[12] Piecuch P, Kucharski S A, Kowalski K and Musiał M 2002 *Comput. Phys. Commun.* **149** 71–96
[13] Nakano H 1993 *J. Chem. Phys.* **99** 7983–92
[14] Kowalski K and Piecuch P 2004 *J. Chem. Phys.* **120** 1715–38
[15] Włoch M, Gour J R, Kowalski K and Piecuch P 2005 *J. Chem. Phys.* **122** 214107
[16] Sutton D 1968 *Electronic Spectra of Transition Metal Complexes* (New York: McGraw-Hill)
[17] Bleuzen A, Foglia F, Furet E, Helm L, Merbach A E and Weber J 1996 *J. Am. Chem. Soc.* **118** 12777–87

IOP Conf. Series: Materials Science and Engineering **773** (2020) 012062 doi:10.1088/1757-899X/773/1/012062

The Reference Interaction Site Model Integrated Calculator (RISMiCal) program package for nano- and biomaterials design

Norio Yoshida*

Department of Chemistry, Graduate School of Science, Kyushu University, 744 Motooka, Nishiku, Fukuoka 819-0395 Japan.

* Corresponding author: Norio Yoshida: noriwo@chem.kyushu-univ.jp

Abstract. Solvent molecules play an essential role in various chemical, physical, and biological processes in solution phase. The reference interaction site model (RISM) and its three-dimensional extension (3D-RISM) are powerful tools for considering the solvation effect on chemical reactions, biological functions, and structure formation upon coupling with the quantum chemistry electronic structure theory and/or molecular dynamics simulation technique. We recently developed the reference interaction site model integrated calculator (RISMiCal) program package based on the RISM and 3D-RISM theories, which has interfaces for external programs such as GAMESS and Tinker. The features of the RISMiCal package are reviewed here.

1. Introduction

Modeling a solvent environment is crucial for nano- and biomaterials design because the solvent plays an essential role in determining the structures, properties, and functions of the materials existing in solution.[1] The solvent models can be classified into several categories. The most popular model is the dielectric continuum model, which treats the solvent environment as a continuum medium characterized by the dielectric constant.[2, 3] Molecular mechanics (MM) is also a widely used approach. In MM, the solvent molecules are explicitly placed around the solute molecule and the coordinates are generated by the molecular dynamics (MD) or Monte Carlo techniques.[4] The integral equation theory (IET) of molecular liquids is another candidate to handle the solvent environment and is based on the statistical mechanics theory of liquids.[5-7] To date, several IETs have been proposed, including the reference interaction site model (RISM),[8-11] and its three-dimensional extension, the 3D-RISM theory.[12-14] The RISM and 3D-RISM theories have been used in a number of studies due to their high applicability involving low computational cost and high accuracy. Based on the RISM and 3D-RISM theories, many theoretical and computational developments have been carried out.

We recently developed the reference interaction site model integrated calculator (RISMiCal) program package, which has interfaces for external programs such as GAMESS (the quantum chemistry software) and Tinker (the MD software). In this paper, we review the features and specifications of the RISMiCal package.

2. RISM and 3D-RISM Theories

The RISM equation is given by

$$h_{\alpha\gamma}(r) = \omega_{\alpha\alpha'} * c_{\alpha'\gamma'} * \left(\omega_{\gamma'\gamma} + \rho_{\gamma'} h_{\gamma'\gamma}\right) \quad (1)$$

and the Kovalenko–Hirata (KH) closure relation is given by[15]

$$g_{\alpha\gamma}(r) = \begin{cases} \exp\left[d_{\alpha\gamma}(r)\right] & \text{for } d_{\alpha\gamma}(r) < 1 \\ d_{\alpha\gamma}(r) + 1 & \text{for } d_{\alpha\gamma}(r) \geq 1 \end{cases} \quad (2a)$$

$$d_{\alpha\gamma}(r) = -\beta u_{\alpha\gamma}(r) + h_{\alpha\gamma}(r) - c_{\alpha\gamma}(r) \quad (2b)$$

where $g_{\alpha\gamma}(r) = h_{\alpha\gamma}(r) + 1$ is a radial distribution function between interaction sites α and γ; $\omega_{\alpha\alpha'}$, $c_{\alpha\gamma}$, β, and ρ denote intramolecular correlation function, direct correlation function, inverse temperature, and number density, respectively, and $u_{\alpha\gamma}$ is a pair interaction potential between sites α and γ. The symbol * indicates that a convolution integral and the summation should be taken for repeated indices. By solving the RISM equation and KH closure under a given $u_{\alpha\gamma}$ iteratively, one can obtain the distribution and direct correlation functions.

Similarly, the 3D-RISM equation and the 3D-KH closure are given by

$$h_{\alpha}(\boldsymbol{r}) = c_{\gamma} * \left[\omega_{\gamma\alpha} + \rho_{\gamma} h_{\gamma\alpha}\right] \quad (3)$$

$$g_{\alpha}(\boldsymbol{r}) = \begin{cases} \exp[d_{\alpha}(\boldsymbol{r})] & \text{for } d_{\alpha}(\boldsymbol{r}) < 1 \\ d_{\alpha}(\boldsymbol{r}) + 1 & \text{for } d_{\alpha}(\boldsymbol{r}) \geq 1 \end{cases} \quad (4a)$$

$$d_{\alpha}(\boldsymbol{r}) = -\beta u_{\alpha}(\boldsymbol{r}) + h_{\alpha}(\boldsymbol{r}) - c_{\alpha}(\boldsymbol{r}) \quad (4b)$$

In the 3D-RISM case, one can obtain the spatial distribution function, which is suitable for describing the solvation structure of complex systems such as nano- and biomaterials.

3. Hybrid Multiscale Implementation

3.1. GAMESS

The RISMiCal package has an interface for the GAMESS quantum chemistry electronic structure program package.[16, 17] RISMiCal with GAMESS realizes the RISM self-consistent field (RISM-SCF),[18] 3D-RISM-SCF,[19] and DFT/3D-RISM computation.[12] An analytical free energy gradient for efficient geometry optimization of solvated molecules is also incremented.[20]

In the RISM/3D-RISM-SCF, the solvated Hamiltonian, $\widehat{\mathrm{H}}_{\mathrm{solv}}$, is given by

$$\widehat{\mathrm{H}}_{\mathrm{solv}} = \widehat{\mathrm{H}}_{\mathrm{iso}} + \widehat{V}_{\mathrm{solv}} \quad (5)$$

where $\widehat{\mathrm{H}}_{\mathrm{iso}}$ and $\widehat{V}_{\mathrm{solv}}$ are the Hamiltonian for isolated molecules and an electrostatic potential due to the solvents, respectively. In the 3D-RISM case, $\widehat{V}_{\mathrm{solv}}$ is given by

$$\widehat{V}_{\mathrm{solv}}(\boldsymbol{r}') = \sum_{\alpha} \rho_{\alpha} \int g_{\alpha}(\boldsymbol{r}) \frac{q_{\alpha} |\Psi(\boldsymbol{r}')|^2}{|\boldsymbol{r} - \boldsymbol{r}'|} d\boldsymbol{r} \quad (6)$$

where $g_{\alpha}(\boldsymbol{r})$ is obtained by solving the 3D-RISM equation. The wave function of solute molecules is obtained by solving the Schrödinger equation for the solvated Hamiltonian. The interaction potential for solving the 3D-RISM is

$$u_{\alpha}(\boldsymbol{r}) = \sum_{i} 4\varepsilon_{i\alpha} \left\{ \left(\frac{\sigma_{i\alpha}}{|\boldsymbol{r} - \boldsymbol{r}_i|} \right)^{12} + \left(\frac{\sigma_{i\alpha}}{|\boldsymbol{r} - \boldsymbol{r}_i|} \right)^{6} \right\} + \int \frac{q_{\alpha} |\Psi(\boldsymbol{r}')|^2}{|\boldsymbol{r} - \boldsymbol{r}'|} d\boldsymbol{r}' \quad (7).$$

Therefore, by solving the Schrödinger equation and the 3D-RISM equation through $\widehat{V}_{\mathrm{solv}}$ and $u_{\alpha}(\boldsymbol{r})$, iteratively, one can obtain the electronic structure Ψ and solvation structure $\{g_{\alpha}\}$, simultaneously.

3.2. *Tinker*

The RISMiCal package has an interface for the Tinker molecular modeling package.[21] RISMiCal/Tinker can perform the MD simulation and structure optimization based on the MM.[22, 23] The solute molecules are treated by Tinker in the usual manner, immersed in solvent, where the solvent distribution is evaluated by 3D-RISM. The force acting on solute atoms due to the solvent is given by

$$F_{\text{solv},i} = \sum_{\alpha} \rho_\alpha \int \frac{\partial u_\alpha(\boldsymbol{r})}{\partial R_i} g_\alpha(\boldsymbol{r}) d\boldsymbol{r} \tag{8}$$

where R_i is a coordinate of solute atom i. The computation of this additional force term to the solute atoms is implemented in the "gradient" module in Tinker.

4. Miscellaneous Implemented Features

The RISMiCal package includes many functions for analyzing solvation thermodynamics based on the RISM and 3D-RISM theories. The implemented functions in the RISMiCal package are summarized in Table 1.

Table 1. Implemented functions in the RISMiCal package ver. 1.0.

Function	**Description**	**Ref.**
QM / MM / RISM	Hybrid method of QM / MM and RISM / 3D-RISM	[24]
FMO / RISM	Hybrid method of the fragment molecular orbital and 3D-RISM	[25]
MOZ-SCF	Hybrid method of the molecular Ornstein–Zernike and electronic structure theory	[26]
RismPath / SM	Method for finding the minimum free energy pathway of ions through protein based on 3D-RISM and the string method	[27]
Solvent placement	Solvent placement algorithms based on the distribution function evaluated by 3D-RISM	[28, 29]
Dielectric consistent RISM / D-Stell dielectric correction	Correction methods to reproduce the dielectric constant of solvent	[30, 31]
Pressure correction / universal correction / Repulsive bridge correction / Partial wave expansion	Correction schemes of solvation free energy	[32-35]

5. Summary

We have developed the RISM/3D-RISM program package, RISMiCal, for use in the design of nano- and biomaterials. RISMiCal has interfaces for the GAMESS and Tinker packages. In addition, it has various functions for analyzing the solvation thermodynamics of nano- and biomaterials based on the RISM and 3D-RISM theories. We are currently preparing this program for public release.

6. References

[1.] Feig M, editor. Modeling Solvent Environments. Weinheim: WILLEY-VCH; 2010.
[2.] Tomasi J, Mennucci B and Cammi R 2005 *Chem. Rev.* **105** 2999-3093.
[3.] Tomasi J and Persico M 1994 *Chem. Rev.* **94** 2027-94.
[4.] Frenkel D and Smit B. Understanding Molecular Simulation: From Algorithms to Applications. 2nd ed. San Diego, CA: Academic Press; 2001.
[5.] Hirata F, editor. Molecular Theory of Solvation. Dordrecht: Kluwer; 2003.
[6.] Gray C G and Gubbins K E. Theory of molecular fluids Volume 1: Fundamentals. Oxford: Clarendon Press; 1984.
[7.] Hansen J P and McDonald I R. Theory of Simple Liquids. 3rd ed. Amsterdam: Academic Press; 2006.
[8.] Chandler D and Andersen H C 1972 *J. Chem. Phys.* **57** 1930-7.
[9.] Andersen H, Chandler D and Weeks J 1972 *J. Chem. Phys.* **57** 2626-31.
[10.] Andersen H and Chandler D 1972 *J. Chem. Phys.* **57** 1918-29.
[11.] Hirata F and Rossky P 1981 *Chem. Phys. Lett.* **83** 329-34.
[12.] Kovalenko A and Hirata F 1999 *J. Chem. Phys.* **110** 10095-112.
[13.] Beglov D and Roux B 1997 *J. Phys. Chem. B* **101** 7821-6.
[14.] Beglov D and Roux B 1996 *J. Chem. Phys.* **104** 8678-89.
[15.] Kovalenko A and Hirata F 2001 *Chem. Phys. Lett.* **349** 496-502.
[16.] Schmidt M W, Baldridge K K, Boatz J A, Elbert S T, Gordon M S, Jensen J H, Koseki S, Matsunaga N, Nguyen K A, Su S J, Windus T L, Dupuis M and Montgomery J A 1993 *J. Comput. Chem.* **14** 1347-63.
[17.] Yoshida N and Hirata F 2006 *J. Comput. Chem.* **27** 453-62.
[18.] Ten-No S, Hirata F and Kato S 1993 *Chem. Phys. Lett.* **214** 391-6.
[19.] Sato H, Kovalenko A and Hirata F 2000 *J. Chem. Phys.* **112** 9463-8.
[20.] Sato H, Hirata F and Kato S 1996 *J. Chem. Phys.* **105** 1546-51.
[21.] Ren P, Wu C and Ponder J W 2011 *J Chem Theory Comput* **7** 3143-61.
[22.] Kiyota Y, Hiraoka R, Yoshida N, Maruyama Y, Imai T and Hirata F 2009 *J. Am. Chem. Soc.* **131** 3852-3.
[23.] Miyata T and Hirata F 2008 *J. Comput. Chem.* **29** 871-82.
[24.] Yoshida N, Kiyota Y and Hirata F 2011 *J. Mol. Liq.* **159** 83-92.
[25.] Yoshida N 2014 *J. Chem. Phys.* **140** 214118.
[26.] Yoshida N and Kato S 2000 *J. Chem. Phys.* **113** 4974-84.
[27.] Yoshida N 2018 *Chem. Phys. Lett.* **699** 22-7.
[28.] Sindhikara D J, Yoshida N and Hirata F 2012 *J Comput Chem* **33** 1536-43.
[29.] Kiyota Y, Yoshida N and Hirata F 2011 *J. Chem. Theory Comput.* **7** 3803-15.
[30.] Yamaguchi T and Koda S 2015 *Bull. Chem. Soc. Jpn.* **88** 804-13.
[31.] Perkyns J S and Pettitt B M 1992 *Chem. Phys. Lett.* **190** 626-30.
[32.] Palmer D S, Frolov A I, Ratkova E L and Fedorov M V 2010 *J Phys-Condens Mat* **22** 492101.
[33.] Kovalenko A and Hirata F 2000 *J. Chem. Phys.* **113** 2793-805.
[34.] Sergiievskyi V, Jeanmairet G, Levesque M and Borgis D 2015 *J Chem Phys* **143** 184116.
[35.] Tanimoto S, Yoshida N, Yamaguchi T, Ten-No S L and Nakano H 2019 *J. Chem. Info. Model* **59** 3770-81.

Impact absorbing kneepad prepared from natural rubber

P Kaewpradit[1]*, A Kongchoo[1], P Chonlathan[1], N Lehman[2], and E Kalkornsurapranee[2]

[1] Department of Chemical Engineering, Faculty of Engineering, Prince of Songkla University, Songkhla 90110, Thailand.
[2] Department of Materials Science and Technology, Faculty of Science, Prince of Songkla University, Songkhla 90110, Thailand.

*Corresponding author: pornsiri.k@psu.ac.th

Abstract. This research aims to develop an impact absorbing material prepared from natural rubber, in order to encourage the competition of Thailand in the global market as well as add the rubber values. The material has been prepared by impregnating natural rubber foam with the dilatant fluid. Experimental results show that the prototype of knee guard impregnated with the proposed dilatant (NRI) gives good shock impact absorbing performance (higher than 5,000 N). This shows 20 – 30% higher performance than the prototype with the conventional dilatant. In other words, the NRI and the commercial knee guard have transmission force < 2,300 N with CV 1.1% and 1.5% respectively. Considering international impact protection standards, the test results have shown that the NRI is corresponding with level 2 class C of the protection standard for roller sports (EN 14120:2003) with the transmission force < 6,000 N. It is also coincident with the protection standard for Motorcyclist' protective clothing against mechanical impact (EN 1621-1:2010) with the transmission force < 35,000 N. In addition, the impact absorbing material prepared from natural rubber has performance within level 1 of EN 1621-2:2003 standard (peak transmission force < 18,000 N), but the commercial knee guard passes level 2 of EN 1621-2:2003 standard (peak transmission force < 9,000 N). It can be seen that the performance of the natural material is comparable with the commercial one, it also has lower manufacturing cost than 300 THB/piece while the commercial has sale cost of 2,500 THB/piece.

1. Introduction

Global status report on road safety in 2018 by World Health Organization (WHO) shows that the number of annual road traffic deaths ranks Thailand as the first country in Asia. Particularly ratio of the motorcycle accidents in Thailand is the third most lethal in the world (74.4% of the road traffic deaths in Thailand) [1], the government and the relating parties will then give especially care. Nowadays personal protective equipment (PPE), with impact absorption or energy dissipation characteristics, must be imported and has high costs. Many patents [2-5] and researches [6-7] have focused on a development of composite material based synthesis polymer to absorb the impact energy for PPE. The materials have been prepared from polymer foam impregnated with dilatant fluid which has special property for promoting impact force absorption. However, the preparation of dilatant fluid, which is a dispersion mixture of nanoparticles requires a special technique since of high viscosity and its energy resistance property.

This research aims to study and develop an impact absorbing material prepared from natural rubber, in order to encourage the competition of Thailand in the global market as well as add the rubber values. The development of dilatant fluid with additive has been initially focused elsewhere for more practical in manufacturing point of view providing lower viscosity (8,000 cP) compared to the conventional dilatant fluid (110,000 cP). In this research, the impact absorption property of the proposed material has been investigated by considering effect of the fluid types, and effect of the initial impact energy. The efficiency of impact energy absorption has also been compared with the commercial knee guard.

2. Experimental

In this research, impact absorbing kneepad based natural rubber (NR) has been prepared by impregnating NR foam with dilatant fluid. The impact absorption property has been investigated by considering effect of the fluid types, and effect of the initial impact energy. Two types of the fluid have been prepared by homogenizing nanoparticles in carrier.

Impact absorption test has been done by using drop test machine with free falling impact mass. For low initial impact force 7,500 N or 10.3 J, the test has been achieved at Department of Mechanical Engineering, Prince of Songkla University with impact mass 4.96 kg at drop height 21.5 cm. For higher initial impact energy, the test has been done at Thailand Automotive Institute, Bang Poo Industrial Estate, Samutprakan province according to EN14120:2003 (Protection for roller sports), EN1621-1:2010 (Part 1: Requirement for impact protectors) and EN1621-2:2003 (Part 2: Motorcyclist' back protectors).

3. Results and discussion

Table 1 shows the effect of dilatant fluid on impact absorption efficiency at initial impact force 7,500 N (or energy of 10.3 J). Drop test has been achieved with impact mass 4.96 kg at drop height 21.5 cm. The results have shown that the absorption efficiency of NR kneepad is increased 97% and 105% by increasing the amount of the conventional dilatant fluid (NRD) 10% and 12%vol, respectively. Furthermore, NRI at 12%vol. of the fluid has about 27% higher efficiency than NRD with only 1.1% of variation coefficient. It has also shown that NRI is comparable with the commercial knee guard (D3o), which has shear thickening behaviour, by considering at 95% confidential interval.

Table 1. Effect of dilatant fluid at initial impact force 7,500 Newton (or 10.3 Joules)

Sample		Avg. absorbed force, N	Standard deviation, N	Coefficient of variation (CV), %	Avg. increment factor
NR kneepad		2,050	N/A	N/A	1.00
NRD	10%	4,072	± 586	14.0	1.97
	12%	4,208	± 212	5.0	2.05
NRI	10%	4,683	± 106	3.0	2.28
	12%	5,329	± 59	1.1	2.60
Commercial (D3o)		5,216	± 78	1.5	2.54

Note: NRD = NR foam with dilatant, NRI = NR foam with the proposed dilatant

From Figure 1, the impact absorbing kneepad has also been tested under initial impact energy of 25 and 50 Joules at Thailand automotive institute according to EN14120:2003 and EN1621, respectively. The results have shown that NRI has maximum peak of transmission force only 3,933 N which could be considered in level 2 class C (highest level) of EN14120:2003 standard. Considering to EN1621 standard, the maximum transmission force of NRI kneepad is 14,476 N, which is lower than 35,000 N according to EN1621-1 (part 1) and 18,000 N (level 1) according to EN1621-2 (part 2). However, the commercial kneepad has transmission force lower than 9,000 N which is considered according to level 2 (highest level) of EN1621-2.

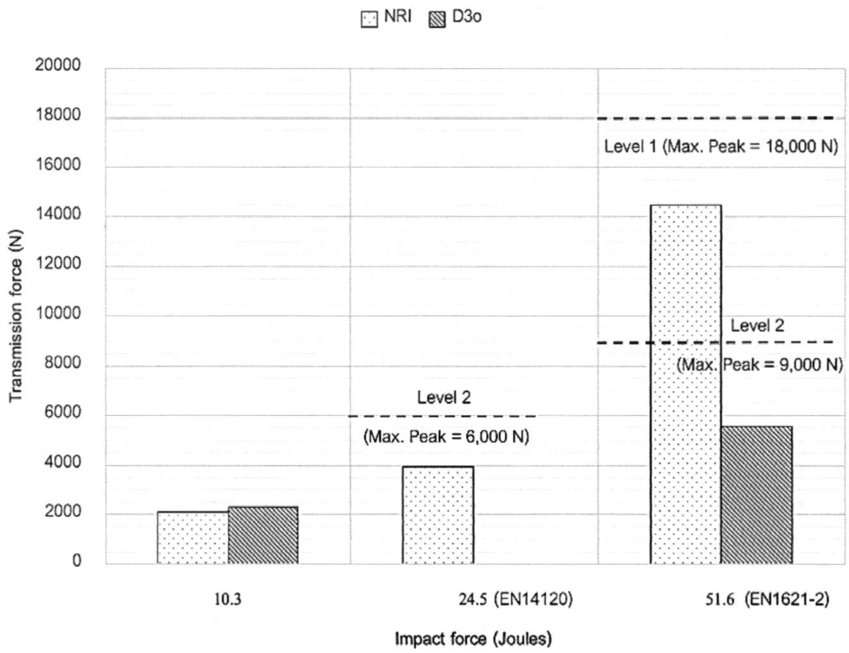

Figure 1. Comparison to the commercial kneepad under different initial impact energy.

4. Conclusions

In this research, impact absorbing kneepad based natural rubber (NR) has been developed. The kneepad has been prepared by impregnating NR foam with the proposed dilatant fluid. Drop test results have shown that the proposed fluid promotes the impact absorption property of the kneepad providing about 27% higher absorption efficiency compared to the conventional fluid. Furthermore, the proposed kneepad has higher stability of the impact absorption property, which is comparable with the commercial one. The results at high initial impact energy have also shown that the impact absorbing kneepad could be considered as protection equipment according to level 2 class C of EN14120 for roller sports, and EN1621-1 for motorcyclist.

5. References

[1] https://www.who.int/thailand/news/detail/19-12-2018-launch-of-the-global-status-report-on-road-safety -2018-in-thailand; December 2018.
[2] Wagner N and Wetzel Eric D. 2004 WO2004/103231.
[3] Francisco Jose G R and Laura Campo D. 2016 WO 2016/051320 A1.
[4] Green Phillip C and Palmer Richard M. 2014 US2013/0269089 A1.
[5] Wagner N, Kirkwood J E and Egres R G Jr 2011 EP 2 319 880 A2.
[6] Majumdar A, Butola B S and Srivastava A 2014 *Materials and Design* **54** 295.
[7] Guida M, Marulo F, Bruno M, Montesarchio B and Orlando S 2018 *Advances in Aircraft and Spacecraft Science* **5** 319.

Acknowledgement

This work was fully financial supported by Thailand Research Fund (TRF). The authors are grateful to Department of Chemical Engineering, Faculty of Engineering and Department of Materials Science and Technology, Faculty of Scinece, Prince of Songkla University for all supports.

Injection-molded polyethylene/natural rubber blends

R Yoksan[1,2,*]

[1] Department of Packaging and Materials Technology, Faculty of Agro-Industry, Kasetsart University, Bangkok 10900, Thailand.
[2] Center for Advanced Studies for Agriculture and Food, Kasetsart University, Bangkok 10900, Thailand.

*Corresponding author: rangrong.y@ku.ac.th

Abstract. The present research aims to develop polyethylene/natural rubber (PE/NR) blends, with a weight proportion of PE:NR of 60:40, for injection-molded articles. Two types of PE, i.e. high density polyethylene (HDPE) and linear low density polyethylene (LLDPE) were used. The blends were fabricated by melt processes and then converted into dumbbell-shaped specimens using an injection molding machine. The effect of NR on properties of PE was investigated. The blends showed lower melt flow ability than the corresponding PE. Although PE/NR blends exhibited lower tensile strength and Young's modulus than PE, their elongation at break and impact strength tended to be higher. In addition, hardness and heat distortion temperature of PE decreased by blending with NR. The obtained PE/NR blends are thermoplastics, which can be further converted into product prototypes using injection molding process.

1. Introduction

Thailand is up to now the largest producer and exporter of natural rubber (NR) in the world accounted for ~40% of global production. However, only around 18% of midstream NR products were consumed in Thailand as raw materials in the manufacture of downstream vulcanized NR products including tires, elastic, latex gloves, rubber bedding and elastic bands, etc., while the rest (82%) is exported for additional processing to produce end products in factories abroad. This makes Thailand loss a chance to enhance its income. Furthermore, NR price decreases should due to its oversupply. Therefore, the development of downstream high value-added NR products is very important at this stage.

Most NR products are fabricated by vulcanization during forming their shapes under applying heat. This process requires many types of chemicals including sulphur to crosslink NR structure, the resulting vulcanized NR products are thus not recyclable. In addition, vulcanization is also time- and energy-consuming process.

Therefore, the present work aims to fabricate thermoplastic natural rubber (TPNR) based on the blending of NR with thermoplastic materials for injection-molded articles. Polyethylene (PE) was chosen as a thermoplastic in this study. Two types of PE, i.e. high-density polyethylene (HDPE) and linear low-density polyethylene (LLDPE), were melt blended with NR at a weight fraction of PE:NR of 60:40. The effect of NR on properties of PE was investigated and discussed.

2. Experimental

Two types of PE, i.e. HDPE (InnoPlus, PTT Global Chemical) and LLDPE (InnoPlus, PTT Global Chemical), were melt blended with NR (STR 5, Natural Art and Technology, Co., Ltd.) in the presence of other additives using a weight fraction of PE:NR of 60:40. The blending was performed in an internal mixer at 160°C for HDPE/NR blend and 140°C for LLDPE/NR blend with a screw speed of 50 rpm until homogeneous mixtures were obtained. The resulting materials were then injection-molded into dumbbell-shaped specimens using an injection molding machine (TUPARL TR80S2A, Sodick Plastech, Japan) at a temperature range of 165–185°C. The processed HDPE and LLDPE were also prepared using the same protocol as the blends, but without NR, and used as controls.

Melt flow index (MFI), tensile properties, impact strength, hardness and heat distortion temperature (HDT) of HDPE, LLDPE, HDPE/NR blend and LLDPE/NR blend were determined. MFI was measured at 190°C with a load of 2.16 kg and a preheating time of 7 min. Tensile strength, Young's modulus and elongation at break were obtained from a Universal testing machine (H50KS, Hounsfield, England) at a crosshead speed of 50 mm/min, a load cell of 50 kN and a gauge length of 109 mm. Impact strength was tested using a Universal impact tester (QC-639, Cometech Testing Machines Co., Ltd., Taiwan) at a load of 12 J. Shore D hardness was measured using a TECLOCK durometer rubber and plastic hardness tester (GS-702N, Japan). HDT was determined by a VICAT AUTO model 69770 (CEAST, USA) using a load cell of 0.455 MPa, a heating rate of 2°C/min and a deformation of 0.25 mm.

3. Results and discussion

Herein, the effect of NR on MFI, tensile properties, impact strength, hardness and HDT of HDPE and LLDPE was investigated. Figure 1 shows that HDPE/NR and LLDPE/NR blends possesses lower MFI than the corresponding HDPE and LLDPE. The result implied that melt flow ability of HDPE and LLDPE decreased by blending with NR or the plastics became more viscous when blended with NR. The reason might be relevant to the higher molecular weight of NR [1] and/or the molecular weight reduction of thermoplastics during the blending process. It should be pointed out that LLDPE had higher melt flow ability than HDPE due to the lower crystallinity or looser packing structure of LLDPE. Similarly, the blend of LLDPE and NR also showed higher MFI than the blend of HDPE and NR.

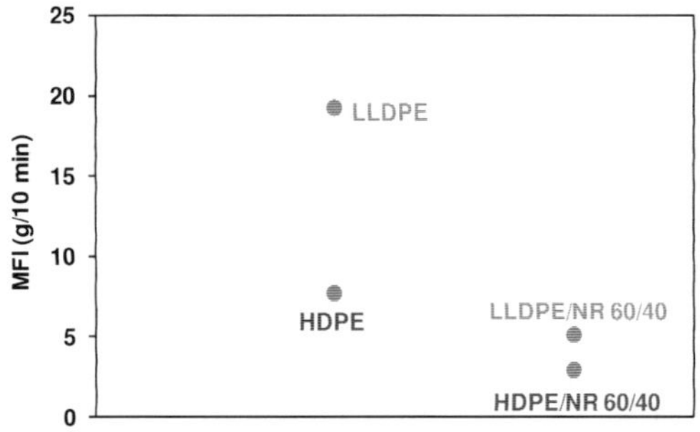

Figure 1. MFI of HDPE, LLDPE, HDPE/NR blend and LLDPE/NR blend.

Concerning to the mechanical properties of the materials, their tensile strength, Young's modulus, elongation at break, impact strength and hardness were reported in Figure 2. HDPE/NR and LLDPE/NR blends exhibited lower tensile strength, Young's modulus and Shore D hardness, but higher elongation at break than HDPE and LLDPE, suggesting that NR reduced tensile strength, stiffness and hardness of the plastics, but provided extensibility. The effect of NR on those mechanical properties was more outstanding in the case of harder plastic like HDPE than that of softer plastic like LLDPE. Similar trend was observed for the impact testing result. It is clearly seen that impact strength of HDPE/NR blend was much higher than that of the corresponding HDPE and the blend showed no break after testing. For the case of LLDPE and its blend, although LLDPE/NR blend had lower impact strength than LLDPE (Figure 2), both materials did not break after receiving the load. The result reflected that NR significantly enhanced toughness of HDPE, but this improvement was not that much pronounced in the case of LLDPE, since the material itself is soft and tough.

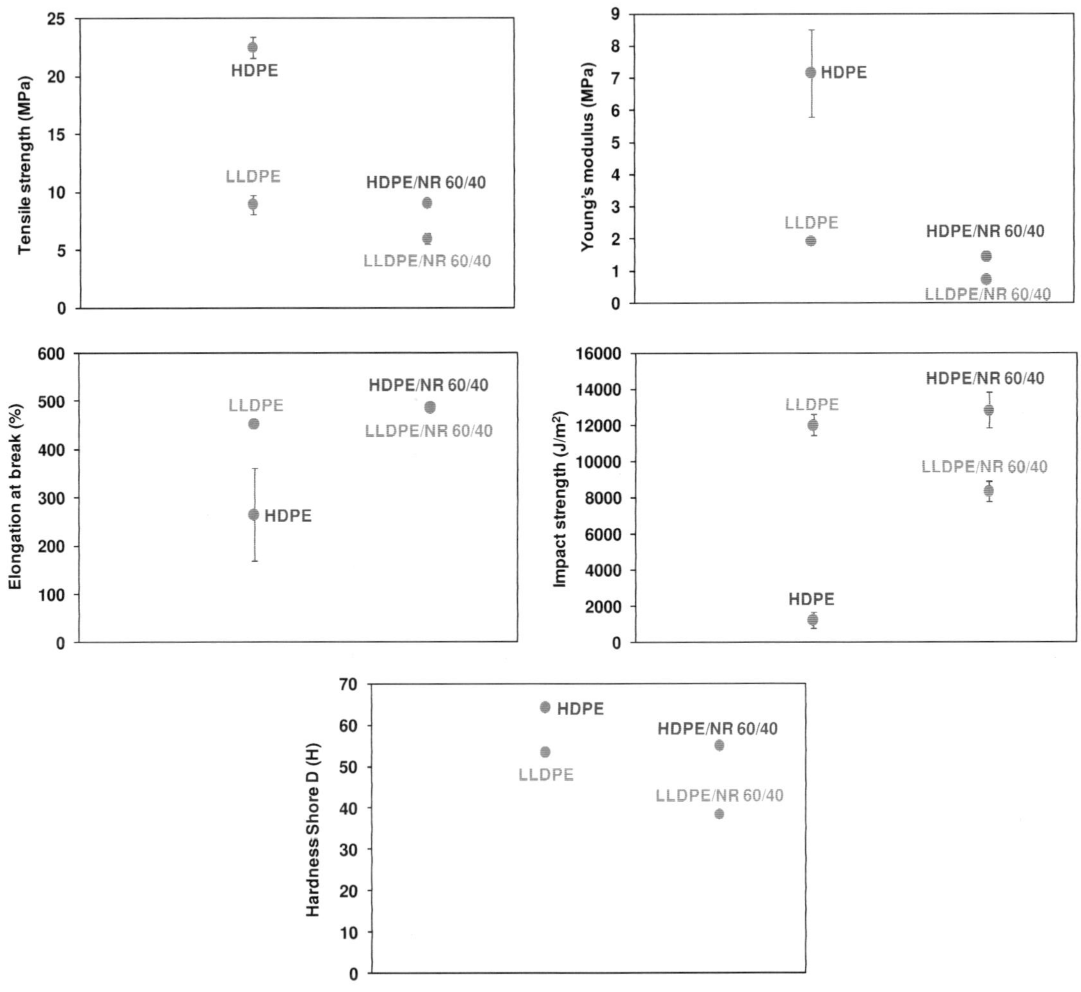

Figure 2. Tensile strength, Young's modulus, elongation at break, impact strength and hardness of HDPE, LLDPE, HDPE/NR blend and LLDPE/NR blend.

HDT of the materials is shown in Figure 3. It was found that HDT of HDPE and LLDPE decreased by blending with NR, implying that HDPE/NR and LLDPE/NR blends distorted at lower temperature than the corresponding plastics. This might be a result of softness of the materials when NR was incorporated. The reduction of HDT was more significant in the case of HDPE than that of LLDPE.

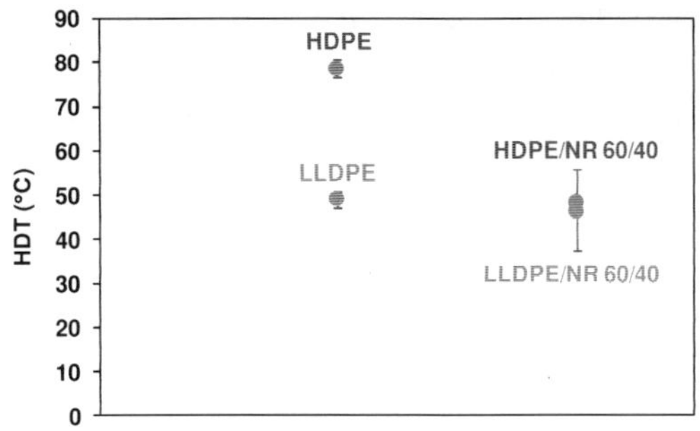

Figure 3. HDT of HDPE, LLDPE, HDPE/NR blend and LLDPE/NR blend.

The resulting HDPE/NR and LLDPE/NR blends could be injection-molded into product prototypes as shown in Figure 4, similar to most thermoplastic materials, however their surfaces had more friction and softer touch similar to elastomers.

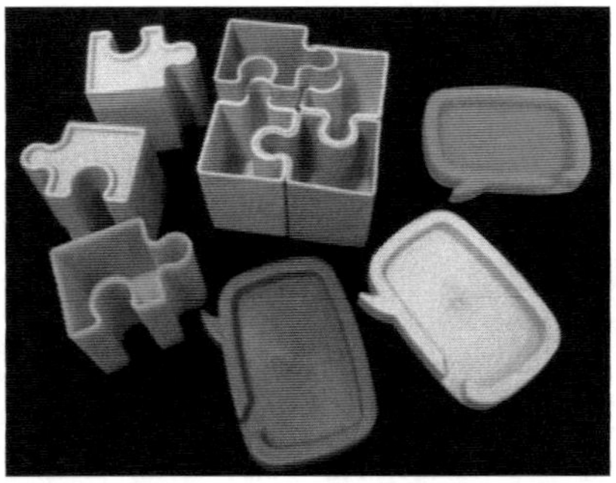

Figure 4. Injection-molded product prototypes of HDPE/NR and LLDPE/NR blends.

4. Conclusions

HDPE/NR and LLDPE/NR blends with a weight proportion of PE:NR of 60:40 were successfully fabricated and could be converted into product prototypes using an injection molding machine. HDPE/NR and LLDPE/NR blends were more extensible and tougher, but possessed lower melt flow ability, tensile strength, stiffness and HDT than the corresponding PE. The effect of NR on the properties of HDPE was more pronounced than that of LLDPE.

5. References

[1] Nakason C, Jamjinno S, Kaesaman A and Kiatkamjornwong S 2008 *Polymers for Advanced Technologies* **19** 85.
[2] Kuriakose B and De S K 1985 *Polymer Engineering and Science* **25** 630.

Acknowledgements

The author thanks the Thailand Research Fund (TRF) for the financial support and SK Polymer Co., Ltd. for the injection molding assistance of the product prototypes.

IOP Conf. Series: Materials Science and Engineering **773** (2020) 012065 doi:10.1088/1757-899X/773/1/012065

Non-destructive assessment of moisture content and modulus of rupture of sawn timber *Hevea* wood using near infrared spectroscopy technique

S Ruangkhasap, S Noypitak*, W Noknoi, and A Terdwongworakul

[1] Department of Agricultural Engineering, Faculty of Engineering at Kamphaengsaen, Kasetsart University, Nakhon Pathom, Thailand 73140.

*Corresponding author: fengsnn@ku.ac.th

Abstract. The sawn timber Para rubber (*Hevea brasiliensis*) wood is an important wood product accounting for the highest export value of Thailand. The objective of this research was to build a prediction model of moisture content and modulus of rupture of sawn timber *Hevea* wood samples using desktop near infrared spectrometer. The timber samples were collected from the southern region and eastern region of Thailand and scanned using Fourier transform near infrared (FT-NIR) spectrometer in a range of 12489–3594 cm^{-1} (800-2700 nm) in diffuse reflectance mode. Then they were determined for moisture content and modulus of rupture (MOR) following ASTM. The predictive models were built by the partial least squares regression (PLSR). The result showed high performance in prediction of moisture content with correlation coefficient of prediction, R_p = 0.89 and root mean square error of prediction; RMSEP = 0.70%db. Regarding a predictive model of modulus of rupture, the results showed fair performance giving R_p = 0.78 and RMSEP = 17.11 MPa. Therefore, using near-infrared spectroscopy technique to predict the moisture content and strength based on the modulus of rupture of timber *Hevea* wood offered a rapid and non-destructive measurement as an alternative to the destructive checking the quality of sawn timber *Hevea* wood.

1. Introduction

In the year 2017, Thailand has the export value of processed rubber wood up to 50,824.29 million baht [1]. Before exporting the wood to foreign countries, not only must the quality of wood be separated into grades such as AB grade, which has damage lower than 80% of length by human eye, but also it must be inspected to ensure that the moisture content is lower than 12% dry basis following Thai Industrial Standards Institute [2]. The sawn timber companies generally use commercially available pin moisture meter and digital moisture meter to determine the moisture content. But there are limits of both meters. For instance, the pin meter leaves the hole damage in the wood. Regarding the digital moisture meter, specific gravity (SG) of wood must be known to program in meter. Besides the strength of timber is a parameter used to indicate the quality of sawn timber. However, the strength of timber has never been tested by any non-destructive method yet. Near infrared (NIR) spectroscopy is an interesting method to verify the quality of agricultural product as it is non-destructive, rapid and friendly with environment. In the past, there were many researches about non-destructive methods which used NIR technique to predict moisture content and modulus of rupture of others wood product [3] with no report in sawn timber *Hevea* wood. The objective of this research was to build a prediction model of moisture content and modulus of rupture of sawn timber *Hevea* wood samples by desktop near infrared spectrometer.

2. Experimental

2.1. Sample preparation
The 216 of sawn timber *Hevea* wood samples were collected from the southern region; Nakhon Si Thammarat and eastern region; Rayong of Thailand in 3 grades as AB grade, C grade and pallet (P) grade. The samples were trimmed to 2.54 x 2.54 x 40 cm_3 then were smooth polished. The samples were stored at room temperature of 25°C for 24 h before measuring near infrared absorption.

2.2. NIRS collection
The samples were scanned using Fourier transform near infrared (FT-NIR) spectrometer (Bruker, MPA, Germany) in a range of 12489–3594 cm_{-1} (800-2700 nm) with an average of 32 scans and a resolution 16 cm_{-1} in diffuse reflectance mode. The spectral data were obtained using the OPUS 6.5 software.

2.3. Determination of reference value
2.3.1. Determination of modulus of rupture. In determination of modulus of rupture of wood, the sample was analysed for modulus of rupture (MOR) by static bending test which is recorded at the rupture point of sawn timber *Hevea* wood with a Universal testing machine (Instron Engineering Co, High Wycombe, England). The timber sample was placed on a span of 37 cm to test compression load using a round head with a constant pressing force speed of 2.5 mm/min at the center of the length of the wood perpendicular to grain according to ASTM D143 [4].

2.3.2. Determination of moisture content. In determination of moisture content of wood, the sample was cut into 2.54 cm x 2.54 cm x 2.54 cm size of cubic at four scanned positions of sample. All samples were dried by hot air oven (FD240, BINDER, Tuttlingen, Germany) at 103±2°C. Then the moisture content (%dry basis; %db) of wood was computed following ASTM D143 [4].

2.4. Data analysis
Reference data, such as moisture content and modulus of rupture, were examined to find the outlier by considering the standard score (Z- score). The samples with standard score greater than 3 or less than -3were considered to be the outlier (Meier, 2014). After that, the mean values of the moisture content and the modulus of rupture were statistically tested for the difference between each grade of the dried sawn timber *Hevea* wood samples using one-way analysis of variance with SPSS v.11.5.

In building the predicting model, the sample data were arranged in ascending order and separated into two sets: a calibration set and a prediction set by ratio of 7:3 ensuring similar variance between the two sets. Partial least squares (PLS) regression models were developed for calibration model of modulus of rupture and moisture content using the Unscrambler v.9.8 (Camo, Oslo, Norway). The spectral pretreatments such as the first derivative (1D), second derivative (2D), multiplicative scatter correction (MSC) and standard normal variate (SNV) were applied to reduce the scattering effect from the spectra in order to improve the model performance.

By considering the performance, the best calibration model should have highest correlation of coefficient of calibration (R_c) and lowest root mean square error of calibration (RMSEC). Similarly, the best prediction model was obtained considering the prediction performance based on the correlation of coefficient of prediction (R_p) with highest value, lowest root means square error of prediction (RMSEP) and residual prediction deviation (RPD). The RPD was calculated from the ratio of the standard deviation (SD) of the prediction group to the standard error of prediction (SEP), which is used to compare the predictive accuracy of the equation [5].

3. Results and discussion

3.1. Properties of reference value

The moisture content and modulus of rupture of dried sawn timber *Hevea* wood was obtained as reference values for prediction. The moisture contents of each grade were found to be not significantly different at the 95% confidence level because the sawn timber companies dried the timber at oven in the same condition. In this study, the moisture content range was 5.73-12.57%db as Table 1.

In case of modulus of rupture of dried sawn timber *Hevea* wood, it was found that the mean of modulus of rupture of AB, C and P the grade were significantly different at the 95% confidence level, with the mean of modulus of rupture ± standard deviation equalling 97.09 ± 15.74 MPa of AB grade, 72.54 ± 21.43 MPa C grade and 37.62 ± 22.66 MPa of P grade, respectively as shown in Table 1. The strength decreased with the lower quality of grades of timber as displayed in Table 1.

Table 1. Descriptive statistics of moisture content and modulus of rupture of sawn timber *Hevea* wood of standard method

Types of wood	Grades	Number of Samples	Min	Max	Mean±SD
Moisture Content (%dry basis)	AB	92	5.73	12.57	$7.92\pm1.44_{a}$
	C	92	7.68	11.98	$10.06\pm1.17_{a}$
	P	32	7.40	11.04	$8.89\pm0.80_{a}$
	Total	216	5.73	12.57	8.98±1.59
Modulus of rupture (MPa)	AB	92	60.54	135.76	$97.09\pm15.74_{c}$
	C	92	16.64	110.94	$72.54\pm21.43_{b}$
	P	30	7.94	87.04	$37.62\pm22.66_{a}$
	Total	214	7.94	135.76	78.20±27.78

Different superscripts in the same column indicate that the values are significantly different ($p < 0.05$) by Duncan's multiple range test.

Table 2. Statistical data of moisture content and modulus of rupture in the calibration and prediction sets.

Parameters	Data set	Number of Samples	Min	Max	Mean ± SD
Moisture Content (%dry basis)	Calibration set	147	5.73	12.57	9.02±1.59
	Prediction set	69	6.17	11.69	8.88±1.60
Modulus of rupture (MPa)	Calibration set	148	7.94	135.76	78.04±28.74
	Prediction set	66	10.07	129.65	78.55±25.72

3.2. Spectral characteristics

Figure 1 showed the domain peak at 8316 cm^{-1}(1203 nm) 6866 cm^{-1}(1457 nm of water) 6287 cm^{-1} (1591 nm of cellulose) 5215 cm^{-1}(1917 nm) 4752cm^{-1}(2104 nm) which were related with vibration bond of OH [6,7] and those at 7344 cm^{-1}(1362 nm) 5832 cm^{-1}(1715 nm of hemicellulose) [6] 4389 cm^{-1}(2278 nm) 4281 cm^{-1}(2336 nm) were related with vibration bond of CH.

Table 3. Statistical data of PLSR models obtained for prediction of moisture content and modulus of rupture.

Models	Pretreatment	LV	Calibration		Prediction	
			R_c	RMSEC	R_p	RMSEP
Moisture content (%dry basis)	2_{nd} derivative	4	0.94	0.53	0.89	0.70
Modulus of rupture (MPa)	Absorbance	9	0.85	15.11	0.78	17.11

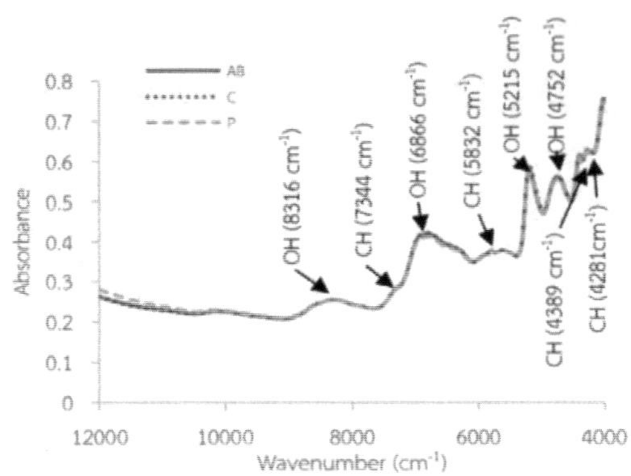

Figure 1. The average absorbance spectra of 3 grades of the oven dried sawn timber *Hevea* wood.

3.3. Assessment of the moisture content and the modulus of rupture using PLS model

The highest accuracy of moisture content model assessed was based on the 2D spectra (Table 3), producing values of $R_p = 0.89$ and RMSEP = 0.70%db with four latent values (LV). The scatter plot displaying the relationship between the actual moisture content and the predicted moisture content of prediction model was shown in Figure 2a). The highest accuracy of the modulus of rupture model assessed was based on the absorbance of spectra as detailed in Table 3, producing values of $R_p = 0.78$ and RMSEP = 17.11 MPa with nine latent values. The scatter plot expressing the relationship between the actual modulus of rupture and the predicted modulus of rupture was shown in Figure 2b).

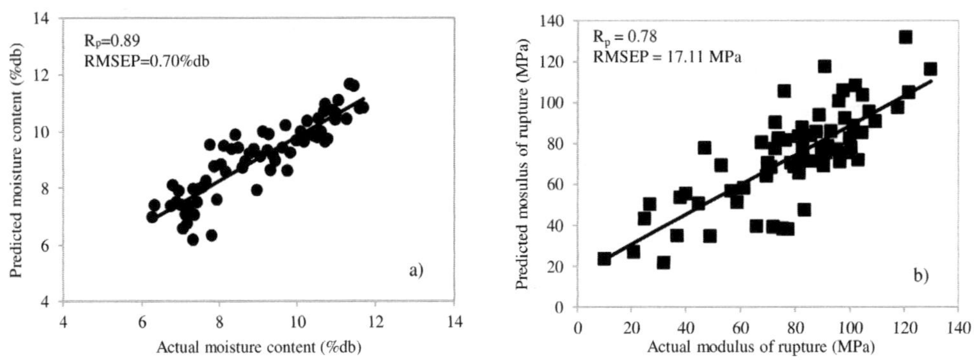

Figure 2. Plot of oven dried wood of reference versus predicted values from prediction set of a) moisture content and b) modulus of rupture.

4. Conclusions

The prediction model of moisture content of sawn timber *Hevea* wood gave high accuracy with fair result based on the R_p and RMSEP which indicated that the model could be applied for rough screening work. And the prediction performance of the model for the modulus of rupture of sawn timber *Hevea* wood indicated that the model could also be applied for rough screening work by the desktop near infrared spectrometer.

5. References

[1] http://rubber.oie.go.th/ImExThaiByProduct.aspx?pt=ex&hgid=4; Febuary 2018.
[2] Thai Industrial Standards Institute. 2009. Timber Hevea Wood. TISI. 2423-2552
[3] Tsuchikawa S 2007 *J. Appl. Spectrosc.* **42** 43–71
[4] American Society for Testing and Materials 2000 Standard Test Methods for Small Clear Specimens of Timber D 143-94

IOP Conf. Series: Materials Science and Engineering **773** (2020) 012065 doi:10.1088/1757-899X/773/1/012065

[5] Williams P 2007 Grains and seeds *Near-Infrared Technology in food science and technology* eds Y Ozaki, W F McClure and A A Christy (New Jersey: John Wiley & Sons, Inc) pp 165-217
[6] Osborne BG, Fearn T Hindle P H 1993 *Practical NIR Spectroscopy with Applications in Food and Beverage Analysis* (U.S.A.: Longman Science & Technology) pp 225
[7] Fujimoto T Kurata, Y Matsumoto K and Tsuchikawa S 2010 *J. Wood. Sci.* **56** 452-459

Acknowledgement
The authors gratefully acknowledge the Thailand Research Fund Grant No. RDG60T0121 and faculty of Kampeangsean campus, Kasesart University for supporting budget for this research.

IOP Conf. Series: Materials Science and Engineering **773** (2020) 012066 doi:10.1088/1757-899X/773/1/012066

Enhancement of the tyre rolling resistance testing machine

S Lakkam*, and A Tempiem

1 Department of Mechanical Engineering, Faculty of Engineering, Rajamangala University of Technology Phra Nakhon, Bang Sue, Bangkok, 10800 Thailand.

*Corresponding author: supachai.l@rmutp.ac.th

Abstract. Rubber tree is an outstanding product in commercial area of Thailand. Tyre is one of the prominent and downstream products as components in the automotive industry with a high production capacity. While tyres are used widely, a regulation for the rolling resistance of tyres has been set up by the European Commission. Therefore, the operators must test their products. By the high costs for the testing before certification, the researchers realize the limitation of the test equipment suffering with the small and medium enterprises that do not have access to check their tyres. This is the weakness of the tyre industry in Thailand. Therefore, the low-cost of tyre rolling resistance testing machine has been built in accordance with the UNECE Regulation No. 117 which is accepted worldwide. As a result, the in-house tyre rolling resistance testing machine is compared with the imported one in terms of production cost. The budget can be reduced up to 50% since the test equipment can be produced with currently technology in the country. Moreover, this is the self-reliance of technological machinery industry. For the working result, it was found that the vibration level of tyre rolling resistance testing machine consented to the standard ISO 10816. Besides, the experimental results of tyre rolling resistance coefficient is compliance with the requirements UNECE Regulation No. 117 Annex 6 and also verified by the international authority (TUV Rheinland Thailand Ltd). It can reduce the limitation in terms of the testing and also support the economic expansion of the production of quality tyres in the country to be accepted in international level.

1. Introduction

Rubber tree is an outstanding product in commercial area of Thailand. Vehicle tyres are the products in the tyre industry that use natural rubber as the number one among all rubber products. The value of exports in the past year is approximately 120 million Baht. The direction of the tyre industry, studied by the International Rubber Study Group or IRSG [1], has predicted that the demand for natural rubber in the world is steadily increasing. By the 2020, the world's demand for rubber will reach 31.7 million tons or approximately value 13 trillion US dollars. The Ministry of Industry has discovered the growth direction of the car industry in China that has become to the No. 1 car of market since 2009. As a results, the potential of Thai tyre production must be complied to the standards relating to tyres as the UN/ECE standards and other international standards. Moreover, adaption of Thailand to be the center of tyre production in ASEAN Economic Community is also a main reason. The development of machinery to support the testing of rubber tyre products has led to such research.

This research aims to establish the tyre friction coefficient testing machine which is compliance to the requirement of UN ECE Regulation No.117 [2] by in-house technology in terms of both controller and mechanic systems.

2. Literature survey

The motion resistance ratio of a high-lug agricultural tyre was investigated by Elwaleed A K. [3] The analysis of covariance (ANCOVA) revealed that both inflation pressure and wheel numeric have significant effects on tyre motion resistance ratio. Regression analysis was also conducted to determine the closeness of fit for Wismer–Luth's and Brixius' equations in predicting the motion resistance ratio of the tested tyre. Finally, three new logarithmic models for tyre motion resistance were formulated. The advantage of reducing tyre inflation pressure from 221 (nominal pressure) to 193 kPa on the motion resistance ratio of the high-lug agricultural tyre was pronounced. However, the tyre's motion resistance ratio deteriorated with further reduction in tyre inflation pressure from 221 (nominal pressure) to 166 kPa. Furthermore, the supervision of vehicle's tyre pressures is a major aspect of improved active car safety. A method for monitoring the tyre pressures is presented [4], using body acceleration signals. Analyzing the frequency spectrum of the virtual transfer function between the body acceleration at the front and rear wheel on one side of the vehicle, characteristic features are generated. Thereby, external interferences on the spectrum and their influences on the symptoms are discussed.

In USA, energy conservation from systematic tire pressure regulation was setup by The 2000 TREAD Act [5]. It requires automakers to gradually provide tire pressure monitoring systems for vehicles sold in the US will correct this problem for new vehicles. This law does not impact the problem in previously deployed vehicles, which have a turnover time of ~20 years. A solution is provided here to address under-inflated tires on the current 220 million vehicles and the concomitant wasted energy due to increased rolling resistance in the US automobile fleet. This communication reports on a preliminary study of tire pressures in randomly chosen vehicles, which were undergoing oil changes at a combined retail/auto-care facility. The study indicates that substantial benefits would accrue if car care facilities systematically offered complimentary tire pressure checks with oil changes.

3. Methodology

From preliminary study, there are 4 defined methods into the UN ECE Regulation No.117 as 1) torque method 2) force method 3) acceleration method and 4) power method. The force method is generally accepted and also popular because it can support twice specimen as the same time for the testing. However, the torque method is most accepted in terms of the accuracy of measurement because the main data can be directly received by machine. Therefore, the tyre rolling resistance testing machine by torque method was designed as the conceptual research and also established to compare the testing results between the torque method and the force method machine. For the main operation of two type of the machine are illustrated in Figure 1 and also calculated by formula 1 for torque method and 2 for force method:

$$F_r = \frac{T_t}{R} \quad (1)$$

$$F_r = Ft[1 + \frac{r_L}{R}] \quad (2)$$

where

Fr is the rolling resistance force
Tt is the input torque
R is the test drum radius
Ft is the tyre spindle force
r_L is the distance from the tyre axis to the drum outer surface under steady-state conditions

According with the conceptual research, the tyre rolling resistance testing machine base on the torque method was designed in terms of the mechanical production. The main sensors and transducers were defined and installed as shown in Figure 2.

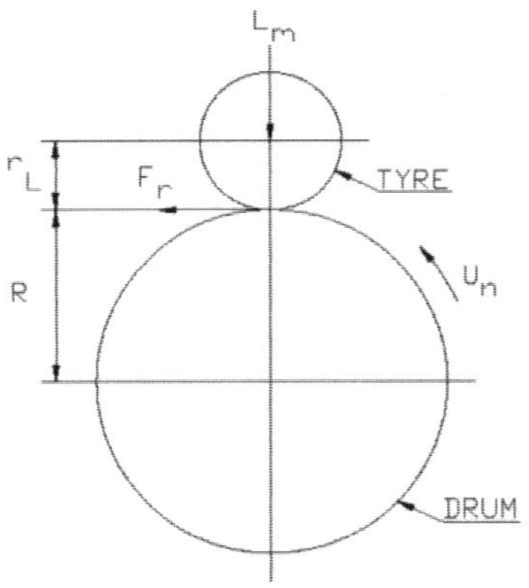

Figure 1. The calculation concept of machine following the UN CEC Regulation No.117 [2].

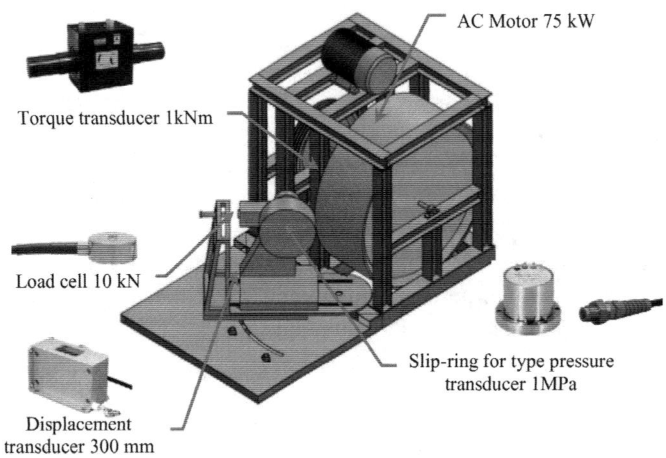

Figure 2. The design of the tyre rolling resistance testing machine base on the torque method.

Since the tyre rolling resistance testing machine base on the torque method has been completely established, the setting- up and calibration process were continuously proceeded. There are both the structural and sensor tolerance calibrated by National Institute of Metrology (Thailand). Furthermore, the machine is certificated that is compliance to requirement of UN CEC Regulation No.117 Annex 6 by TUV rheinland Ltd. Consequently, the 4 tyre specimens were donated by industry company to prove the tyre rolling resistance coefficient by means of the torque and force method following UN ECE Regulation No.117 Annex 6. The experimental work condition is shown in Table 1.

Table 1. Test condition [2]

No.	Tyre type	Test speed (km/h)	Load (N)	Inflation pressure (kPa)	Warm up duration (min)
1	C1	80	80% of maximum load capacity	250	30
2	C1	80		250	30
3	C2	80	85% of maximum load capacity	450	50
4	C2	80		450	50

4. Results and discussion

As a results, the rolling resistance coefficient of the 4 tyre specimens were conducted by means of the torque and force method following UN ECE Regulation No.117 Annex 6. For the force method, the specimens were tested on the standardized testing machine that is used in factory. At the same time, the same tyres were conducted to test on the tyre rolling resistance testing machine by means of torque method. The results of tyre rolling resistance coefficient are presented in the Table 2. The tendency of data show that the torque method machine reports the value slightly higher than the force method machine. There are 6 percentage of maximum difference value of both methods. It should be caused by mechanical loss and sensitivity of measurement system.

Table 2. The results of tyre rolling resistance coefficient

No.	Tyre type	Torque method machine (N/kN)	Force method machine (N/kN)	Difference (%)
1	C1	8.84	8.73	1%
2	C1	9.59	8.99	6%
3	C2	8.83	8.77	1%
4	C2	9.13	8.85	3%

5. Conclusions

In conclusion, the in-house tyre rolling resistance testing machine (torque method machine) is compared with the imported one (force method machine). The test results between the two machine are slightly difference. Simultaneously, the in-house tyre rolling resistance testing machine is accepted that it is compliance with the requirements UNECE Regulation No. 117 Annex 6 by the international authority (TUV Rheinland Thailand Ltd). In terms of production cost, the budget can be reduced up to 50% since the test equipment can be produced with currently technology in the country. Moreover, this is the self-reliance of technological machinery industry. It can reduce the limitation in terms of the testing and also support the economic expansion of the production of quality tyres in the country to be accepted in international level.

6. References

[1] International Rubber Study Group 2017 *statistical report of the global rubber industry.*

[2] The United Nations Economic Commission for Europe The United Nations Economic Commission for Europe Regulation117.

[3] Elwaleed A K, Yahya A, Zohadie M, Ahmad D, and Kheiralla A F 2016 Effect of Inflation Tyre Pressure on Motion Resistance Ratio of A High-lug Agricultural *Journal of Department of Biological and Agricultural Engineering*. Faculty of Engineering, University Putra Malaysia **43** 69.

[4] Halfmann C, Ayoubl M and Hoizmann H 1997 Supervision of Vehicles' Tyre Pressure by Measurement of Body Accelermeters *Institute of Automatic Control*. Darmstadt University of Technology. Laboratory for Control Engineering and Process **5** 1151.

[5] Pearce Joshua M and Hanlon Jason T 2007 Energy Conservation from Systematic Tire Pressure Regulation *Energy Policy* Clarion University of Pennsylvania **35** 2673.

Acknowledgement

The authors are grateful for the financial support from The Thailand Research Fund (TRF) and also Rajamangala University of Technology Phra Nakhon for the experimental facilities.

AUTHOR INDEX